About the Pacific Institute for Studies in Development, Environment, and Security

The Pacific Institute is one of the world's leading research and policy non-profits working to create a healthier planet and sustainable communities, with a focus on freshwater. The mission of the Institute is to create and advance solutions to the world's most pressing water challenges—including unsustainable water management and use, climate change, and environmental degradation; food, fiber, and energy production; and lack of access to fresh water and sanitation. The Institute was cofounded in 1987 by Peter Gleick and three colleagues. Based in Oakland, California, the Institute combines science-based thought leadership with active outreach to influence policies that ensure society, the economy, and the environment have the water needed to thrive.

The Pacific Institute cuts across traditional areas of study and actively collaborates with a diverse set of stakeholders—including leading policy makers, scientists, corporate leaders, international organizations such as the United Nations, advocacy groups, and local communities. This interdisciplinary and independent approach helps bring opposing groups together to forge effective real-world solutions. We have worked to change policy and to focus thinking about water away from narrow approaches and toward more integrated and sustainable water practices and concepts through rigorous independent research, extensive policy engagement, and intensive outreach to the public. The Institute has formulated a new vision for long-term water planning in California and internationally; developed a new approach for valuing well-being in local communities; worked on transborder environment and trade issues; analyzed standards for global water stewardship; clarified key concepts and criteria for sustainable water use; offered recommendations for reducing conflicts over water in the Middle East, Latin America, and Central Asia; championed the human right to water; assessed the impacts of global warming on freshwater resources; and created programs to address environmental justice concerns in low-income communities and communities of color.

Our research has reached tens of millions of people through reports, projects, speeches, testimony, and media; more than 100 research papers, reports, and testimonies are available free on the Pacific Institute websites at http://www.pacinst.org and http://www.worldwater.org. More information about the Pacific Institute and its staff, board of directors, issue areas, and programs can be found at http://www.pacinst.org.

THE WORLD'S WATER

Volume 9

The Report on Freshwater Resources

Peter H. Gleick
with Michael Cohen, Heather Cooley, Kristina Donnelly, Julian Fulton,
Mai-Lan Ha, Jason Morrison, Rapichan Phurisamban, Heather Rippman,
and Stefanie Woodward

Foreword by Alexandra Cousteau

2018

The Pacific Institute for Studies in Development, Environment, and Security
Oakland, California

Copyright © 2018 Pacific Institute for Studies in Development, Environment, and Security

All rights reserved under International and Pan-American Copyright Conventions. No part of this book may be reproduced in any form or by any means without permission in writing from the publisher: Pacific Institute, 654 13th Street, Oakland, California 94612.

ISBN 10: 1983865885
ISBN 13: 978-1983865886

ISSN 15287-7165

Printed on recycled, acid-free paper

Manufactured in the United States of America
10 9 8 7 6 5 4 3 2 1

Keywords: water, freshwater, water quality, corporate water stewardship, water conflict, human right to water, drought, climate change, water footprint, drinking water fountains, cost of water, water trading, water markets, peak water, water use

Contents

Foreword by Alexandra Cousteau xi

Introduction xiii

About the Authors xvii

ONE   The UN Global Compact CEO Water Mandate: History, Objectives, Strategy   1
Heather Rippman and Stefanie Woodward

Introduction: Some History on the Evolution of Corporate Social and Environmental Responsibility   1
The Formation and Role of the UN Global Compact   3
The UN CEO Water Mandate   4
Corporate Water Stewardship   6
Building Consensus: Key Water Stewardship Concepts and Terminology   8
Making Water Stewardship Accessible: Producing and Distributing Tools and Guidance   9
Managing Water-Related Impacts   10
Ensuring Integrity in Water Stewardship Initiatives   14
Making an Impact: Sustainable Management of Water and Sanitation for All   16
Conclusion   16

TWO   A Human Rights Lens for Corporate Water Stewardship: Toward Achievement of the Sustainable Development Goal for Water   21
Mai-Lan Ha

Introduction   21
The Business Case for Action on Water   22
State Recognition of the Human Rights to Water and Sanitation   25
Human Rights Responsibility of the Corporate Sector   28
Support for the Human Rights to Water and Sanitation   31
The Path Forward   35

THREE   Updating Water-Use Trends in the United States   39
Kristina Donnelly and Heather Cooley

Introduction   39
Data Collection and Organization 41
Total Water Use   42
Conclusions   49

FOUR   The Water Footprint of California's Energy System, 1990-2012   55
Julian Fulton and Heather Cooley

Introduction   55
Background   55
Evaluating Water-Energy Links   57
Water for California's Total Energy System   58
Summary   61
Acknowledgements   64

FIVE   The Nature and Impact of the 2012-2016 California Drought   67
Peter H. Gleick

Introduction   67
The Hydrological and Climatological Conditions behind the California Drought   67
Impacts of the California Drought   74
Overall Economic Well-Being   81

SIX   Water Trading in Theory and Practice   87
Michael Cohen

Water Trading, Transfers, Markets, and Banks   88
Water Trading in Theory   88
Water Trading in Practice   90
Environmental, Economic, and Social Performance   95
Necessary, Enabling, and Limiting Conditions for Water Trades   103
Conclusion   107

SEVEN   The Cost of Water Supply and Efficiency Options: A California Case   113
Heather Cooley and Rapichan Phurisamban

Introduction   113
Methods and Approach   114
Storm Water Capture   115
Water Recycling and Reuse   117
Desalination   120
Urban Water Conservation and Efficiency   122
Summary and Conclusions   126

WATER BRIEFS

One	The Human Right to Water and Global Sustainability: Actions of the Vatican 129 *Peter H. Gleick*
Two	Access to Water through Public Drinking Fountains 135 *Rapichan Phurisamban and Peter H. Gleick*
Three	Water and Conflict Update 141 *Peter H. Gleick*

WATER UNITS, DATA CONVERSIONS, AND CONSTANTS   147

COMPREHENSIVE TABLE OF CONTENTS   157

Volume 1: The World's Water 1998–1999: The Biennial Report on Freshwater Resources   157

Volume 2: The World's Water 2000–2001: The Biennial Report on Freshwater Resources   160

Volume 3: The World's Water 2002–2003: The Biennial Report on Freshwater Resources   163

Volume 4: The World's Water 2004–2005: The Biennial Report on Freshwater Resources   166

Volume 5: The World's Water 2006–2007: The Biennial Report on Freshwater Resources   170

Volume 6: The World's Water 2008–2009: The Biennial Report on Freshwater Resources   174

Volume 7: The World's Water Volume 7: The Biennial Report on Freshwater Resources   178

Volume 8: The World's Water Volume 8: The Biennial Report on Freshwater Resources   182

Volume 9: The World's Water Volume 9: The Report on Freshwater Resources   186

INDEX TO VOLUME 9   189

COMPREHENSIVE INDEX TO VOLUMES 1-8   197

LIST OF FIGURES

Figure 1.1: How Key Concepts and Terms Relate to One Another   10
Figure 1.2: The Water Stewardship Progression   11
Figure 1.3: Corporate Water Disclosure Framework   14
Figure 2.1: Relationship between SDG 6 and Other Sustainable Development Goals   23

Figure 2.2: Heat Map for Determining Severity of Human Rights Impacts 33
Figure 3.1: Total Annual and per Capita Water Use (Freshwater and Saline Water), 1900–2010, by Sector 43
Figure 3.2: Economic Productivity of Water, 1900–2010 44
Figure 3.3: Water Use for Thermoelectric Power Generation, 1900–2010, by Type 44
Figure 3.4: Water Use Intensity for Thermoelectric Power Generation, 2010, by State 46
Figure 3.5: Total and per Capita Water Use for the Municipal and Industrial Sector, 1900–2010 47
Figure 3.6: Total and per Capita Water Use for the Residential Sector, 1950–2010 48
Figure 3.7: Residential Per-Capita Water Use in 2010, by State 50
Figure 3.8: Annual Freshwater Use for Irrigation (1900–2010) and Irrigated Area (1950–2010) 51
Figure 3.9: Average Application Depth, 1950–2010 51
Figure 3.10: Irrigated Area in Million Hectares, 1950–2010, by Irrigation Method 52
Figure 4.1: Changes in California GDP, Population, Energy Use, and Energy Greenhouse Gas Inventory, 1990–2012 (Index: 1990=1.0) 56
Figure 4.2: California's Energy–Water Footprint, 1990–2012, by Energy Type 60
Figure 4.3: California's Energy–Water Footprint, 1990–2012, by Type of Water 61
Figure 4.4: California's Energy–Water Footprint, 1990–2012, by Location 62
Figure 5.1: California Water Year Precipitation, 1896–2016 (in millimeters per water year) 69
Figure 5.2: Percentage of California's Area in Drought from January 2000 to October 2016 70
Figure 5.3: California Water Year Temperature, 1896–2016 (in degrees Celsius) 70
Figure 5.4: Unimpaired Runof for the Sacramento–San Joaquin Index, Water Years 1906–2016 71
Figure 5.5: Deviation from 1906–2016 Average Runoff, by Water Year, for the Sacramento–San Joaquin River Systems (in million acre-feet per year) 73
Figure 5.6: Cumulative Storage Capacity behind California Dams (in million acre-feet) 73
Figure 5.7: California Crop Revenue by Crop Type (in billions of 2015 dollars) 76

Figure 5.8: California Gross and Net Farm Income, and Production Expenses (in billions of 2015 dollars)   77

Figure 5.9: Central Valley Groundwater Change over Time   78

Figure 5.10: California In-State Electricity Generation by Source, 2013   79

Figure 5.11: Reductions in Hydroelectricity Generation Due to California Drought, 2001 through September 2015 (in gigawatt-hours per month)   79

Figure 5.12: Gross State Product (GSP) for All 50 U.S. States, 1997–2014 (in millions of 2009 chained dollars)   82

Figure 6.1: Non-Project Water Transfers within the Sacramento-San Joaquin Watersheds in 2013   93

Figure 6.2: The Colorado River Basin   95

Figure 6.3: Water Purchases for the Environment in California, 1982–2011   99

Figure 7.1: Potential Residential Water Savings, by End Use, in California   122

Figure WB-2.1: Typologies of Drinking Water Fountains, Including Information on Features, Location, and Other Fountain Characteristics   137–138

Figure WB-3.1: The Number of Water Conflict Events Reported per Year, 1930–2016   143

LIST OF TABLES

Table 1.1: Integrity Risk Areas   15

Table 2.1: Dimensions of the Human Rights to Water and Sanitation   26

Table 2.2: Examples of Water-Related Impacts Experienced by Affected Stakeholders   30

Table 2.3: Relationship Between UN Guiding Principles and Elements of Corporate Water Management   31

Table 2.4: Elements of Corporate Water Stewardship   32

Table 3.1: USGS Changes in Water-Use Categories   40

Table 4.1: Median Consumptive Water-Use Factors for Electricity Production Technologies Used to Calculate California's Blue EWF   58

Table 4.2: Median Consumptive Water-Use Factors for Liquid Fuel Production Used to Calculate California's Blue and Green EWF   59

Table 5.1: Water-Year Index for the Sacramento and San Joaquin Valleys, 2000–2016   72

Table 5.2: Number and Area of California Wildfires by Agency, 2015   82

Table 7.1:	Storm Water Capture and Reuse Cost	117
Table 7.2:	Water Recycling and Reuse Cost	119
Table 7.3:	Relative Salinity of Water	120
Table 7.4:	Seawater and Brackish Water Desalination Cost	121
Table 7.5:	Residential Water Conservation and Efficiency Measures	123
Table 7.6:	Nonresidential Water Conservation and Efficiency Measures	125

Foreword

I'm delighted to offer this foreword to the ninth volume of what is the single best resource for the public, researchers, and advocates working to protect the world's freshwater resources—*The World's Water*, produced by Peter Gleick and the Pacific Institute. As we tackle the complex water issues of today and anticipate the challenges of the future, we will increasingly rely on the knowledge, analysis, and insight provided by Peter Gleick and the Institute. They have been an invaluable resource for my work to empower people to reclaim and restore the world's water, one community at a time.

I came to freshwater by way of the oceans and as I made my way upstream, I realized how interconnected our oceans and freshwater systems are. It also became clear how water connects us to each other. Our health, our happiness, and our prosperity all depend on an abundance of clean water available to our communities.

We all live on the waterfront. Your waterfront may be the storm drain on your street, the creek in your backyard, the river through your city, or the ocean that borders your town—our relationship with water in all its forms is critical to the health and well-being of our families, our communities, and our water-covered planet. Taking care of water goes beyond what most think of as "environmentalism" and gets to the very heart of how we define healthy communities; how we manage the resources that create jobs and local economies; and how we build local capacity now for the challenges ahead.

There are many challenges: even now, as we approach 2020, nearly a billion people do not have access to safe, affordable freshwater. Water-related diseases kill more people than all forms of violence, including war. This is inexcusable in our day and age—we have the money, technology, and know-how to tackle the problem, but are failing to do so. Conflict over water continues around the world, as growing populations demand more and more water from our vulnerable ecosystems. Climate change threatens our water resources in many ways.

All these problems are tackled in *The World's Water* series. These volumes go back to 1998 and have addressed every freshwater challenge the planet faces, with a fresh eye and a commitment to finding answers to these challenges. The current volume is the first to be available in an entirely digital format (though you can buy a "print-on-demand" copy for your shelves!). It continues to be the go-to resource for information on current problems and effective solutions.

My grandfather Jacques-Yves Cousteau was always baffled when people would ask him why he was such a vocal advocate for protecting water resources. He would usually begin his response with "When you go and see..." and then paint the picture as only he could of the majesty and importance of water.

His advice still rings true today. I challenge you to explore your local waterfront. Take a walk along the creek or river in your city and ask yourself if it's the kind of place where you'd let your children swim. Stop for a moment the next time it rains and consider the water you see running from your property or along a nearby street and ask yourself if you'd eat fish from the waters it drains to. Go and see the places where your drinking water is sourced. And think about those less fortunate who may have to struggle every day to find water for their families to drink, cook, and clean with.

Regardless of background or political philosophy, I believe we all want to live and raise our families in communities where our local water is safe enough for swimming, drinking, and fishing.

Read this volume of *The World's Water*. Read the earlier ones. Engage with scientists, activists, your local politicians, your neighbors. Explore your watershed. Save the planet.

Alexandra Cousteau
Berlin, Germany
Fall 2017

Introduction

Welcome to the newest version of *The World's Water: The Report on Freshwater Resources—Volume 9.* As the world of publishing has changed, so too are we trying to evolve. It has been nearly 20 years since publication of the first volume. Volume 7 in this series marked a shift from the "biennial" scheduling of the book's release and the elimination of the date from the title. Volume 8 marked the last edition to be labeled "biennial" and to be produced by Island Press, our long-time publisher in this effort. This new edition, Volume 9, marks the next stage in the evolution of *The World's Water*, with a purely electronic edition (though readers now have the new option of purchasing an "on-demand" hard copy of the book), and the first to eliminate the data tables that have been a major part of the earlier print editions. We are now moving to post water data tables exclusively on http://www.worldwater.org, where they will continue to be available for free. We are also in the process of updating and modifying that website and developing a more comprehensive and innovative data portal there.

When the first volume of *The World's Water* was published in 1998, the United Nations' Millennium Development Goals (MDGs) for 2015 had not even been established. The concepts of water "footprints," "virtual water," "corporate water stewardship," "peak water," and other now-central topics had not yet been put forward or were mostly unknown. Internet data visualizations and electronic book publishing were unheard of. Yet today, the MDGs have been replaced with a new set of comprehensive environmental and social targets, the Sustainable Development Goals (SDGs) for 2030. A wide variety of research, academic, advocacy, and policy groups are addressing water problems in new and innovative ways. And the demand for good water analysis is higher than ever.

In this new volume, we continue to offer insights into critical global water problems, overviews of data and analysis around water use and management, and case studies of some of the greatest water challenges around the world. *The World's Water*, however, has always been about more than just bad news. There is plenty of good news and many innovative efforts to identify and implement sustainable solutions, and we include many of them here. There is no shortage of topics to address, and as always it is a challenge to try to choose among them for inclusion in the books. In this latest volume, we tackle some new topics and revisit and update some older ones.

Chapter 1 looks at the broad effort that has developed around the issue of corporate water stewardship, with a summary of the history, objectives, and strategies behind the efforts of the United Nations Global Compact, focusing on the CEO Water Mandate. The Pacific Institute has been a leader in helping to define and coordinate work around corporate water issues, and we publish extensively in this area.

Chapter 2 expands on previous work by the Institute on the human right to water and sanitation, and looks at how corporate water stewardship must integrate this formal right into private sector efforts to more sustainably manage water resources. What are the rights and responsibilities of corporations in meeting the right to water? How can the concept of the human right to water and sanitation be used to improve corporate water management?

Chapter 3 offers a comprehensive look at the critical issue of water use, with a focus on the data sets on water use collected in the United States by the U.S. Geological Survey. Water-use data are among the least collected in the world, and even the U.S. efforts in this area are incomplete. Nevertheless, the available data offer some key insights into trends in how people, agriculture, and industry are using water. The Pacific Institute has been at the forefront of advancing the discussion about smart and efficient water use.

Chapter 4 expands work done in recent years on "water footprints," including work we've pursued at the Pacific Institute on California's water footprint. In this new chapter, we summarize research into the water footprint of energy use in the California context. Policy makers have often failed to consider the implications of energy policies on water resources, and this chapter uses the case of California's energy system from 1990 to 2012 to examine how energy policies have affected demands on water resources and provides insights into potential climate mitigation policies.

Chapter 5 summarizes some of the key impacts and implications of the severe five-year drought that afflicted California through 2016. The Pacific Institute has regularly analyzed and published research on extreme hydrologic events in California, and the current chapter offers an overview of the hydrologic conditions behind the recent drought and offers insights into the impacts on agriculture, ecosystems, hydropower production, and urban centers.

Chapter 6 summarizes a set of tools that are increasingly thought to be vital for more sustainable management of water resources: market-based water reallocation mechanisms, often known as water trading or water markets. Water markets have received increased attention and support in recent years because of their perceived adaptability and ability to meet changing water needs, especially in places where other strategies, such as pre-assigned water rights, are under new stress. This chapter discusses water trading in theory and practice, including its environmental, economic, and social performance, and the conditions needed for implementing different market mechanisms.

The final chapter, Chapter 7, also addresses a key economic issue associated with water management—the question of the cost of water alternatives. The cost of water supply and demand options is key to determining which water strategies to pursue. Yet determining these costs has been limited by data and methodological challenges. A groundbreaking Pacific Institute study, summarized in this chapter, examined the cost of a range of efficiency and alternative supply options in urban areas for the state of California: storm water capture, water reuse, brackish and seawater desalination, and a range of urban water conservation and efficiency measures. There is a growing recognition that, while these factors are hard to quantify, improving economic assessments is vital.

As always, the chapters in *The World's Water* are supplemented with shorter "Water Brief" reports on items of interest. The current volume includes the regular update on our unique Water Conflict Chronology, with historical examples of conflicts related to water going back to 2500 BC and new entries through early 2017. The Chronology is also available as maps, data, and timelines at the website at http://www.worldwater.org. Other Water Briefs include a summary of a meeting held at the Pontifical Academy of Sciences in the Vatican on the human right to water, with the text of Pope Francis' statement on this issue, and a review of critical issues around public access to water through drinking water fountains.

Thanks, and acknowledgements to all of my coauthors; my copyeditor Alison S. Britton and designer Michael Mott who helped produce the electronic version of the book; the former publisher, Island Press, for their long support of our efforts (and indeed, you can still get the hard copies of Volumes 1 through 8 from them!); and the David and Lucille Packard Foundation for financial support of the transition to the new formats.

Peter H. Gleick
Oakland, California
Fall 2017

About the Authors

Peter H. Gleick is president emeritus and cofounder of the Pacific Institute. He is a hydroclimatologist working on global freshwater challenges and solutions. He created *The World's Water* book series and continues to serve as editor and author. He is a member of the US National Academy of Sciences and a MacArthur Foundation Fellow; obtained a B.S. in Engineering and Applied Sciences from Yale University and an M.S. and Ph.D. in Energy and Resources from the University of California, Berkeley.

Michael Cohen is a Senior Research Associate at the Pacific Institute, where he has written many articles and reports on water, water use, water trading, and environmental protection in California and the binational Colorado River basin. Mr. Cohen has a Master's degree in Geography, with a concentration in Resources and Environmental Quality from San Diego State University, and a B.A. in Government from Cornell University.

Heather Cooley is Director of Research at the Pacific Institute and conducts and oversees research on an array of water issues. Prior to joining the Institute in 2004, she worked at Lawrence Berkeley National Laboratory studying climate and land use change and carbon cycling. She has a B.S. in Molecular Environmental Biology and a Master's degree in Energy and Resources from the University of California, Berkeley.

Kristina Donnelly was a research associate with the Pacific Institute's Water and Sustainability Program from 2011 to 2016. Prior to joining the Institute, Kristina organized educational programs and published reports with the Arava Institute for Environmental Studies. She received a B.S. in Mathematics from American University and an M.S. in Natural Resources and Environment from the University of Michigan.

Julian Fulton is an Assistant Professor of Environmental Studies at California State University Sacramento, where he teaches and conducts research on issues related to water sustainability. He is a former Research Affiliate of the Pacific Institute and has contributed to several institute publications on the water footprint and water–energy nexus. Dr. Fulton holds a Ph.D. in Energy and Resources from the University of California, Berkeley.

Mai-Lan Ha is a Senior Research Associate focusing on corporate water stewardship and voluntary sustainability standards to improve company practice on water issues from human rights to environmental protection. Prior to joining the Pacific Institute, Mai-Lan worked on trade, sustainable development, and human rights issues in Southeast Asia.

Jason Morrison is president of the Pacific Institute and Head of the CEO Water Mandate, an initiative of the United Nations Global Compact. He cofounded the Alliance for Water

Stewardship, which had developed and is implementing a freshwater certification program to advance water stewardship. He studies the policy implications of private sector sustainability initiatives, freshwater-related business risks, and sustainable water management in the business community. He received a B.A. in Philosophy from the University of California, San Diego, and a Master's Degree in Energy and Environmental Studies from Boston University.

Rapichan Phurisamban worked as a Research Associate at the Pacific Institute from 2014 to 2017. Her research interests span various topics, including water resources management, climate change adaptation, and environmental justice. She holds a Master of Public Policy degree from the University of California, Berkeley, and a Bachelor's degree in Economics from the University of British Columbia.

Heather Rippman was a Senior Research Associate at the Pacific Institute and Senior Advisor to the UN Global Compact CEO Water Mandate from 2015 to 2017. Her professional interests include freshwater and marine stewardship, corporate sustainability, and collective action to support effective management of shared natural resources.

Stefanie Woodward is a consultant at WSP USA, where she specializes in corporate water stewardship and supports clients with water strategy and reporting. She was formerly a Research Associate at the Pacific Institute, where she served as an advisor to the UN CEO Water Mandate. Stefanie holds a Master of Environmental Management degree from the Yale School of Forestry and Environmental Studies.

CHAPTER 1

The UN Global Compact CEO Water Mandate: History, Objectives, Strategy

Heather Rippman and Stefanie Woodward

Introduction: Some History on the Evolution of Corporate Social and Environmental Responsibility

The last few decades have seen a shift in the way modern corporations perceive their role in society, especially in social and environmental responsibility. In regard to water, new principles and coalitions are being organized around the concept of water stewardship and how to manage operations under increasingly challenging water conditions. This chapter addresses a major focus of these efforts: the United Nations CEO Water Mandate, which has played a leading role in developing the theoretical and practical underpinnings of the sustainable management and use of water by the private sector.

Business management theorists began to debate the social role of corporations in a modern, interconnected society during and after the worldwide economic depression of the 1930s. Stanford professor Thomas Kreps, known as "the conscience of the business school," introduced a course in 1931 entitled *Business Activity and Public Welfare*, and first published Measurement of the *Social Performance of Business* in 1940 (Kreps 1962). In 1953, American economist and academic Howard R. Bowen published Social Responsibilities of the Businessman, appealing to corporate executives to make decisions based on both business objectives and social values, and earning him the nickname, "the father of corporate social responsibility." In contrast, economist Milton Friedman (1962) famously developed and publicized an opposing point of view that corporations exist only to generate profits and reward shareholders for their investments. Friedman anticipated that the rule of law—particularly property and liability law—would protect other interests, and that shareholders themselves should engage as individuals in social initiatives of their own choosing. In this narrow interpretation of corporate responsibility, Friedman said:

There is one and only one social responsibility of business—to use its resources and engage in activities designed to increase its profits so long as it stays within the rules of the game, which is to say, engages in open and free competition without deception or fraud.

At that time, the world's population was 3.1 billion and total economic activity was a fraction of the US $77 trillion global economy that would emerge over the next 50 years. With the end of the Cold War came a period of optimism and expanded international cooperation and trade. Innovations in agriculture, industry, transportation, and communications were accompanied by the development of global finance and international trade agreements and "globalization."

Proponents of globalization believed that unrestricted international trade and investment would lead to unprecedented prosperity for all. Companies pursuing growth in revenue, profits, and share value obtained inexpensive labor and raw materials far from corporate headquarters, setting off trends toward industrialization and urbanization, and contributing to the development of new markets worldwide. But extraordinary economic growth often came with extraordinary exploitation of human and natural resources. Countries eager for economic development and lacking strong labor and environmental protections became home to sweatshops and industrial pollution. Large-scale environmental disasters with their roots in industrial activities, like Bhopal (1984), Chernobyl (1986), and Exxon Valdez (1999), undermined public trust in both business and government. Chronic unjust and unethical business practices were increasingly exposed and publicized, leading to lost revenue, damaged company reputations, and reduced share value. Long-term trends like population growth and climate change began to call into question the possibility and appropriateness of unlimited growth.

By 1999, public opposition to globalization culminated during World Trade Organization (WTO) negotiations in Seattle, where massive protests by labor unions, human rights activists, and environmental organizations brought the negative consequences of free-trade policies into the mainstream media spotlight. In this context, leading companies, seeking to restore the trust of consumers, investors, and shareholders, began to take voluntary steps to improve labor conditions, manage environmental impacts, and increase transparency about their practices. Over time, these efforts coalesced into a broader movement called—variously—corporate responsibility (CR), corporate social responsibility (CSR), corporate citizenship, corporate sustainability, corporate stewardship, and so forth. The key concept that has emerged in mainstream business is that corporations can and should not only take responsibility for their own environmental and social impacts but also act voluntarily in the absence of effective governmental policy, oversight, and enforcement. Today, almost all major international corporations have some form of sustainability officer and strategy. New efforts are underway to develop standard reporting tools and metrics. And there is a growing understanding of the diverse risks to companies that fail to evaluate and tackle corporate stewardship challenges.

The Formation and Role of the UN Global Compact

Unless globalization works for all, it will work for nobody. I propose that you, the business leaders, and we, the United Nations, initiate a global compact of shared values and principles, which will give a human face to the global market.
—Kofi Annan, UN Secretary-General
January 31, 1999, at the World Economic Forum in Davos

The United Nations Global Compact was launched in 2000 in recognition of the connections between sustainable development and sustainable business. Core to the Global Compact, the United Nations issued a call to action to voluntarily align private sector operations and strategies with ten universally accepted principles in the areas of human rights, labor, environment, and anti-corruption (Box 1.1), and to take action in support of the newly developed UN Millennium Development Goals (MDGs), which were focused on the multiple dimensions of extreme poverty.

BOX 1.1 UN Global Compact Principles

HUMAN RIGHTS

1. Businesses should support and respect the protection of internationally proclaimed human rights; and
2. make sure that they are not complicit in human rights abuses.

LABOUR

3. Businesses should uphold the freedom of association and the effective recognition of the right to collective bargaining;
4. the elimination of all forms of forced and compulsory labour;
5. the effective abolition of child labour; and
6. the elimination of discrimination in respect of employment and occupation.

ENVIRONMENT

7. Businesses should support a precautionary approach to environmental challenges;
8. undertake initiatives to promote greater environmental responsibility; and
9. encourage the development and diffusion of environmentally friendly technologies.

ANTI-CORRUPTION

10. Businesses should work against corruption in all its forms, including extortion and bribery.

Source: UN Global Compact 2016.

The Global Compact has since grown to become the largest corporate responsibility initiative in the world, with over 8,000 corporate signatories based in more than 135 countries (UN Global Compact 2016). UN Global Compact member companies commit to the ten principles and to communicate goals, actions, and progress toward meeting these principles on an annual basis.

In practice, making a positive contribution to a social or environmental outcome beyond a company's own operations requires a nuanced understanding of local context and conditions, a locally appropriate solution or portfolio of solutions, and often a coalition of local partners with a shared definition of success and a commitment to take action. Today, the UN Global Compact supports the development of collaborative solutions through more than fifty Local Networks worldwide.[1]

To help facilitate achievement of social and environmental commitments that demand specialized expertise, tools, and guidance, the UN Global Compact's activities are also organized by issue—including, for example, Business for Peace, Women's Empowerment Principles, and Caring for Climate. For freshwater resources, the CEO Water Mandate is the UN Global Compact's platform for corporate environmental responsibility, focused on water scarcity, pollution prevention, access to water and sanitation, and meeting the challenges initially set by the Millennium Development Goals, now superseded by the Sustainable Development Goals (SDGs).

The UN CEO Water Mandate

Historically, water has been plentiful and cheap in the temperate and tropical regions that are home to most of the world's human population. Early settlements and entire civilizations alike thrived based on their proximity and access to sufficient water resources. Advances in engineering and the development of public institutions made it possible to build and operate large infrastructure projects to deliver water resources beyond the capacity of natural systems and over large distances. Massive dams and reservoirs and the ability to tap into large volumes of groundwater have helped mitigate droughts and floods, brought more land under agricultural cultivation, extended growing seasons, and supported food production at a scale that wouldn't be possible with rainfed agriculture alone. Water has been diverted over large distances to support cities and industries that would not be able to survive on limited local water resources.

However, as populations and economies have continued to grow, decreasing per capita water availability, declining water quality, and a systemic failure to fulfill the human rights to water and sanitation increasingly affect the well-being of workers and communities, threaten the long-term viability of farms and factories, and pose risks to consumers, investors, and shareholders. At the UN Global Compact Leaders' Summit in 2007, a group of six companies—including The Coca-Cola Company, Levi Strauss & Co., Läckeby Water Group, Nestlé S.A., SAB Miller, and Suez—announced the creation of the CEO Water Mandate, a voluntary initiative focused on engaging the private sector in sustainable water management (UN Global Compact 2007). Similar to the UN Global Compact, companies endorse the CEO Water Mandate with a commitment to action on a set of six key elements of water stewardship (Box 1.2).

---

1. https://www.unglobalcompact.org/engage-locally/about-local-networks.

BOX 1.2 CEO Water Mandate: Six Key Elements

Direct Operations
Assess water use, set targets for water conservation and wastewater treatment, and invest in new technologies to achieve these goals. Raise awareness of water impacts, risks, and opportunities within corporate culture and include water sustainability in business decisions.

Supply Chain and Watershed Management
Share water stewardship best practices with suppliers, and encourage them to assess and improve water efficiency, manage wastewater quality, and increase water reuse. Build capacities to analyze and respond to watershed risk and encourage major suppliers to report regularly on progress.

Collective Action
Build relationships and work with local and regional civil society organizations, governments, and public authorities on water sustainability issues, policies, and innovations. Support the work of other private sector water initiatives and collaborate with relevant UN bodies and intergovernmental organizations, especially including the UN Global Compact's Local Networks.

Public Policy
Exercise business statesmanship by participating in global and local policy discussions, recommending and supporting regulation and market mechanisms that drive water sustainability, and expanding the role of the private sector in supporting integrated water resource management. Partner with governments, businesses, civil society, and other stakeholders to advance the body of water stewardship knowledge, guidance, and tools.

Community Engagement
Understand water and sanitation impacts and challenges, advance water and sanitation education and awareness, and support local government and other initiatives in the development of adequate water and sanitation infrastructure.

Transparency and Disclosure
Be transparent in dealings with governments and others on water issues. Publish and share water strategies, targets, progress, and areas for improvement in relevant corporate reports. Communicate progress to the UN Global Compact and the CEO Water Mandate.

Source: Adapted from CEO Water Mandate 2011.

The CEO Water Mandate provides leadership, enhances understanding, and contributes to the advancement of corporate water stewardship practice. The Mandate develops and distributes guidance to fill gaps in knowledge, make complex concepts accessible, and expedites a transition from an emerging field of expertise to a mainstream practice with many informed and capable leaders and practitioners, widely distributed across a diverse set of companies, industry sectors, and geographies. Finally, the Mandate facilitates, builds, and maintains partnerships to address the world's most pressing water issues through corporate water stewardship.

Corporate Water Stewardship

The CEO Water Mandate's primary objective is to mobilize a critical mass of business leaders to address global water challenges through corporate water stewardship, in partnership with the United Nations, civil society organizations, governments, and other stakeholders. Generally, water stewardship refers to responsible management and future planning of water resources. The concept is rooted in the belief that all water users have a role to play in the sustainable management of shared freshwater resources. Jones et al. (2015) suggested that there seems to be no agreed-upon definition of water stewardship, but it is now increasingly common to describe corporate engagement with water use. The Alliance for Water Stewardship (AWS) defines water stewardship as **use of water that is socially equitable, environmentally sustainable, and economically beneficial, achieved through stakeholder-inclusive processes that involve site- and catchment-based activities** (Alliance for Water Stewardship 2013).

Importantly, the concept of corporate water stewardship addresses three main aspects of water stress: water scarcity, water quality, and access to water, sanitation, and hygiene (WASH). To effectively address the drivers of water-related business risk, water stewardship requires organizations to take shared responsibility for meaningful individual and collective actions that benefit people and nature (CEO Water Mandate 2015a).

Ultimately, corporate water stewardship is a comprehensive method of addressing critical water challenges and driving sustainable water management. In the early years of stewardship activities, specific activities consisted of:

1. measuring current water use;
2. assessing water landscape and water risks;
3. consulting stakeholders;
4. engaging supply chain;
5. establishing a water policy and setting corollary goals and targets;
6. implementing Best Available Technology;
7. factoring water risk into relevant business decisions;
8. measuring and reporting performance;
9. forming strategic partnerships; and
10. helping remediate any negative impacts a business causes or contributes to (UN Office of the High Commissioner 2011).

(Morrison and Gleick 2004; Gleick and Morrison 2006)

More recent Mandate language and stewardship principles are organized around the following objectives:

1. providing adequate water, sanitation, and hygiene for all employees;
2. increasing efficiency and reducing pollution in owned operations;
3. facilitating improved water performance in value chains;
4. advancing collective action and sustainable water management in river basins; and
5. achieving continuous dialogue with stakeholders.

An annual water questionnaire is prepared by CDP, a nonprofit organization formerly known as the Carbon Disclosure Project, which aims to reveal water-related risk in institutional investment portfolios and to reflect the effectiveness of corporate water stewardship strategies. In their 2013 annual report, *From Water Management to Water Stewardship*, CDP recognizes that

> Companies with robust water stewardship strategies are typically characterized by having a comprehensive knowledge of water use across their value chain and the impact (current and projected) that water-related issues have on their business and vice versa. More importantly, they have appropriate plans and procedures in place to mitigate risks that give adequate consideration to priorities of the local watershed in which they operate.

CDP's 2014 Global Water Report revealed that, of nearly 1,100 responding companies, 74 percent had evaluated how water quantity and quality could affect their growth strategy. However, of these, only 38 percent assessed water-related risk in both directly owned operations and their supply chain, and only 25 percent conducted detailed water risk assessment at the watershed level (CDP 2014).

In fewer than 10 years, the CEO Water Mandate has grown from its six founding members to include around 150 companies.[2] The UN Global Compact's assessment of impact found that 60 percent of Mandate-endorsing companies report on water use and 53 percent recognize and report on water scarcity in areas where they have operations or supply chain facilities (DNV GL and UN Global Compact 2015). Although it is not possible to conclude that improvements in water efficiency and reductions in water use are necessarily taking place in the most at-risk watersheds, Mandate-endorsing companies report saving an estimated 12.7 billion m^3 of water since they joined the initiative (DNV GL and UN Global Compact 2015).

To date, the CEO Water Mandate's activities have been largely focused on and supported by leading companies testing and implementing advanced water stewardship practices. To scale up the impact of corporate water stewardship practices globally and achieve an objective of a critical mass of companies practicing effective water stewardship, the Mandate—and stewardship activities overall—must convince multinational corporations, small and medium enterprises, and suppliers of all sizes, at all stages of development and in diverse cultures and geographies to understand, prioritize, and implement elements of corporate water stewardship (CEO Water Mandate 2015b).

Through the UN Global Compact's Local Networks and Mandate-endorsing company

---

2. https://ceowatermandate.org/about/endorsing-companies/.

supply chains, the Mandate stands to increase its reach substantially. It can continue to provide leading companies with cutting-edge tools and guidance to predict and overcome obstacles associated with innovative and inclusive water stewardship strategies. It can also empower companies that are committed to improving operations to take their first steps toward water stewardship in their own direct activities and key supply chains by simplifying existing guidance and making best practices accessible and central to operations.

Building Consensus: Key Water Stewardship Concepts and Terminology

Corporate water stewardship is an emerging discipline that demands collaboration, cooperation, and collective action. To prioritize, plan, and implement watershed-scale collective action projects, diverse stakeholders require a common language of key concepts and terminology to communicate with each other, operations managers and suppliers, and communities, non-governmental organization (NGO) partners, governments, investors, and consumers.

Companies typically come to understand their relationship with water in terms of their water footprint and water-related business risk. A water footprint assessment—which estimates the volume of water consumed and polluted in the production of a material or a product, or in the operation of an entire business, industry, or nation—can help to express the nature and extent of a company's dependence and impact on water resources (see, for example, Hoekstra 2008 and Hoekstra et al. 2011). It is also appealing as a basis for setting targets to reduce water use related to manufacturing processes or production of agricultural raw materials. For example, some companies are beginning to set a goal of offsetting their water use, or even seeking water "neutrality" similar to carbon neutrality or offsets. Such a target implies that a company can compensate for the negative impacts of its water footprint. However, there is no accepted standard for measuring negative impacts or defining which types and how much of any given activity is sufficient compensation. While a water footprint assessment can inform a risk assessment, a simple volumetric footprint measurement omits the local context necessary to characterize the risks related to water use, and obscures the difference in impact between using water from a source that's plentiful and using the same volume of water from a source that's overexploited or not readily replenished.

Water-related business risks generally fall into three broad and interrelated categories (Gleick and Morrison 2006; Morrison and Gleick 2004):

- *Physical risks* include scarcity, degraded source water quality, and flooding.
- *Regulatory risks* relate to inconsistent, ineffective, or poorly enforced public policy, particularly when a change in regulation or enforcement could disrupt production or lead to an unexpected cost of compliance.
- *Reputational risks* are faced by companies that overexploit or are perceived to overexploit water resources—including inefficient use, water pollution, excessive withdrawal, competition with other users, or other negligent water-related activities.

All three categories of risk can lead to business and financial impacts from increased operating costs, fines or unplanned capital expenditures, supply chain disruptions, damage to the value of a brand, or lost access to markets. If effective water strategy depends on a nuanced understanding of local watershed context, then a proliferation of seemingly interchangeable terms such as "water scarcity," "water stress," and "water risk" could be especially problematic for companies seeking to interpret geographic assessments and develop effective water initiatives.

In 2013, the CEO Water Mandate initiated a dialogue among organizations developing corporate water tools to see if a shared understanding could be reached on a number of key issues. The Alliance for Water Stewardship, Ceres, CDP, The Nature Conservancy, the Pacific Institute, Water Footprint Network, World Resources Institute, WWF, Global Reporting Initiative (GRI), PricewaterhouseCoopers, corporate water stewardship practitioners, water resource managers, and others in the scientific community provided expertise and insights. The paper resulting from this collaborative effort, *Driving Harmonization of Water Related Terminology*, describes critical distinctions between key terms such as water withdrawal and consumption, for example. It also explains that when assessing the nature and severity of water-related challenges, "water scarcity," "water stress," and "water risk" refer to three distinct concepts and should not be used interchangeably (Figure 1.1). The next step is to incorporate the resulting definitions into organizational efforts wherever possible (CEO Water Mandate 2014a).

Making Water Stewardship Accessible: Producing and Distributing Tools and Guidance

Water stewardship requires specialized capabilities—such as watershed assessment and collective action—beyond those that commonly exist on corporate environment, safety, and health teams. In addition to developing clear terminology and definitions, the CEO Water Mandate produces tools and guidance with contributions by Mandate-endorsing companies and expert advisors, and helps to promote water stewardship tools produced by other leading organizations in the field. The Mandate works not only to put stewardship concepts into practice, but also to introduce complex concepts, provide access to simplified or introductory guidance, and drive adoption of best practice at the facility level.

Many existing product and material standards and certifications address water together with other social and environmental impacts, but are not necessarily aligned with best practices for water stewardship. In contrast, the AWS standard does incorporate special expertise in corporate water stewardship, but in practice it does not explicitly address tradeoffs with other environmental priorities.

At Stockholm World Water Week in 2015, the Mandate introduced a Water Stewardship Toolbox.[3] The Toolbox is organized around the Mandate's Water Stewardship Progression, making guidance readily available for corporate entities working on water efficiency, water quality, and water and sanitation in the workplace, and for advanced leaders of complex multi-stakeholder water stewardship initiatives (Figure 1.2).

---

3. http://www.ceowatermandate.org/toolbox.

FIGURE 1.1 HOW KEY CONCEPTS AND TERMS RELATE TO ONE ANOTHER.
Source: CEO Water Mandate 2014a.

Managing Water-Related Impacts

There are many ways to improve the environmental performance of companies producing goods and services, from voluntary sustainability standards for raw material production to corporate codes of conduct governing processing and manufacturing facilities. The Mandate's toolbox contains a growing collection of resources that support companies that have not yet fully addressed issues around access to safe water and sanitation, treating wastewater, or improving water efficiency in their direct operations. These are first steps that position companies for more advanced water stewardship and external engagement.

Supply Chain Water Stewardship

In many industries, water-related business risks and impacts in supply chains are more substantial than those in their direct operations. For example, in the apparel sector, cotton cultivation and dyeing textiles represent the largest water footprint and the most pressing water-related issues, but these impacts occur outside the direct operational control of most brands and retailers. Some companies rely on supplier codes of conduct and systems of audits, rewards, and sanctions to manage the social and environmental performance of suppliers. Codes of conduct are becoming more common and more complex. Changing regulations and consumer preferences create incentives to increase standards, track new metrics, and set more aggressive social and environmental targets, but a condition sometimes called "audit fatigue" can occur when a supplier has to comply with more than one client company's standards. Such conflicting priorities and stan-

Figure 1.2 The Water Stewardship Progression.
Source: UN Global Compact 2017.

dards are driving greater harmonization across industry sectors when there are opportunities to make a common standard possible.

Codes of conduct can only be reliably enforced under the terms of a contract with a direct supplier, so many only apply to first-tier suppliers. Meanwhile, actors in more extended supply chains may not be obligated to meet any such standards. As a result, some companies rely on voluntary sustainability standards—such as Forest Stewardship Council (FSC) or Marine Stewardship Council (MSC) certifications—to manage sustainability issues related to high-risk or high-impact raw materials or industrial processes like forest products or commercial fishing.

For companies with complex extended supply chains, limited traceability, poor understanding of the nature and location of diverse supply chain products and processes, failure to evaluate the water-related risks that affect suppliers' operations can increase costs, lead to fines and penalties, and limit or disrupt production. Furthermore, inadequate or inequitable access to water, sanitation, and hygiene (WASH) in the workplace or lack of access to WASH services in communities where workers and their families reside can reduce productivity, increase absenteeism or turnover, worsen the spread of preventable waterborne illnesses, and create other threats to human health and well-being. These, in turn, can affect corporate reputation and profitability, which adds to the incentive to develop stewardship standards and practices.

Watersheds

The private sector increasingly recognizes the need to evaluate site-level water use in the context of local water conditions in order to inform and prioritize efficiency targets for different locations. For example, companies can manage risk more effectively in direct operations and supply chains by giving higher priority to efficiency improvements for water-intensive activities in drought-prone locations than for similar operations where water resources are more plentiful.

Tools like the World Resources Institute (WRI) Aqueduct Water Risk Atlas[4] and Worldwide Fund for Nature (WWF) Water Risk Filter[5] can provide information on where supplier facilities may face the most severe water-related risks. Once a geographic area has been identified as a high priority for corporate water stewardship, local team members, decision makers, and suppliers can develop a better understanding of the physical conditions and sociopolitical forces shaping the water management decisions that affect specific locations.

For directly owned and operated facilities and for supplier locations alike, water-related risks sometimes originate not from on-site activities that farms or manufacturing facilities themselves control, but rather from physical or political conditions outside the direct influence of both brands and suppliers. For owned operations, companies can and should assess watershed context in detail (using a tool like GEMI water management risk questionnaire[6]) and take steps to participate in integrated resource management as water users, rate payers, and members of their communities.

Collective Action

Companies wishing to operate sustainably must participate in the stewardship of common resources, especially in stressed watersheds where owned operations or strategic suppliers are located. Until they assess local watershed context, companies primarily act alone, often focused on reducing water use at direct operations or key suppliers. However, until sustainable water management is achieved in the watersheds where they do business, companies can continue to face water-related risks.

Forward-thinking companies understand that working with other stakeholders at the watershed scale, outside the fence lines of direct operations or supply chain farms or factories, may be required to address root causes of resource scarcity, accessibility, or source water contamination, which can increase costs or disrupt operations. For example, the Beverage Industry Environmental Roundtable (BIER), a coalition of business leaders in an industry that faces substantial water-related risks, has acknowledged that in some locations, watershed-level interventions may in fact be more effective at mitigating water-related risk than facility-level water use efficiency or other activities (BIER 2015). To assist companies in prioritizing their efforts, BIER has proposed developing a decision support tool that would give higher priority to interventions outside the fence line than to internal efficiency or water-quality improvements in certain circumstances.

---

4. http://www.wri.org/applications/maps/aqueduct-atlas.
5. http://waterriskfilter.panda.org/.
6. http://waterplanner.gemi.org/questionnaire.asp.

To improve the likelihood and effectiveness of collective action, the CEO Water Mandate has helped develop water stewardship initiatives that bring together the private sector, governments, and communities in support of sustainable water management for shared benefits (CEO Water Mandate 2015c). The Mandate defines collective action as coordinated engagement between interested parties within an agreed-upon process in support of common objectives (CEO Water Mandate 2015c, p. 7).

A key enabling function of the CEO Water Mandate is the Water Action Hub (the Hub),[7] a web-based tool that originated from the Mandate's Collective Action work in 2012. Envisioned as a matchmaking platform for prospective participants in regional water stewardship initiatives, the Hub now contains information for around 400 organizations with more than 200 projects around the world. It promotes collaboration among groups of companies and/or external stakeholders to address local water challenges, helping potential collaborators to find each other and to join forces on water-related collective action projects that improve water management in regions of critical interest.

The CEO Water Mandate Guide to Water-Related Collective Action (2013) provides detailed explanations and best practices for five elements of collective action:

1. scoping water challenges and action areas that collective action will address;
2. identifying and characterizing the interested parties with the potential to influence key problems;
3. embedding the challenges, action areas, and interested parties in a level of engagement that will optimize the effort and shared benefits of participants;
4. designing the collective action engagement; and
5. structuring and managing the collective action.

Importantly, the CEO Water Mandate's Guide to Responsible Business Engagement with Water Policy (2010) outlines principles that are needed to maintain integrity at all stages of collective action initiatives, not limited to those involving policy engagement. These principles include striving for inclusiveness and integrated approaches, setting clear objectives to advance sustainable water management for shared benefits, and maintaining transparency (CEO Water Mandate 2010).

Corporate Water Disclosure

A core concept in water stewardship is data sharing and transparency. Such transparency contributes to the credibility of the CEO Water Mandate and endorsing companies' water stewardship efforts, helps to mainstream adoption of best practices, and keeps stakeholders informed of strategies, progress, and opportunities for improvement. In fact, transparency is itself one of the six core elements defined by the Mandate (see Box 1.2). CEO Water Mandate guidance on transparency includes an early summary of corporate water accounting methods and tools published in 2010 (CEO Water Mandate 2010). More recently, the Mandate's Corporate Water Disclosure Guidelines (2014b) highlight

---

7. http://www.wateractionhub.org.

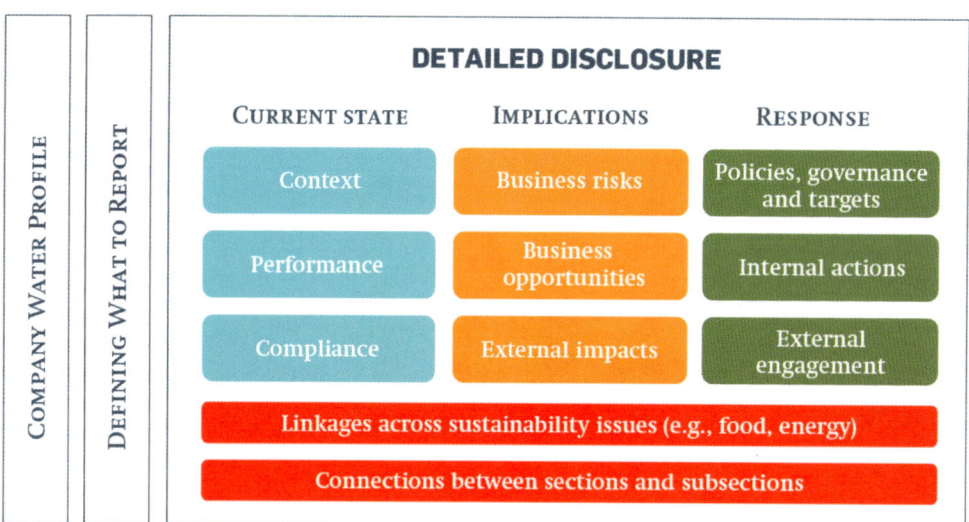

FIGURE 1.3 CORPORATE WATER DISCLOSURE FRAMEWORK.
Source: CEO Water Mandate 2014b.

the misspent resources resulting from the proliferation of assessment tools and sustainability questionnaires and promote a common approach to reporting on water (Figure 1.3).

A robust, holistic approach to corporate water accounting that measures the effectiveness of water stewardship initiatives in terms of sustainable water management outcomes at the watershed, national, and global scale would help farms and facilities understand and manage risks and opportunities, reduce the reporting burden on companies, and reinforce the potential of the private sector—and the CEO Water Mandate itself—to contribute to sustainable water resource management.

Ensuring Integrity in Water Stewardship Initiatives

Corporate water stewardship necessarily brings together a diverse range of companies, NGOs, governments, and communities. If environmental and social benefits are perceived to be in conflict with business objectives, companies might choose to engage with government to instead pursue short-term benefits, opaque deals, or special treatment. However, when water risk is understood in relation to sustainable economic development, improved supply-chain capacity, and emerging market opportunities, a strong business case in favor of sustainable water management and innovative public–private sector partnerships emerges (CEO Water Mandate 2015c).

To promote effective, transparent, and mutually beneficial corporate water stewardship initiatives that serve public as well as private interests, the CEO Water Mandate's 2015 Guide for Managing Integrity in Water Stewardship raises awareness of the potential pitfalls of such collective action (Table 1.1). The Guide also provides a framework to help emerging partnerships proceed with high levels of accountability and transparency and ensure that all stakeholders truly benefit, including the most vulnerable populations (CEO Water Mandate 2015a).

TABLE 1.1 Integrity Risk Areas

	Risk	Description
Risks related to participants	Track record	Ideal participants have good reputations for acting with integrity, including compliance with policy and regulation, transparency, professional behavior, ethics, and values. A poor track record may have a negative impact on the credibility of the initiative and its participants.
	Representation	The selection of participants and representatives should include all stakeholders affected by the initiative and those influential to the attainment of its objectives. Proxies should possess the mandate, legitimacy, and authority to meaningfully represent stakeholders. Otherwise, others may pursue vested interests or undermine informed decision making, accountability, credibility, inclusiveness, responsiveness, and ultimately the delivery of beneficial outcomes.
	Intent and incentives	Participant motivations should be aligned with long-term goals, addressing shared water risks and advancing sustainable water management, to prevent misuse of the partnership in pursuit of self-interest or short-term benefits.
	Capability	Participants require the capacity to carry out key functions in a partnership, including implementing projects, monitoring progress, controlling processes, and/or holding other participants accountable. Unless participants engage meaningfully, initiatives are susceptible to manipulation, and poorly conceived and executed projects may result.
	Conduct	Participants should be expected to demonstrate commitment to the initiative and follow agreed-upon procedures. Superficial engagement or non-constructive conduct of participants jeopardizes fair process and outcomes.
	Continuous engagement	Participants should maintain long-term commitment and engagement. Without ongoing accountability, effective implementation is undermined, jeopardizing positive outcomes.
Risks related to governance	Planning and design	Rationale, focus, content, and governance of the initiative should be well-defined. Inadequate, incomplete, or inappropriate planning processes can discourage participant engagement, support weak governance structures, risk ineffective collective action outcomes, and create opportunities for unethical behavior.
	Stakeholder engagement	Exclusion or omission of affected stakeholders negatively affects decision-making processes, biases objectives, and undermines the credibility, accountability, and responsiveness of partnerships.
	Responsibilities, decision making, and communication	Poorly informed participants, weak reporting mechanisms, unclear responsibilities, lack of oversight, and collusion among key participants also undermining the accountability and outcomes of water stewardship initiatives.
	Financial management	Effective financial planning, allocations, arrangements, and transactions are essential. Lack of transparency and mistrust in the financial management of the partnership can enable misuse of funds, nurture corruption, or allow abuse of influence in service of self-interest.
	Monitoring, evaluation, and learning	Sufficient and transparent monitoring, evaluation, and learning systems improve the effectiveness of initiatives, and are necessary to prevent dishonest claims of positive outcomes and failure to honor commitments.
Risks Related to Context and Outcomes	Capture: organizational resources and investment	Water stewardship initiatives should be analyzed and aligned with public policy priorities and targets, to avoid diverting organizational resources and public funds away from issues of greatest local priority and societal benefit, and toward addressing the priorities of private or foreign entities.
	Capture: regulatory action, policy, and water	Government institutions are mandated to serve the public interest and should fairly balance legitimate interests. The types of government institutions and the specific representatives that engage in a water stewardship initiative have varying degrees of influence on policy and regulatory processes, creating risks related to policy and regulatory capture.
	Perverse outcomes	Social and environmental impacts must be adequately evaluated and safeguards established to prevent harm. Perverse outcomes may arise from poorly informed water stewardship initiatives, damaging social equity or environmental assets, or undermining institutional performance.
	Limited contribution to SWM	Water stewardship initiatives should seek to affect the root causes of water challenges, not only the symptoms of poor water management. Furthermore, partnerships with benefits conceived as "offsets" for a company's negative impacts on society and the environment are sometimes criticized as greenwashing. In cases where negative individual impacts continue unabated or root causes of water stress go unaddressed, some participants may use the partnership to disguise the pursuit of self-interest to the detriment of other stakeholders.

Source: Adapted from CEO Water Mandate 2015a.

Making an Impact: Sustainable Management of Water and Sanitation for All

When the UN agreed to its eight Millennium Development Goals for the period between 2000 and 2015, the relationship between environmental issues and sustainable development was not prominently featured. Today, a wider group of stakeholders understands the vital role water and sanitation play in the economy, society, and the environment. The process of defining the recently updated 2030 Sustainable Development Goals (SDGs) was more inclusive, and the resulting goals are relevant to development concerns facing all nations, including the developed world. The adoption of the SDGs introduces a compelling framework for collective action by the private sector, government, and civil society through which it becomes possible to address social and environmental issues that inhibit economic development and shared prosperity (United Nations 2015).

Of the 17 new goals, SDG 6 is dedicated exclusively to ensuring availability and sustainable management of water and sanitation for all. Box 1.3 outlines the targets underpinning SDG 6, which represent a shift to a more holistic approach including issues associated with water scarcity, quality, management, and ecosystems.

Companies seeking to manage water-related business risks can and should contribute to improved water and sanitation management and governance that is also in the public interest. If done responsibly, integrating private sector action into global policy frameworks and local implementation practices makes it possible for companies to contribute considerable resources and expertise to the achievement of SDG 6. In keeping with its role in the UN Global Compact, the CEO Water Mandate is well positioned to build consensus within the water stewardship community around metrics and indicators of progress, and to orient corporate water stewardship initiatives toward the achievement of the SDG 6 targets. CEO Water Mandate tools and guidance can inform the development of corporate water strategies; and the Mandate's network of companies, expert partners, and UN Local Networks can accelerate positive outcomes to achieve sustainable management of water and sanitation for all (CEO Water Mandate 2015a).

Conclusion

In the long run, those who do not use power in a manner that society considers responsible will tend to lose it.
—Keith Davis, Management Theorist
1971

Wherever economic growth outpaces the capacity of government to balance development with protection of shared natural resources, companies may face threats to their long-term sustainability. As the roles of governments, companies, and civil society continue to evolve in response to changing climate, resource availability, and stakeholder expectations, larger segments of the private sector are working to align corporate water stewardship initiatives with both local priorities and global goals. The CEO Water Mandate is facilitating this opportunity to bring corporate capabilities and resources to water

BOX 1.3 UN Sustainable Development Goal 6

Ensure availability and sustainable management of water and sanitation for all

6.1 – Access to water
By 2030, achieve universal and equitable access to safe and affordable drinking water for all.

6.2 – Access to sanitation
By 2030, achieve access to adequate and equitable sanitation and hygiene for all and end open defecation, paying special attention to the needs of women and girls and those in vulnerable situations.

6.3 – Pollution prevention
By 2030, improve water quality by reducing pollution, eliminating dumping and minimizing release of hazardous chemicals and materials, halving the proportion of untreated wastewater and substantially increasing recycling and safe reuse globally.

6.4 – Sustainable withdrawals and efficiency
By 2030, substantially increase water-use efficiency across all sectors and ensure sustainable withdrawals and supply of freshwater to address water scarcity, and substantially reduce the number of people suffering from water scarcity.

6.5 – Integrated water resource management
By 2030, implement integrated water resources management at all levels, including through transboundary cooperation as appropriate.

6.6 – Ecosystem health
By 2020 protect and restore water-related ecosystems, including mountains, forests, wetlands, rivers, aquifers and lakes.

6.a – International cooperation
By 2030, expand international cooperation and capacity-building support to developing countries in water- and sanitation-related activities and programs, including water harvesting, desalination, water efficiency, wastewater treatment, recycling and reuse technologies.

6.b – Community participation
Support and strengthen the participation of local communities for improving water and sanitation management.

Source: United Nations 2015. Resolution adopted by the General Assembly on 25 September 2015.

stewardship by developing and disseminating tools and guidance, deepening understanding and engagement at the local level, and promoting collective action to achieve sustainable management of water resources.

Companies can no longer deny responsibility for their own water impacts or those of their suppliers. In the face of increasing risks related to climate change, groundwater depletion, extreme events and disasters, and the impacts of inadequate water and sanitation on human health and well-being, consumers and investors increasingly expect companies to take a more active approach to environmental sustainability. In the coming years, the strategies and methods to do so will gain more traction as commitments from the private sector expand.

References

Alliance for Water Stewardship (AWS). 2013. The AWS International Water Stewardship Standard. http://a4ws.org/our-work/aws-system/the-aws-standard/.

Beverage Industry Environmental Roundtable (BIER). 2015. Performance in Watershed Context: Concept Paper. http://www.bieroundtable.com/#!/c16n5.

Bowen, H. R. 1953. *Social Responsibilities of the Businessman*. New York: Harper.

CDP. 2013. From Water Management to Water Stewardship: Annual Report. https://www2.deloitte.com/us/en/pages/operations/articles/cdp-us-water-report.html.

———. 2014. From Water Risk to Value Creation: CDP Global Water Report. https://b8f65cb373b1b7b15feb-c70d8ead6ced550b4d987d7c03fcdd1d.ssl.cf3.rackcdn.com/cms/reports/documents/000/000/646/original/CDP-Global-Water-Report-2014.pdf?1470394078.

CEO Water Mandate. 2010 (November). Guide to Responsible Business Engagement with Water Policy. https://ceowatermandate.org/files/Guide_Responsible_Business_Engagement_Water_Policy.pdf.

CEO Water Mandate. 2011. The CEO Water Mandate: An Initiative by Business Leaders in Partnership with the International Community. http://ceowatermandate.org/files/CEO_Water_Mandate.pdf.

CEO Water Mandate, Ross Strategic, Pegasys Consulting and Development, Pacific Institute, and Water Futures Partnership. 2013 (September). Guide to Water-Related Collective Action. https://www.unglobalcompact.org/docs/issues_doc/Environment/ceo_water_mandate/Water_Guide_Collective_Action.pdf.

CEO Water Mandate, Alliance for Water Stewardship, Ceres, CDP, The Nature Conservancy, Pacific Institute, Water Footprint Network, World Resources Institute, and WWF. 2014a (September). Driving Harmonization of Water-Related Terminology: Discussion Paper. http://ceowatermandate.org/files/MandateTerminology.pdf.

CEO Water Mandate, Pacific Institute, CDP, World Resources Institute, PricewaterhouseCoopers. 2014b (September). Corporate Water Disclosure Guidelines: Toward a Common Approach to Reporting Water Issues. http://ceowatermandate.org/files/Disclosure2014.pdf.

CEO Water Mandate and Water Integrity Network (WIN). 2015a (August). Guide for Managing Integrity in Water Stewardship Initiatives: A Framework for Improving Effectiveness and Transparency. http://ceowatermandate.org/files/integrity.pdf.

CEO Water Mandate. 2015b (October). UN Global Compact CEO Water Mandate: 2016–2018 Strategic Plan. https://ceowatermandate.org/files/CEO_Water_Mandate_2016-2018_Strategic_Plan-FINAL.pdf.

CEO Water Mandate, WWF, and WaterAid, 2015c (September). Serving the Public Interest: Corporate Water Stewardship and Sustainable Development. http://pacinst.org/wp-content/uploads/2015/09/Corporate-Water-Stewardship-and-SDG-6-on-Water-and-Sanitation.pdf.

Davis, K., and R. Blomstrom. 1971. *Business, Society, and Environment: Social Power and Social Response*. New York: McGraw-Hill.

DNV GL and UN Global Compact. 2015. Impact: Transforming Business, Changing the World. http://globalcompact15.org/report.

Friedman, M. 1962. *Capitalism and Freedom*. Chicago, IL: University of Chicago Press.

Gleick, P. H., and J. Morrison, 2006. Water Risks that Face Business and Industry. In *The World's Water 2006–2007*, pp. 145–174. Washington, DC: Island Press.

Hoekstra, A. Y. 2008. Water Neutral: Reducing and Offsetting the Impacts of Water Footprints. Value of Water: Research Report Series No. 28. http://waterfootprint.org/media/downloads/Report28-WaterNeutral_1.pdf.

Hoekstra, A. Y., A. K. Chapagain, M. M. Aldaya, and M. M. Mekonnen, 2011. The Water Footprint Assessment Manual: Setting the Global Standard. London: Earthscan. http://waterfootprint.org/media/downloads/TheWaterFootprintAssessmentManual_2.pdf.

Jones, P., D. Hillier, and D. Comfort. 2015. Corporate Water Stewardship. *Journal of Environmental Studies & Sciences* 5(3): 272–276.

Kreps, T. 1962. Measurement of the Social Performance of Business. *The Annals of the American Academy of Political and Social Science* 343(1): 20–31. http://ann.sagepub.com/content/343/1/20.full.pdf.

Morrison, J., and P. H. Gleick, 2004. Freshwater Resources: Managing the Risks Facing the Private Sector. http://pacinst.org/wp-content/uploads/2013/02/business_risks_of_water3.pdf.

UN Global Compact. 2007. The Global Compact Leaders' Summit 2007. Facing Realities: Getting Down to Business. Palais des Nations, Geneva, 5-6 July 2007. https://www.unglobalcompact.org/docs/news_events/8.1/GC_Summit_Report_07.pdf.

———. 2017. Establishing Your Water Stewardship Journey. 2017. https://www.unglobalcompact.org/take-action/action/water-stewardship-journey.

———. 2016. Who We Are. https://www.unglobalcompact.org/what-is-gc/mission/principles.

United Nations (UN). 1999 (February). Secretary-General Proposes Global Compact on Human Rights, Labour, Environment, in Address to World Economic Forum in Davos. Press Release. SG/SM/6881. http://www.un.org/press/en/1999/19990201.sgsm6881.html.

———. 2015. Transforming Our World: The 2030 Agenda for Sustainable Development. Resolution adopted by the General Assembly on 25 September 2015. http://www.un.org/ga/search/view_doc.asp?symbol=A/RES/70/1&Lang=E.

CHAPTER 2

A Human Rights Lens for Corporate Water Stewardship: Toward Achievement of the Sustainable Development Goal for Water

The Sustainable Development Goals—Ensure Availability and Sustainable Management of Water and Sanitation for All

Mai-Lan Ha

Introduction

In September 2015, the UN General Assembly adopted a new set of international development objectives called the Sustainable Development Goals (SDGs) to guide the implementation of development priorities through 2030. With 17 goals and 169 targets, the SDGs are more complex than the Millennium Development Goals (MDGs) that they replace. Although the MDGs provided a starting point for action, they are generally recognized to be incomplete because they focus predominantly on issues facing developing countries—such as eradicating extreme poverty and achieving universal primary education—while not providing avenues or priorities for substantive action by developed countries. The new SDGs offer a more coherent framework that takes into account both the complexity and interlinkages inherent in sustainable development and the opportunity for action by all countries and sectors. This chapter offers a review of the role that the human rights to water and sanitation play in the water-related targets of the SDGs and ways in which the business community can integrate these rights into their larger water stewardship efforts.

The sustainable management of water resources and the goal of ensuring water and sanitation for all are central to the achievement of a number of the SDGs, including those related to eradicating hunger, improving child mortality, and ensuring environmental sustainability. SDG 6 acknowledges the interlinked nature of water to other SDGs—including increasing access to water, sanitation, and hygiene for populations currently not served or underserved; and addressing issues of water stress, water quality, integrated water management, and ecosystems. There is also recognition that meeting an SDG on water requires that all societal actors take action by committing resources, skills, and expertise. Two of the targets of SDG 6 focus on the means of implementing the SDG, including increasing cooperation and capacity building, as well as improving the ability of local communities to participate in water management planning and decisions. The diagram in Figure 2.1 shows how SDG 6 supports a number of other SDG goals.

While all sectors of society will have to be engaged to meet the SDG targets, businesses have a clear role to play given their dependency on water and their impacts on water supplies and quality. Many companies are already expanding their engagement in water issues through a variety of corporate water stewardship practices. Considerable effort in recent years in defining and codifying such practices has already been coordinated by UN agencies and partnerships—such as by the CEO Water Mandate, under the UN Global Compact (see, for example, Chapter 1). Many of these efforts directly align with the water-related objectives and means of implementation of the SDGs. These practices can be further strengthened by integrating business responsibility for the human rights to water and sanitation into corporate water stewardship practices, thereby enhancing the social dimensions of stewardship.

The Business Case for Action on Water

Wherever we look, businesses today touch upon aspects of water, either through their direct operations or in their supply chains that rely on water or produce wastewater, or in their role as water service providers. Given the importance of water, the business case for corporate action is generally based upon a number of factors, as well as the characteristics of fresh water.

Water Is a Non-Substitutable Resource

Water or the services it provides or enables is an indispensable input for most businesses. Managing secure access to water in the quantities needed, of the quality required, and at the right time and place is often essential for economic viability. This becomes increasingly important as "peak water" pressures on the finite quantities of water available increase in many regions (Gleick and Palaniappan 2010).

Corporate Water Stewardship 23

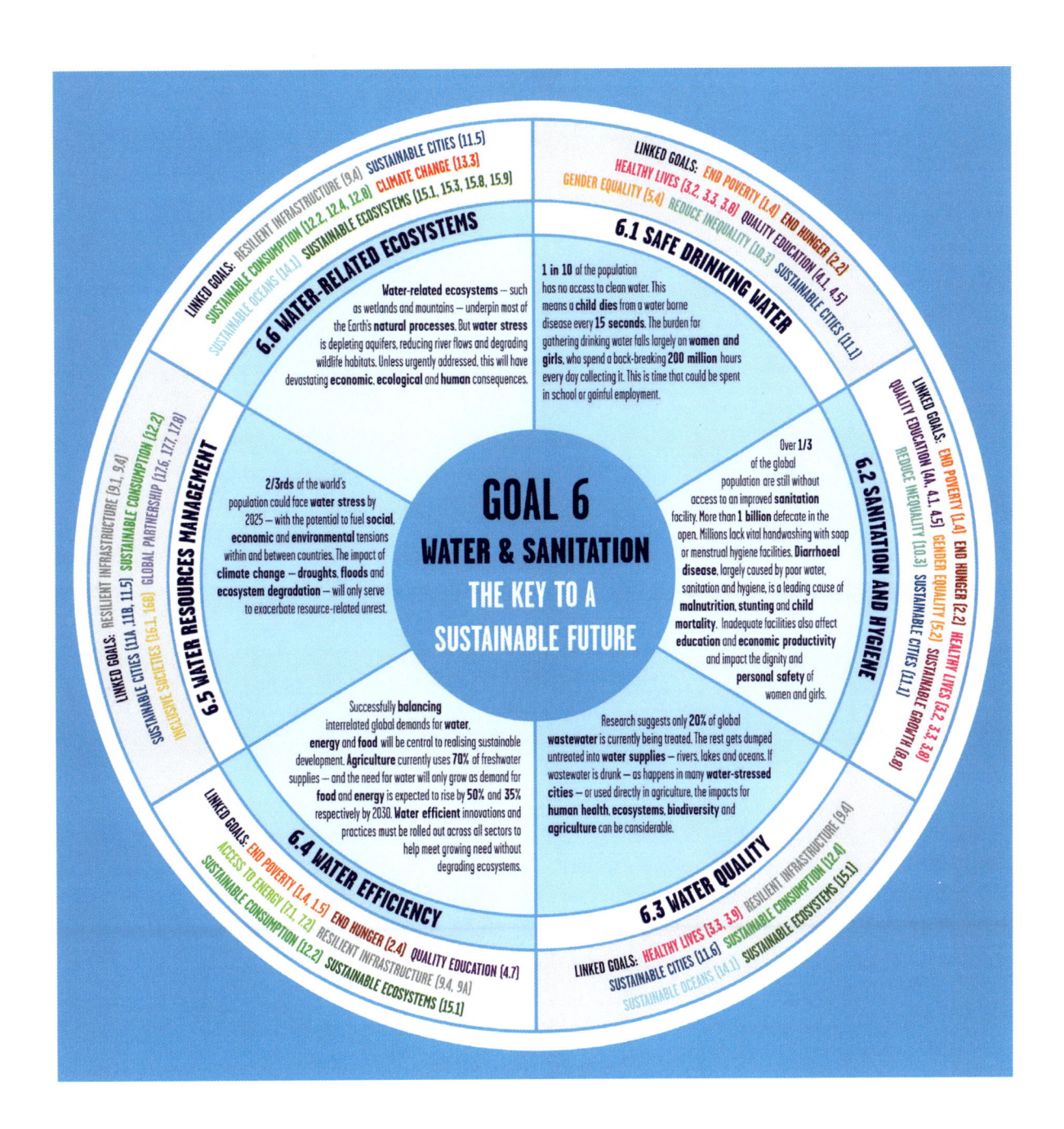

FIGURE 2.1 RELATIONSHIP BETWEEN SDG 6 AND OTHER SUSTAINABLE DEVELOPMENT GOALS.
Source: WaterAid and Unilever 2015.

Water in the Value Chain

Water plays an important role throughout the whole value chain of industrial production and commercial activity, as well as in multiple interactions with communities and stakeholders. Businesses have an interest in and responsibility to understand these complex relationships and conduct their activities accordingly.

Water and the License to Operate

Ensuring the company's local legal and social license to operate in a specific location increasingly depends on how communities understand and view local business water behavior.

Business Operations Depend on Water

Preventing or reacting to operational crises resulting from the inadequate availability, supply, or quality of water or water-dependent inputs in a specific location is an increasing challenge.

- *Competitive advantages:* Companies can gain an advantage over competitors because of stakeholder perception that businesses are implementing effective water stewardship practices.
- *Financial advantages:* Sustainable water use and management can assure investors and markets that business operations will continue to be profitable by securing water availability for operations and reducing water-related costs.
- *Corporate values around equity and sustainable development:* Upholding corporate values based on sustainable and equitable development can contribute to the well-being of the catchments, ecosystems, and communities in which the company operates.

Businesses have a central role to play in ensuring sustainable development policies are implemented because of their critical and active role in transforming resources into products and services required by societies. This case is further strengthened by the realization that business efforts toward sustainable development can influence their long-term survival and success. The case revolves around a number of areas:

- *Ensuring good water governance:* Businesses that depend upon water realize that meeting development goals necessitates addressing aspects of water sustainability more broadly than simply ensuring access to supply—including improving water governance systems and addressing water security and water quality—all issues of importance for addressing water-related business risk.
- *Healthier employees:* Business actions to ensure adequate water and sanitation in the workplace provide the opportunity for companies to ensure their employees are sufficiently cared for. Healthier employees contribute to overall long-term company productivity through less frequent sick days and absence of costs associated with the need to replace or train new employees (CEO Water Mandate et al. 2014).

- *Vibrant communities:* Beyond their employees, businesses also realize that healthy communities have a positive impact on their business. Businesses are engaging in activities that focus not only on employees, but increasingly on the families of their employees and communities at large. Healthy families ensure a high level of productivity in the workplace, while vibrant communities often serve to bolster a company's social license to operate and a healthy customer base.
- *Triple bottom line:* Businesses realize that a strong case can be made that helping to achieve sustainable development goals offers opportunities to create innovative new products and markets.

These elements make it clear that ensuring adequate water for employees, communities, and society is needed for the long-term well-being of businesses. Not taking action, on the other hand, is increasingly untenable, leading to the potential for greater conflict over water resources, decreased social license to operate, and increased reputational risks.

State Recognition of the Human Rights to Water and Sanitation

Underpinning the achievement of SDG 6 on water and sanitation is the recognition of the importance of the human rights to water and sanitation (HRWS). In 2010, the UN General Assembly officially recognized the rights to water and sanitation as fundamental human rights (UN Global Compact and Deloitte 2010). Table 2.1 defines these rights.

With the recognition of water and sanitation as human rights, governments across the world are now tasked with meeting their obligations and responsibilities under the UN declarations. Today, over 80 countries have recognized either explicitly or implicitly the rights to water and sanitation for their citizens through constitutional amendments and national legislation, or implicitly through interpretations of provisions such as those related to the right to life, the right to health, or the right to a safe environment (CEO Water Mandate et al. 2012). Regionally, countries in Africa and South America have been at the forefront of adopting such legislation. It should also be noted that in the majority of cases, legislation has focused on access to drinking water, with less recognition being given to the right to sanitation.[1]

---

1. The following sections draw on three bodies of work. The first is a sourcebook of national laws and policies relating to water, *The Human Right to Safe Drinking Water and Sanitation in Law and Policy: A Sourcebook* (WASH United et al. 2012). The second is an analysis prepared by the Pacific Institute and the UN CEO Mandate of national legislation in countries that have explicitly or implicitly recognized the HRWS, including South Africa, Kenya, Indonesia, Costa Rica, India, and Belgium. South Africa, Kenya, and Belgium explicitly recognize the HRWS, while Indonesia, India, and Costa Rica implicitly recognize the right. The countries chosen include countries recognized for their progressive water laws and those which may provide an indicator of regional trends. Finally, it includes an examination of recent national jurisprudence relating to the HRWS. These are drawn from case examples of jurisprudence accumulated by The Center on Housing Rights and Evictions in *Legal Resources for the Right to Water and Sanitation: International and National Standards*, 2nd Edition, January 2008.

TABLE 2.1 Dimensions of the Human Rights to Water and Sanitation

Dimension	Definition
Availability	Water and sanitation facilities must be present in order to meet peoples' basic needs. This means a supply of water that is sufficient and continuous for personal and domestic uses, which ordinarily include drinking and food preparation, personal hygiene, washing of clothes, cleaning, and other aspects of domestic hygiene, as well as facilities and services for the safe disposal of human excreta (i.e., urine and feces).
Accessibility	Water and sanitation facilities must be located or constructed in such a way that they are accessible to all at all times, including to people with particular needs (such as women, children, older persons, or persons with disabilities). Accessibility is particularly important with regard to sanitation, as facilities that are not easily accessible are unlikely to be used and may raise safety risks for some users, especially women and girls.
Quality and safety	Water must be of a quality that is safe for human consumption (i.e., drinking and food preparation) and for personal and domestic hygiene. This means it must be free from microorganisms, chemical substances, heavy metals, and radiological hazards that constitute a threat to a person's health over a lifetime of consumption. Sanitation facilities must be safe to use and prevent contact between people and human excreta.
Acceptability	Water and sanitation facilities must meet social or cultural norms from a user's perspective; for example, regarding the odor or color of drinking water, or the privacy of sanitation facilities. In most cultures, gender-specific sanitation facilities will be required in public spaces and institutions.
Affordability	Individual and household expenditure on water and sanitation services, as well as associated hygiene, must be affordable for people without forcing them to resort to other unsafe alternatives and/or limiting their capacity to acquire other basic goods and services (such as food, housing, or education) guaranteed by other human rights.

Source: CEO Water Mandate et al. 2017.

Regardless of how various countries have come to formally recognize the HRWS, there have been clear trends in what this adoption means for water-using companies (CEO Water Mandate et al. 2012), including public trusteeship of water resources, prioritization of water uses that emphasize meeting human needs before other needs, the protection of water resources leading to increased regulations to limit water resource degradation, and increased participation in water resource management.

Public Trusteeship of Water Resources

Governmental recognition that every individual must have access to safe water in order to survive and thrive has led countries to designate water as a public good under public control in order to ensure that it is managed in an equitable and sustainable manner for all. This has been codified in water laws, constitutions, or judicial decisions in many countries. Water use (including withdrawals, diversions, and discharge) is managed through a wide range of institutional systems that differ depending on societal sectors, governmental structures, and industry makeup.

Prioritization in Water Use

One of the most crucial implications of state adoption of the HRWS is recognition that water must first be used to meet basic human or domestic needs, prior to it being made available for other uses (such as for agribusiness or industry). Governments have codified this through legislation that explicitly prioritizes water for human needs or through the creation of a system that tasks water authorities with determining a "reserve requirement" to ensure adequate water is set aside for human and ecological needs. In turn, this has led to systems that require permits for all uses outside of those to meet basic human needs, coupled with the ability of governing authorities to amend or cancel these water use permits in times of water scarcity, drought, and emergencies (such as the Kenya Water Act of 2002). This shift toward explicit prioritization has also resulted in case law requiring water authorities to change water allocations to meet human needs before providing water for businesses, as well as the suspension of company activities for fear that the company's water use would affect communities' ability to access water. For example, Pakistan's High Court in Karachi found that Nestle's proposed bottling plant would diminish underground aquifers affecting local communities' water needs (High Court of Sindh at Karachi 2004).

> "The future development agenda must aim at universal enjoyment of the human right to water and sanitation by every single human being."
>
> —*Former Special Rapporteur on the Human Right to Safe Drinking Water and Sanitation, Catarina de Albuquerque*

Protection of Water Resources

Legal recognition of the HRWS has put the onus on governments to better protect water resources in order to ensure that adequate water is available for all segments of the population. To do so, some countries have adopted precautionary and "polluter pays" principles, increased regulations aimed at preventing water resource degradation, adopted legislation calling on those who do pollute to bear the costs of remediation, and fines or imprisonment for those found guilty of purposefully polluting water resources. For example, South Africa requires any person or entity engaged in an activity that may cause pollution to take "reasonable measures" to prevent pollution from occurring, continuing, or reoccurring. Once pollution manifests, the polluter is responsible for all clean-up, even if the entity is no longer engaged in the activity. In 2012, the North Gauteng High Court in Pretoria utilized the anti-pollution clause of the National Water Act—the cornerstone of 1994 legislation to implement South Africa's adoption of the human right to water—to rule that the Harmony Gold Mining Company must continue to pay for the pumping and treatment of acid mine water around the Orkney Mine, even though it had sold the mine in 2007. The court ruled that Harmony must bear the costs of remediation for activities that caused pollution before the sale (Sapa 2012).

Increased Participation in Water Resource Management

National water laws and policies are increasingly recognizing the importance of public participation in water resource management. Kenya offers the most progressive example of this trend. The result of Kenya's water sector reform, its Water Act of 2002, explicitly calls for greater public participation in many aspects of water service provision and resource management. The Act led to the creation of water resource user associations and catchment area advisory committees. Both these types of organizations require participation from not only local government and businesses, but also local community groups, NGOs, and individuals with knowledge of local water issues. The groups are tasked with a range of activities, including collaborating on catchment-level allocation and management decisions, monitoring of water use and quality, and advising the Water Resource Management Authority on permits for water use.

Taken together, these trends indicate that in the future, businesses may face more robust (and in some cases complicated) water governance systems. In sum, the interest of countries in meeting their responsibility to protect water resources for human needs changes how they will approach water resource oversight and allocation for commercial and industrial purposes. At the same time, governments are increasingly creating legal controls to ensure that companies' actions do not adversely impact available water resources. Finally, the introduction of more actors into water governance processes will increase the number of groups with whom companies will need to engage in order to assure continuity of supply.

Human Rights Responsibility of the Corporate Sector

In 2011, the UN General Assembly and the UN Human Rights Council in tandem adopted the UN Guiding Principles for Business and Human Rights for implementation of the UN "Protect, Respect, and Remedy Framework," making them the authoritative framework for business responsibility toward human rights, including the rights to water and sanitation (United Nations 2008). The Protect, Respect, and Remedy Framework lays out the three basic responsibilities of states and businesses:

1. The State duty to protect against human rights abuses by third parties, including businesses, through appropriate policies, regulation, and adjudication.

2. The corporate responsibility to respect human rights, which means to avoid infringing on the rights of others and to address adverse impacts with which a business is involved.

3. The need for greater access for victims to effective judicial and non-judicial remedies (United Nations 2008).

The Guiding Principles look to help implement this framework by enabling businesses to develop policies and practices to show that they are respecting human rights. These include:

1. developing and articulating a human rights policy;
2. assessing the company's actual and potential impacts;

3. integrating findings from such assessments into the company's decision making and taking actions to address them;

4. tracking how effectively the company is managing to address its impacts;

5. communicating with stakeholders about how the company addresses impacts; and

6. helping remediate any negative impacts a business causes or contributes to (UN Office of the High Commissioner 2011).

Together, the recognition of water and sanitation as human rights and the adoption of the UN Guiding Principles set baseline expectations for companies.

Corporate Water Stewardship and Business Respect for the Human Rights to Water and Sanitation

Over the past decade, a growing number of companies have recognized that increasing water stress poses significant risks to their operations. They also increasingly recognize that negative impacts on communities, particularly on issues related to human rights, also may detrimentally affect their business. In response, a number of companies have adopted a range of corporate water stewardship practices. Corporate water stewardship (CWS) has been defined as the process of a company's progression from understanding environmental and social water risks, to improving water management in operations and supply chains, to working collaboratively with other water users and water managers to improve governance of shared water resources. Companies that commit to water stewardship broadly understand that there are two sets of risks that need attention: company-related risks that require individual company actions, and river basin-related risks that require collective action with diverse stakeholders. A foundational premise of corporate water stewardship is that businesses can take positive action to mitigate adverse impacts on communities and ecosystems and thereby manage water-related business risks—including physical, reputation, and regulatory risks (CEO Water Mandate et al. 2015b).

Generally, companies manage and implement their stewardship practices and policies through corporate water management cycles that vary from company to company. A typical process, which has been adapted from the UN Global Compact Management Model for water-related management, is outlined below:

Commit—Commit to drive sustainable water management.

Account—Collect data on internal water performance and the condition of the basins in which the company operates.

Assess—Use the data generated in the Account phase to identify water-related business risks, opportunities, and negative impacts.

Define—Define and refine corporate water policy, strategies, and performance targets that drive performance improvements and address risks and negative impacts.

Implement—Implement water strategies and policies throughout the company and across the company's value chain.

TABLE 2.2 Examples of Water-Related Impacts Experienced by Affected Stakeholders

Impacts	Description
Lack of access to water and/or sanitation services in the workplace	Some workplaces lack adequate sanitation facilities or access to potable water. This can lead to more severe impacts on migrant or other workers who live on-site in company dormitories. A lack of safe and adequate sanitation facilities may particularly affect women.
Scarcity of water	Community members may be concerned that a company's water use will put additional stress on local water resources. For example, a large agricultural company or a mining operation can draw large quantities of water from an aquifer, affecting the local communities' shallow wells.
Pollution of water	Certain kinds of industrial processing, industrial effluents, or agricultural practices can contaminate local water resources.
Physical barriers to water access	Community members' access to water is affected by business activities that divert a watercourse or block an access route to a water source (e.g., when exclusion zones associated with a hydroelectric dam or intake pumping station inhibit traditional access routes, or land is sold by the government to a private owner who blocks access to traditional sources of water). Community members subsequently may need to travel a significant distance to access clean water (a task that is borne disproportionately by women and girls).
Inequitable access to water or economic constraints on access	A government authority upgrades the water-supply system specifically to encourage a company to expand its operations in an area. It increases the rates charged for connections and/or use for all users without regard to the effect it may have on peoples' ability to pay. The new charges are too high to be affordable for poorer community members, some of whom are also members of potentially vulnerable or marginalized groups (e.g., women). The state does not provide subsidies or other programs to ensure access to water for those who now can't afford it.

Source: CEO Water Mandate et al. 2015a.

Monitor—Monitor progress and changes in performance and basin conditions.

Communicate—Communicate progress and strategies and engage with stakeholders for continuous improvement by means of corporate water disclosure (UN Global Compact and Deloitte 2010).

The human rights to water and sanitation influence all companies' water stewardship practices. By applying a human rights lens to water stewardship, a new focus on the social dimension of water is added by moving action away from a limited focus on the most pressing economic water-related risks for companies toward addressing pressing impacts on humans. Examples of impacts on the HRWS are highlighted in Table 2.2.

However, taking action on human rights is not so different from existing stewardship practices. In fact, the due diligence elements of the UN Guiding Principles outlined above align well with elements of some company corporate water management practices, as shown in Table 2.3.

Companies that look to respect the human rights to water and sanitation will often need to build upon the work and competencies already present in their water and human rights teams. At a practical level, this may mean integrating elements of water or human rights into existing systems, structures, and policies. For example, companies

TABLE 2.3 Relationship Between UN Guiding Principles and Elements of Corporate Water Management

UN Guiding Principles Element		Corporate Water Management Elements
Policy Commitment and Embedding Respect	*Is similar to*	Commit; Define
Assessing Impacts	*Is similar to*	Account; Assess
Integrating and Taking Action	*Is similar to*	Implement
Tracking Performance	*Is similar to*	Monitor
Communicating Performance	*Is similar to*	Communicate
Remediation	*No clear match but*	Elements of Implement are relevant

Source: CEO Water Mandate et al. 2015a.

may have both standalone water management and human rights policies. When they look to make a public commitment to the rights to water and sanitation, they can look to integrate water and sanitation into human rights policies, or vice versa (CEO Water Mandate et al. 2015a). A key here, however, is ensuring that human rights are preserved by calling out how companies are meeting their responsibilities to respect the rights to water and sanitation in both operations and in business relationships, as well as expectations for entities within their value chain.

In many cases, companies meeting their responsibility to respect the human rights to water and sanitation will likely undertake a range of activities that also fall under existing corporate water stewardship practice, described in Table 2.4. Fundamental to any action related to respecting human rights includes ensuring appropriate and ongoing stakeholder engagement to develop policies and respond to identified impacts in a way that ensures they are in line with stakeholders' expectations and needs.

Examples of Applying a Human Rights Lens to Aspects of Corporate Water Stewardship

Assessing Risks and Impacts

Companies undertaking effective water stewardship activities are already taking action to understand their basin contexts, as well as their impacts on ecosystems. This provides them with a concrete starting point from which to assess impacts on communities. In many cases, impacts on the human rights to water and sanitation will depend on a variety of actions—including companies' (or their suppliers') own water use, how that affects local ecosystems, and how that in turn affects communities. To meet their responsibilities, companies may conduct further standalone human rights impact assessments or utilize revised water risk and assessment processes that integrate aspects of the human rights to water and sanitation. Once companies understand their impacts, how they are involved, and prioritize the most pressing human rights impacts, they can take appropri-

TABLE 2.4 Elements of Corporate Water Stewardship

Key Elements	Description of Activities
Addressing operational issues	Technical and management changes that improve water efficiency, wastewater treatment, and employee access to water, sanitation, and hygiene (WASH).
Understanding basin, contexts, and impact	Awareness of how the company interacts with surrounding basin(s), including the nature and extent of local water stress and local regulation; and the company's impacts on ecosystems and communities, including any potential impacts on the human rights to water and sanitation.
Developing a water strategy and raising awareness internally	Developing goals, strategies, and policies that integrate water risks and impacts into core business processes and decision making; raising awareness of the company's water impacts and stewardship strategy throughout the business—from the CEO and leadership team, to facility managers, to suppliers.
Leveraging improvements in value chain	Managing water-related risks and impacts throughout the value chain from raw materials to consumers—including water use, water quality, access to WASH services in the supply chain, and other social and environmental impacts outside the company's direct operations.
Advancing water sustainability via collective action	Actions that address basin-related risks or identified collective impacts, which require proactive collaboration with others to improve local conditions and reduce water stress in the basin.
Advancing water sustainability via public policy engagement	Responsible engagement by the private sector, which improves public sector capacity and advances better water governance.
Communicating with external stakeholders	Ongoing transparent reporting, disclosure, and dialogue with diverse stakeholders about corporate water stewardship strategy, policies, activities, baseline conditions, and progress toward targets.

Source: CEO Water Mandate et al. n.d.

ate action. While stewardship does not provide guidance in regard to which impacts to prioritize, a human rights lens provides specific guidance for this based on severity and likelihood of impacts. Severity is based upon three factors:

1. Scale—how grave is the impact
2. Scope—how many are affected
3. Irremediability—how difficult it is to restore the situation (such as contaminated water resources)

Companies will also need to determine how likely it is that an impact will occur. The human rights lens necessitates that companies strive to address the worst impacts from the viewpoint of affected stakeholders first (i.e., those in the top right quadrant of Figure 2.2 with both high likelihood and severity).

Once companies have identified impacts, depending on how they are involved in the impact (either causing, contributing, or linked to it) they can take a range of actions (CEO Water Mandate et al. 2015a; 2015c). Often these actions are directly related to operational performance (such as limiting water use, increasing efficiency, implementing improved wastewater treatment processes), remedial actions (ceasing actions leading to

Severity

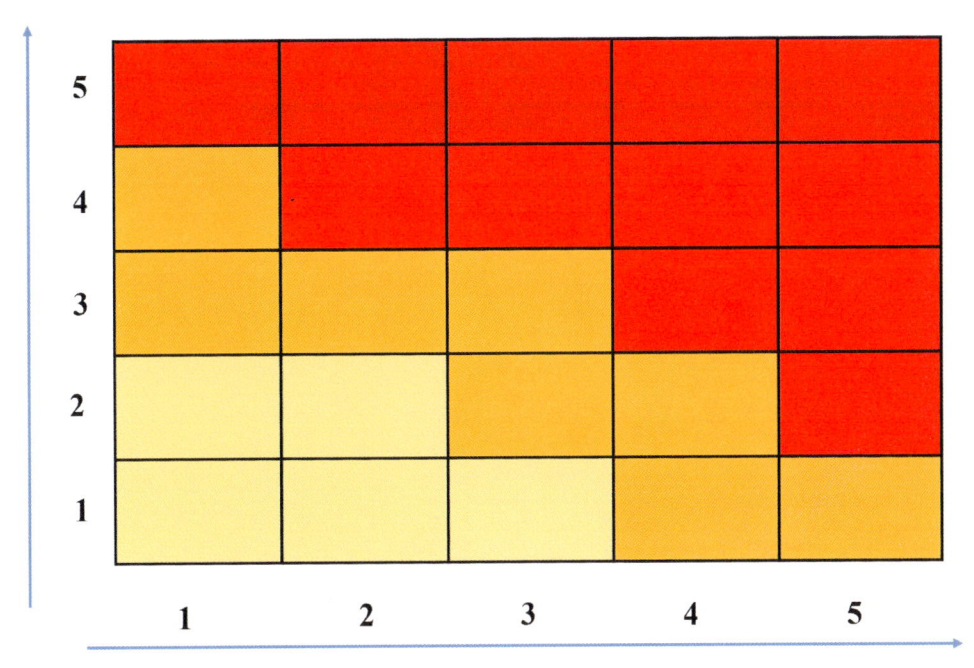

FIGURE 2.2 HEAT MAP FOR DETERMINING SEVERITY OF HUMAN RIGHTS IMPACTS.
Source: CEO Water Mandate et al. 2015a.

negative impacts and providing alternative water resources), or working with others to improve water performance through collective action or using their leverage to bring about improvements by their supply chain actors.

Addressing Cumulative Impacts

In many cases, impacts on the right to water and sanitation are cumulative, resulting from the actions of a variety of actors operating in a basin. Together, these actors' water use might lead to unsustainable use of local water resources or alteration in water quality to an extent that it affects local communities' rights to water and sanitation. In order to both identify these impacts and take appropriate action, companies will need to work with other stakeholders in the basin. Corporate water stewardship's strong emphasis on collective action enables exactly this type of analysis and action via joint monitoring, and local projects that leverage the resources of the private sector or promote engagement with policy makers.

Leveraging Improvements in the Supply Chain

In many cases, a company's greatest water-related risks do not lie in direct operations but rather in its supply chains (CEO Water Mandate et al. 2015c). Similarly, the greatest impacts on the rights to water and sanitation often lie in a company's supply chains. Companies that recognize both their increased water risks and water impacts, and work to bring about better water performance in their supply chains, are better able to meet their responsibilities in both areas.

BOX 2.1 Case Example: Company Action to Identify and Respond to Human Rights Impacts

A company in the food and beverage industry regularly conducts human-rights impact assessments in high-risk countries and has begun incorporating impacts on the HRWS into its assessments. In one country where it has a plant, the company's assessment highlighted local community members' concerns that they were experiencing reduced access to safe water and associated health problems. Local stakeholders expressed the view that the irrigation practices of local farmers (responsible for 96% of the water use in the country) and the activities of the various companies located in the watershed area were responsible for using the majority of available groundwater. This input helped the company evaluate the nature of its own involvement in the negative HRWS impacts on local communities. Following the human rights impact assessment, an independent third party-verified review was completed, which concluded that the company's operations were not causing or contributing to depletion of water in the region and that the company's approach to water stewardship, and wastewater treatment in particular, was effective. But the assessment also suggested that the negative HRWS impacts were nonetheless directly linked to the company's operations through its business relationships, since some of the local farmers were supplying milk to the company. In response to the linkage situation, the company committed to strengthen its engagement with local farmers about more effective use of water for irrigation purposes and responsible water stewardship, thereby using its leverage to try to mitigate the risk of the impact continuing.

To help mitigate the risk that the company's own activities might contribute in the future to negative HRWS impacts, the company also took some additional steps. The company committed to regular consultations with local NGOs, water experts, environmental groups and other companies located in the area about access to water issues to help evaluate whether local approaches prove effective over time. The company signed a memorandum of understanding with a major environmental NGO in order to improve water usage within the company's operations, including its supply chain, and to further implement a standard developed by the Alliance for Water Stewardship in the region and, ultimately, in the whole country.

Source: CEO Water Mandate et al. 2015a.

Support for the Human Rights to Water and Sanitation

For some companies, particularly those who are UN Global Compact endorsers, there is an additional expectation that they might go beyond minimum efforts toward more active support of the rights to water and sanitation, which can be supported through a number of different means, including:

1. providing core efforts through innovation and services rendered;
2. through social investment or philanthropy;
3. engaging in collective action and public-policy; and
4. developing partnerships.

In many cases, businesses that take steps to respect the HRWS have positioned themselves to be able to effectively support these rights. Some of the key obstacles to increased private-sector engagement for activities that support access to water, sanitation, and hygiene are concerns about the long-term sustainability of such projects, as well as lack of clarity regarding government versus company roles. Often, these projects require an array of competencies that go beyond the company's core expertise. A strong focus on effective stakeholder engagement enables companies to determine what type of support would be most appropriate to local circumstances, thereby increasing the likelihood of long-term success. In addition, new guidance related to managing the integrity of multi-stakeholder water stewardship initiatives, which would cover a number of partnerships, social investments, and collective action that support the HRWS, also provides guidance on how to undertake projects in a way that meets local needs and respects the role of government (CEO Water Mandate et al. 2015a).

Other companies are taking a different approach, by using their core businesses to directly contribute to supporting the human rights to water and sanitation, and achievement of WASH targets. For example, Unilever works on changing consumer behavior

---

BOX 2.2 Case Example: Respect as a Basis for Support

A company that is reviewing how to strengthen increased access to WASH in its own facilities may learn from its workers that there is a poor understanding of sanitation in the local community that may hamper the company's efforts within its factories. Via engagement with workers and others it also learns that there are existing government-led programs to increase awareness around WASH in the local community. The company can then decide to invest in specific initiatives to both ensure that it meets its own responsibilities within its factories but also contributes to the broader goal of meeting the right to sanitation in the local community.

Source: CEO Water Mandate et al. 2015a.

and promoting greater access through WASH with specific products aimed at not only improving local communities' access to sanitation and hygiene but also focused on improving the effectiveness of such interventions.

The Path Forward

Meeting the long-term objectives of the UN's Sustainable Development Goals for water will require broad efforts by all actors. The private sector has a unique role to play. Central to these efforts will be an alignment between companies' broader water stewardship practices with public efforts to satisfy the formal human rights to water and sanitation. Leading companies have already taken action to do exactly this, though given the extent of the challenge many more need to develop and implement integrated strategies. By making these efforts, businesses will not only help reduce their own water risks and improve their long-term viability but they can also play a significant role in reaching larger public goals related to the sustainability of this life-sustaining resource.

References

Center on Housing Rights and Evictions. 2008. *Legal Resources for the Right to Water and Sanitation: International and National Standards.* 2nd ed. http://www.worldwatercouncil.org/fileadmin/wwc/Programs/Right_to_Water/Pdf_doct/RWP-Legal_Res_1st_Draft_web.pdf.

CEO Water Mandate, Shift, and the Pacific Institute. 2012 (August). Bringing a Human Rights Lens to Corporate Water Stewardship: Results of Initial Research. http://ceowatermandate.org/files/HumanRightsLens2012.pdf.

CEO Water Mandate, UN Global Compact, and the Pacific Institute. 2014 (September). Exploring the Business Case for Corporate Engagement on Sanitation: White Paper. http://ceowatermandate.org/files/Sanitation.pdf.

CEO Water Mandate, Shift, and the Pacific Institute. 2015a (January). Guidance for Companies on Respecting the Human Rights to Water and Sanitation: Bringing a Human Rights Lens to Corporate Water Stewardship. http://ceowatermandate.org/files/business-hrws-guidance.pdf.

CEO Water Mandate, UN Global Compact, and the Pacific Institute. 2015b. Stewardship is Good for Business. http://ceowatermandate.org/why-stewardship/stewardship-is-good-for-business/.

CEO Water Mandate, Water Integrity Network, and the Pacific Institute. 2015c (August). Guide for Managing Integrity in Water Stewardship Initiatives: A Framework for Improving Effectiveness and Transparency. http://ceowatermandate.org/files/integrity.pdf.

CEO Water Mandate, Shift, and the Pacific Institute. 2017. Respecting Human Rights to Water and Sanitation. http://ceowatermandate.org/humanrights/understanding-impacts/hrws/.

CEO Water Mandate, UN Global Compact, and the Pacific Institute. No date. Serving the Public Interest: Corporate Water Stewardship and Sustainable Development. https://ceowatermandate.org/files/Stockholm/Corporate%20Water%20Stewardship%20and%20the%20SDGs.pdf.

Gleick, P. H., and M. Palaniappan. 2010. Peak Water: Conceptual and Practical Limits to Freshwater Withdrawal and Use. *Proceedings of the National Academy of Sciences (PNAS)* 107(25): 11155–11162. http://www.pnas.org/cgi/doi/10.1073/pnas.1004812107.

High Court of Sindh at Karachi. 2004. Agha Khan University and Sindh Institution of Urology and Transplantation vs. Nestle Milkpak Lt. Suit No. 567. https://www.informea.org/sites/default/files/court-decisions/COU-159317.pdf.

Kenya Water Act. 2002. http://www.kenyalaw.org/lex/rest/db/kenyalex/Kenya/Legislation/English/Amendment%20Acts/No.%208%20of%202002.pdf.

Sapa. 2012 (February 27). Court: Harmony Must Pay for Acid Water. http://m.fin24.com/fin24/Companies/Mining/Court-Harmony-must-pay-for-acid-water-20120702.

United Nations. 2008 (June). The UN "Protect, Respect and Remedy" Framework for Business and

Human Rights. https://business-humanrights.org/sites/default/files/reports-and-materials/Ruggie-report-7-Apr-2008.pdf.

UN Global Compact and Deloitte. 2010. UN Global Compact Management Model: Framework for Implementation. https://www.unglobalcompact.org/docs/news_events/9.1_news_archives/2010_06_17/UN_Global_Compact_Management_Model.pdf.

UN Office of the High Commissioner for Human Rights. 2011. Guiding Principles on Business and Human Rights: Implementing the United Nations "Protect, Respect, and Remedy" Framework. http://www.ohchr.org/Documents/Publications/GuidingPrinciplesBusinessHR_EN.pdf.

WASH United, Freshwater Action Network, and Waterlex. 2012 (March). The Human Right to Safe Drinking Water and Sanitation in Law and Policy: A Sourcebook. http://www.waterlex.org/resources/documents/RTWS-sourcebook.pdf.

WaterAid and Unilever. 2015. The Global Goals Toolkit: Ensure Availability and Sustainable Management of Water and Sanitation for All. http://www.wateraid.org/uk/what-we-do/policy-practice-and-advocacy/research-and-publications/view-publication?id=1d43a620-c800-41d4-9dee-0a3171ced923.

CHAPTER 3

Updating Water-Use Trends in the United States

Kristina Donnelly and Heather Cooley

Introduction

The United States Geological Survey (USGS) has estimated and published data on national and statewide water use approximately every five years since 1950. These data identify the total amount of water used by state, source (i.e., ground or surface water), and type (i.e., fresh or saline) for broad categories of water use. Although the categories have changed somewhat over time (see Table 3.1), the most recent data from 2010 include the following sectors: public supply, domestic, industrial, irrigation, livestock, mining, and thermoelectric.

These data serve a variety of purposes. They provide a means for understanding how water use is changing over time and how it varies nationally, regionally, and on a state level.[1] The data help researchers and government agencies estimate important quantities, such as interbasin transfers, water availability, and other components of and changes to the water cycle (USGS 2002). And they offer insights into trends in industrial, environmental, and agricultural water use, including growing gaps between demand and water availability on a regional basis.

In this chapter, we discuss the USGS national water use data set, describe how the data are collected and organized, and evaluate some of the national trends that the data suggest. Our analysis finds there have been important structural changes in the U.S. economy and we have made considerable progress in managing the nation's water, with total water use less than it was in 1970, despite continued population and economic growth. Indeed, every sector, from agriculture to thermoelectric power generation, shows reductions in water use. National water use, however, remains high compared to other industrialized Western nations, and many freshwater systems are under stress from overuse. Moreover, there is growing evidence that climate change will worsen existing water resource challenges, affecting the supply, demand, and quality of the nation's

---

1. Given the complexity and uncertainty inherent in the data collection, it is not advisable to try to analyze local trends from these data.

TABLE 3.1 USGS Changes in Water-Use Categories

1950	1955	1960	1965–1980	1985	1990–1995	2000–2010
Municipal	Public Supply	Public Supply	Public Supply	Public Supply	Public Supply	Public Supply
Irrigation	Irrigation	Irrigation	Irrigation	Irrigation	Irrigation	Irrigation
Rural	Rural	Rural Domestic	Rural Domestic	Domestic	Domestic	Domestic
		Livestock	Livestock	Livestock	Livestock	Livestock
						Aquaculture (incl. fish farming & hatcheries)
	Other Industrial	Other Industrial	Other Industrial	Commercial	Commercial (incl. off-stream fish hatcheries)	Commercial not estimated
				Industrial	Industrial	Industrial
				Mining	Mining	Mining
Self-supplied Industrial	Condenser Cooling	Condenser Cooling	Condenser Cooling		Animal Specialties (incl. fish farming)	
	Other	Other	Other			
	Fuel-Electric Power	Fuel-Electric Power	Fuel-Electric Power	Thermoelectric Power (Fossil Fuel / Geothermal / Nuclear)	Thermoelectric Power (Fossil Fuel / Geothermal / Nuclear)	Thermoelectric Power (Once-through Cooling / Closed-loop Cooling)
Water Power	Water Power	Water Power	Hydroelectric Power	Hydroelectric Power	Hydroelectric Power	Hydroelectric Power not estimated
				Sewage Treatment (releases)	Wastewater Treatment (releases)	Wastewater Treatment not estimated

Source: USGS 2016.

water resources (Walsh et al. 2014). To address these challenges, we must continue and even expand efforts to improve water use efficiency in our homes, businesses, industries, and on our farms.

Data Collection and Organization

The first collection of historical national water use data was produced by the U.S. Department of Commerce in 1948 and provided estimated water use as far back as 1900 (U.S. Bureau of the Census 1975). In 1965, Congress established the U.S. Water Resources Council to conduct a comprehensive study of the nation's water resources, which was published in 1968 and again in 1978. The USGS also began publishing national water-use data in 1950, centralizing this data collection effort in 1978 through the establishment of the National Water-Use Information Program (NWUIP) (Hutson et al. 2004). The reports, now published as part of that data collection effort, are among the most cited of all USGS products (USGS 2014c).

Although data collection was fragmented for many years, coordination of that effort and the accuracy of the data have improved markedly since the establishment of the NWUIP. In 2002, the National Academy of Sciences, an independent scientific advisory group, evaluated the NWUIP and recommended a number of tasks that would improve the program (Committee on USGS Water Resources Research 2002). Today, national data are collected from or calculated using a variety of sources, including national data sets, state agencies, questionnaires, and local contacts (Maupin et al. 2014). USGS regional and state offices typically submit data representing their region to USGS headquarters, which then compiles and standardizes it, filling in any gaps using statistical analysis or data from other federal agencies (Maupin et al. 2014).

The USGS reports water use for human purposes as withdrawals from water bodies—be it a lake, river, estuary, or aquifer. When water is discharged back to a water body after it has been withdrawn, it is considered a "non-consumptive use." Most indoor residential water use, for example, would be considered non-consumptive because after the water is used, it is discharged to sewers or septic tanks, treated, and then returned to the environment. In some cases, water is either evaporated or consumed through use, or incorporated into a product; when water is not returned to the system, it is called a "consumptive use." Until 2000, the USGS calculated the amount of water that was consumed by each sector, but this effort stopped primarily because of resource and data constraints (Maupin et al. 2014). There are also non-withdrawal uses of water, where the water is used *in situ* and not diverted; these uses (also called "instream uses") include navigation, ecosystem protection, recreation, and waste disposal, among others (MacKichan 1957).

Substantial improvements in water use data are still needed. Despite the advances in collecting water use data, a great deal remains unknown. Much of the data consist of estimates by expert analysts, rather than actual measurements. More detailed and more accurate measurement of groundwater use is required, especially in Western states reliant on irrigation. Data on the penetration of water-efficient appliances and technologies in the municipal and industrial (M&I) sectors are missing, and would help us understand the additional potential for improvements. State methods for collecting data remain inconsistent, making comparisons over time and over regions unreliable. The distinction between withdrawals and consumptive use is important, but the last several reports have not included consumptive use estimates—it would be valuable to reinstate this metric. Furthermore, a comprehensive census and inventory of U.S. water use as requested by Congress in the Omnibus Public Land Management Act of 2009 (Public Law 111–11), also known as the Secure Water Act, should be fully funded and completed. Addressing these limitations to improve the quality and timeliness of water-use data would ultimately help land-use managers, water utilities, and local communities to better plan, develop, and manage their water resources sustainably.

Total Water Use

National water use has declined over the last three decades and experienced a major drop between 2005 and 2010.[2] These trends have been evident for a while (see Gleick 2003), and they continue today. Total water use, which includes both freshwater and saline water, peaked in 1980 at 610 km^3 before falling to 550 km^3 in 1985 (Figure 3.1). Between 1985 and 2005, water use remained relatively flat, but by 2010 total water use declined to 490 km^3—lower than it was in 1970.

Total water use declined at the same time as the population and economy grew. As a result, daily per capita water use has also been falling since reaching a peak of 7.37 m^3 in 1980. In 2010, per capita use was 4.33 m^3 per day, down 17 percent from 2005 levels and the single largest decline in any five-year period. Figure 3.2 shows the "economic productivity" of water in the United States from 1900 to 2010; i.e., the inflation-adjusted gross domestic product (GDP) for every cubic meter of water used. Between 1900 and 1980, the United States experienced only a modest increase in the economic productivity of water, and by 1980, $1.50 of GDP was produced per m^3 of water used. Since that time, economic productivity has increased dramatically. Indeed, during the most recent period (2005–2010), economic productivity increased by 20 percent to $4.30 per m^3 of water. These results show that the United States now produces far more wealth with far less water than at any time in the past.

The USGS makes a distinction between saline water and freshwater. Saline water, which includes seawater, brackish water from estuaries, and salty groundwater, has a higher concentration of salts, containing about 1,000 mg/L or more of total dissolved solids (TDS). Throughout the period of record (1955–2010), freshwater has constituted the majority (85 percent) of national water withdrawals and use. In 2010, agriculture and thermoelectric power were each about 40 percent of freshwater use, with the remaining 20 percent withdrawn by the M&I sector. Freshwater use, however, has changed dramatically over time, with particularly large increases in water withdrawals for thermoelectric power. For example, in 1955, total freshwater use was 310 km^3, of which 27 percent (83 km^3) was for thermoelectric power, 49 percent (154 km^3) for agriculture, and 23 percent (71 km^3) for the M&I sector. Between 1955 and 1980 (when U.S. freshwater use peaked), agricultural and M&I water use increased 37 percent and 45 percent, at about the same rate as the population (a 38 percent increase), while water use for thermoelectric power increased by 150 percent. Freshwater use remained relatively constant over the next two

---

2. Unless otherwise specified, the geographic extent of the data is as follows: 1950 represents the lower 48 states, DC, and Hawaii; 1955 represents the lower 48 states and DC; 1960 and 1975–2010 represent all 50 states, DC, Puerto Rico, and U.S. Virgin Islands; 1965–1970 represent all 50 states, DC, and Puerto Rico.

Water Use Trends in the U.S. 43

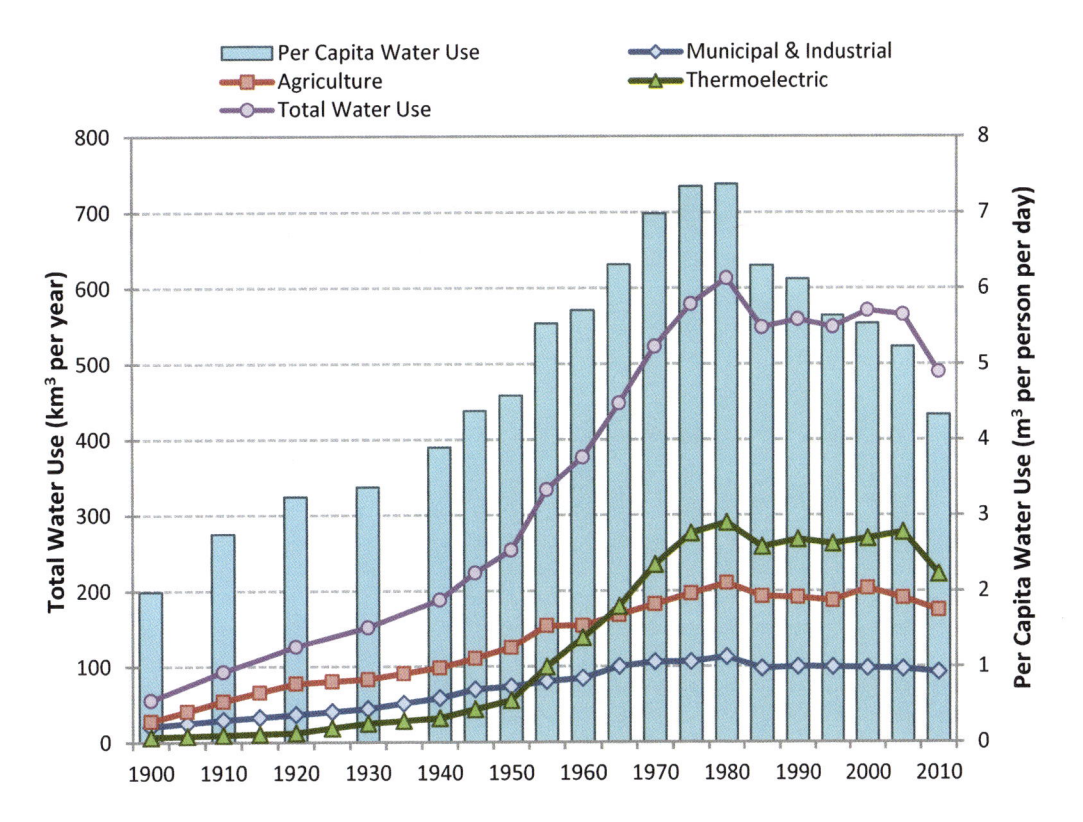

FIGURE 3.1 TOTAL ANNUAL AND PER CAPITA WATER USE (FRESHWATER AND SALINE WATER), 1900–2010, BY SECTOR.

Notes: Municipal and Industrial (M&I) includes public supply, self-supplied residential, self-supplied industrial, mining, and self-supplied commercial (self-supplied commercial was not calculated in 2000–2010). Agriculture includes aquaculture (1985–2010 only), livestock, and irrigation. Between 1900 and 1945, the M&I category includes water for livestock and dairy.

Sources: Data for 1900–1945 from the Council on Environmental Quality (CEQ) 1991. Data for 1950–2010 from USGS 2014a. Population data from Williamson 2015.

decades, but between 2005 and 2010, freshwater use dropped by 13 percent. Thermoelectric power, which represented about one-third of total freshwater use in 2010, was responsible for nearly two-thirds of the overall reductions. We explore these trends in the following sections.

Water Use for Thermoelectric Power Generation

Water requirements for thermoelectric power production are substantial, representing the single largest withdrawal of water—both fresh and saline—in the United States. Thermoelectric power plants, typically powered by fossil, geothermal, nuclear, and biomass fuels, use water for cooling purposes and to replenish boiler water lost through evaporation. In 2010, thermoelectric power plants withdrew 220 km^3, nearly all of which was surface water (Figure 3.3). Nearly three-quarters of the total amount of water withdrawn by thermoelectric power plants in 2010 is freshwater. The use of saline water is largely confined to coastal regions with access to the ocean.

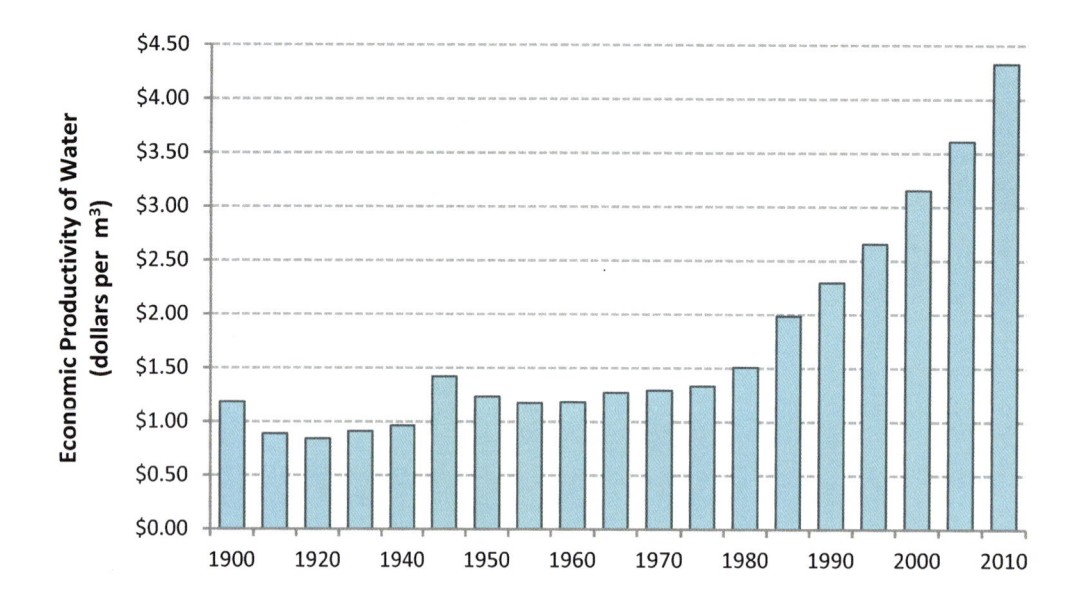

FIGURE 3.2 ECONOMIC PRODUCTIVITY OF WATER, 1900–2010.
Note: All estimates have been adjusted for inflation and are reported in year 2009 dollars.
Sources: Updated from Gleick (2003). Data for 1900–1945 from CEQ (1991). Data for 1950–2010 from USGS (2014a). Population data from Williamson (2015).

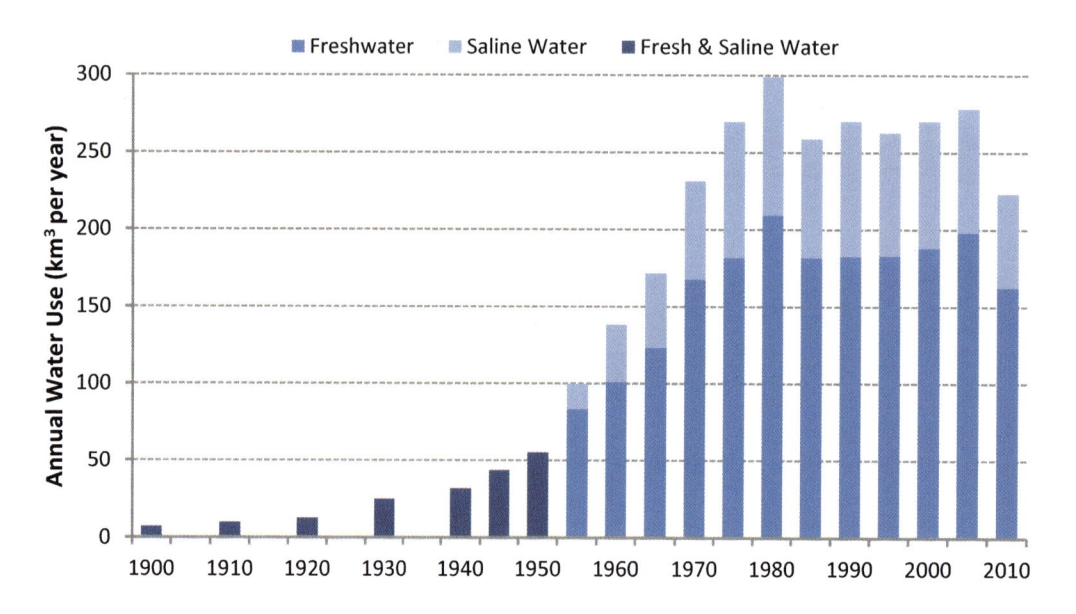

FIGURE 3.3 WATER USE FOR THERMOELECTRIC POWER GENERATION, 1900–2010, BY TYPE.
Sources: Data for 1900–1950 from CEQ 1991. Data for 1955–1980 from the USGS water use data companion publications, *Estimated Use of Water in the United States,* which are published along with each data release: MacKichan 1957; MacKichan and Kammerer 1961; Murray 1968; Murray and Reeves 1972 and 1977; Solley et al. 1983. Data for 1985–2010 from USGS 2014b.

Both total water withdrawals and freshwater withdrawals for thermoelectric power plants are lower than they were in 1970. This represents an important reversal of a 25-year trend of increasing water use for producing energy. Total and freshwater use for thermoelectric power plants peaked in 1980 (Solley et al. 1983) at 300 and 210 km^3, respectively. In 1985, water use declined but then increased in nearly every five-year period through 2005. The 20 percent reduction during the most recent period (2005–2010) represents a significant shift in national water use for thermoelectric power plants, which the USGS attributes to upgrades to intakes and cooling systems, especially a reduction in the use of water-intensive once-through cooling systems (Maupin et al. 2014). Once-through cooling systems can cause harm through "thermal pollution," altering ecosystems and killing aquatic life. To address this, states like California have begun to phase out the use of once-through cooling systems, arguing that it no longer represents "best available technology" as required by the federal Clean Water Act (CEC 2016). However, federal regulators, following a 20-year long rulemaking process, decided to allow the practice to continue in 2014, requiring only that plan operators must take steps to decrease the number of fish killed by cooling systems (U.S. EPA 2014).

On average, thermoelectric power plants in the United States withdraw 0.07 m^3 of water (both fresh and saline) for every kWh generated in 2010. The water intensity of thermoelectric power production, however, varies tremendously across the United States, ranging from 0.002 m^3 per kWh in Arizona to 0.28 m^3 per kWh in Rhode Island (Figure 3.4). This variation is primarily driven by the type of cooling system employed, with states that rely on once-through cooling using far more water per unit of energy produced than states using recirculating or dry cooling. Overall, by 2010, the United States reduced the water use intensity of thermoelectric power production by 41 percent since 1985 and 18 percent since 2005, with the largest reductions in the northwest and southwest. Despite these improvements, thermoelectric power plants still represent the single largest use of water in the United States. Water use could be further reduced by accelerating water and energy efficiency improvements, the development and deployment of less water-intensive renewable energy systems, and the adoption of recirculating- and dry-cooling systems (Cooley et al. 2011).

Water Use for the Municipal and Industrial Sector

Municipal and industrial water use represents the amount of water withdrawn to meet the needs of cities, towns, and small communities. This includes water used in homes for both indoor and outdoor needs (i.e., cleaning, bathing, cooking, and maintaining gardens and landscapes), as well as water used in the commercial, industrial, and mining sectors to produce goods and services. M&I water use also includes water used by institutions, such as schools, cities, prisons, and government agencies, as well as water losses due to system leakage, firefighting, theft, hydrant flushing, and unmetered connections.

In 2010, M&I water withdrawals in the United States totaled 92 km^3, or 19 percent of total national water use. During much of the twentieth century, M&I water use increased as the population grew, reaching a record high of 112 km^3 in 1980 (Figure 3.5). This trend reversed in 1985, after which total water use for M&I began to level off and then decline. During the most recent period (2005–2010), M&I water use decreased by 4 percent, despite a 4 percent increase in both population and GDP. As a result, per capita water use has declined in every five-year period over the last three decades, from 1.35 m^3 per capita

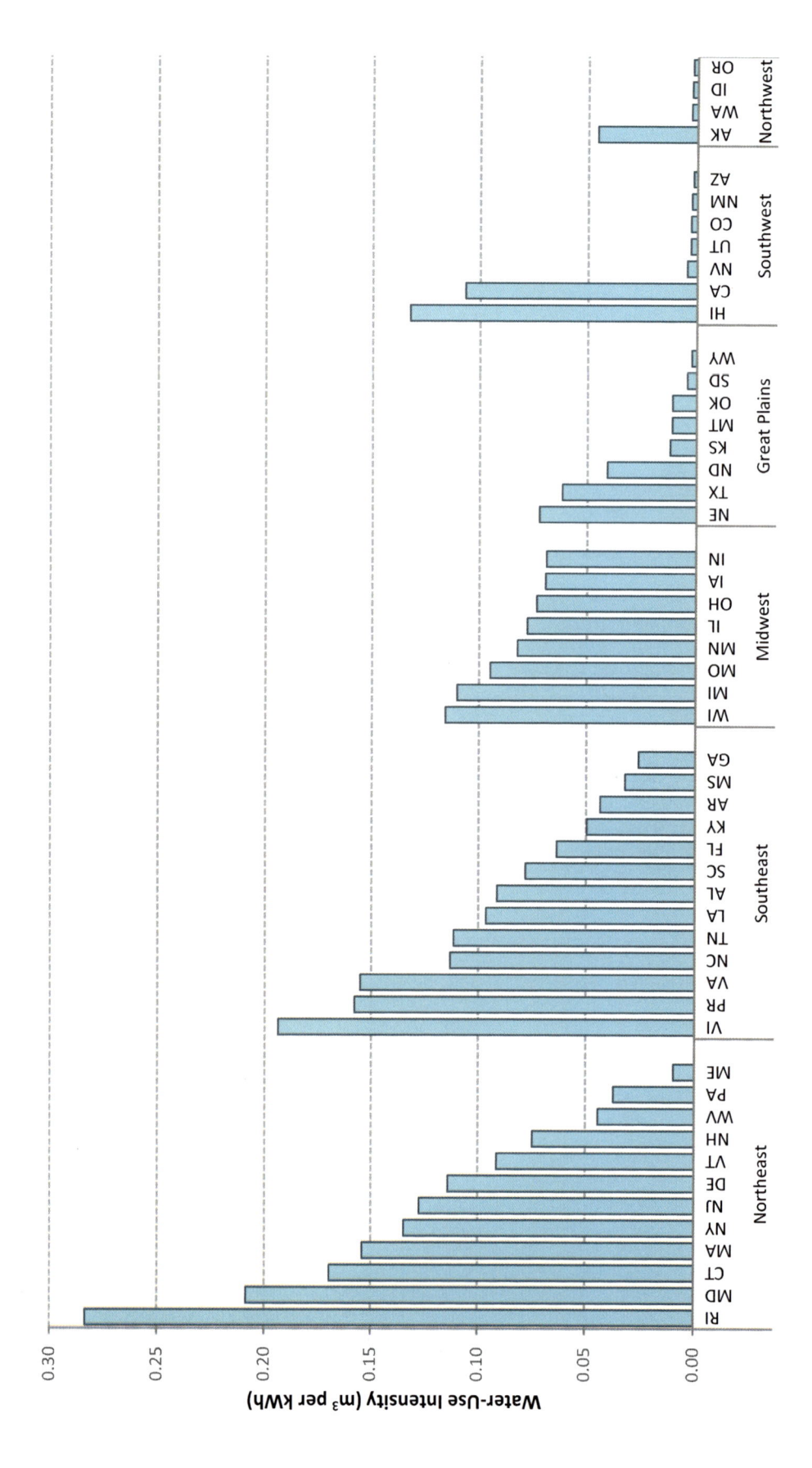

FIGURE 3.4 WATER USE INTENSITY FOR THERMOELECTRIC POWER GENERATION, 2010, BY STATE.
Source: UUSGS 2014b.

Water Use Trends in the U.S. 47

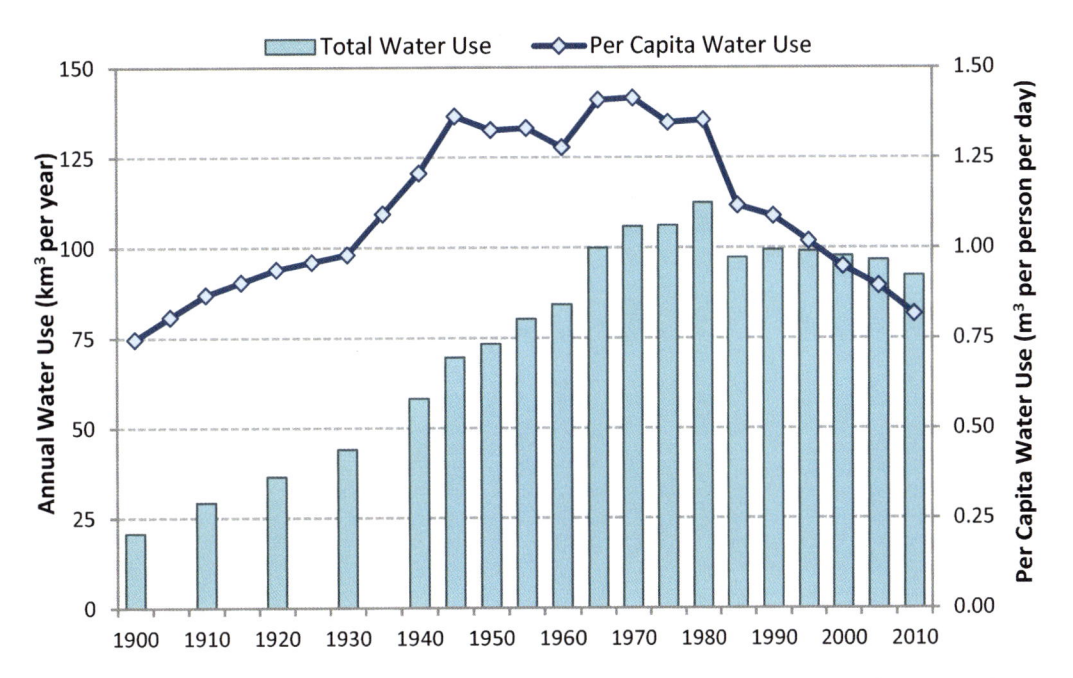

FIGURE 3.5 TOTAL AND PER CAPITA WATER USE FOR THE MUNICIPAL AND INDUSTRIAL SECTOR, 1900–2010.

Notes: Self-supplied commercial was not calculated in 2000, 2005, or 2010, which would account for some of the reduction in use that occurred during that period. In addition, USGS documentation notes that water use estimates for self-supplied industrial use were more realistic in 1985 than in 1980 and would account for some of the reduction between these years (Solley et al. 1988). M&I water use from 1900–1945 also includes water for livestock and dairies. Some years include public supply deliveries to thermoelectric; although it was not possible to exclude these deliveries for all years, the years for which data are available suggest that this use was relatively very small. Washington DC was excluded from the analysis due to lack of data.

Sources: 1900–1945 data from CEQ 1991; 1950–2010 data from USGS 2014a; population data from Williamson 2015.

per day in 1980 to 0.82 m³ per capita per day in 2010.

Reductions in M&I per capita water demand were driven by two major factors. First, the economy shifted from one dominated by water-intensive manufacturing to a less water-intensive service-oriented economy. Second, numerous federal, state, and local policies and actions have resulted in extensive water efficiency improvements. For example, the National Energy Policy Act of 1992 established efficiency standards for all toilets, urinals, kitchen and lavatory faucets, and showerheads manufactured after January 1, 1994 (for a longer discussion, see Gleick 2012). Subsequent legislation established additional standards for products not included in the original act—including clothes washers, dishwashers, and several commercial products. More recently, the Environmental Protection Agency (EPA) developed the WaterSense program, a voluntary labeling program inspired by the Energy Star program, to help customers identify and purchase water-efficient appliances. Unlike Energy Star, which relies on manufacturers to report the energy use for their products, WaterSense fixtures are tested and certified by an independent third party, guaranteeing that they meet the EPA's specifications for water efficiency and product performance.

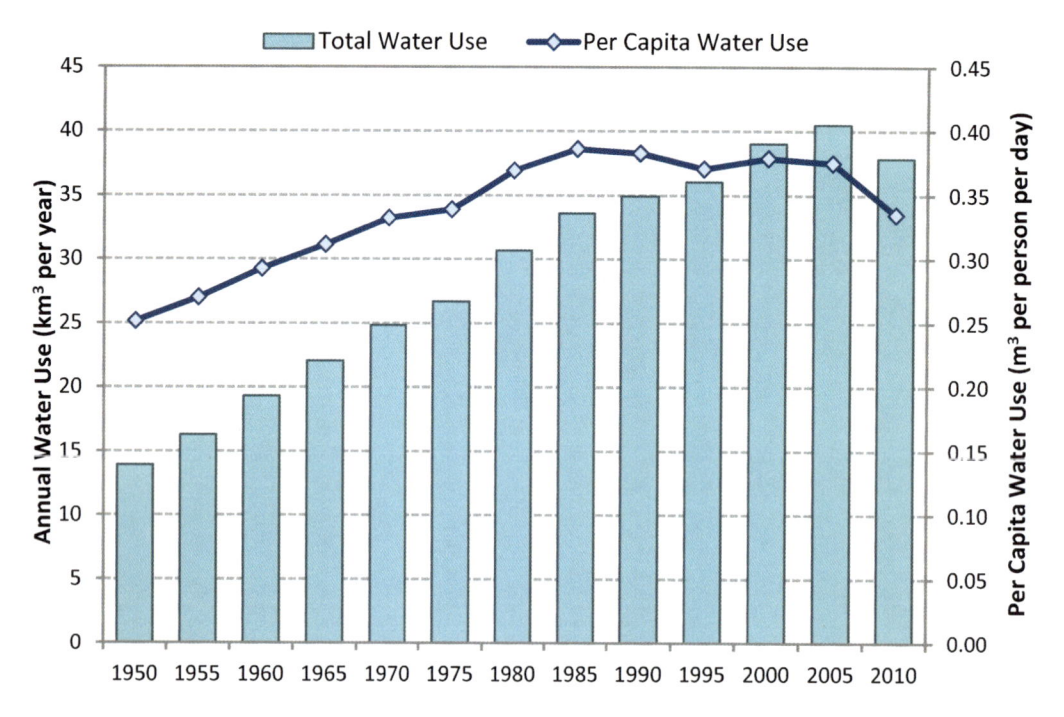

FIGURE 3.6 TOTAL AND PER CAPITA WATER USE FOR THE RESIDENTIAL SECTOR, 1950–2010.
Notes: The publicly available USGS data only estimate residential water use for 1985–2010 (excluding 2000). Residential water use for 1960–1980 included public use and losses. In the years available, about 57 percent of the public supply went to residential use. For the years in which residential water use data were not separately reported (1950–1980 & 2000), we multiplied the total public supply by 57 percent and added it to self-supplied residential. Washington DC was excluded from the analysis due to lack of data.
Sources: Data for 1950–1980 from the USGS water use data companion publications, *Estimated Use of Water in the United States*, which are published along with each data release: MacKichan 1951 and 1957; MacKichan and Kammerer 1961; Murray 1968; Murray and Reeves 1972 and 1977; Solley et al. 1983. Data for 1985–2010 from USGS 2014b. Population data from Williamson 2015.

Water Use for the Residential Sector

Residential water use is a subset of M&I water use that includes household water use—including for drinking, bathing, washing clothes and dishes, flushing toilets, and landscaping. Residential water can be supplied by private well or spring or delivered by a public supplier. Between 1950 and 2005, total residential water use in the United States steadily increased, reaching 40 km³ in 2005 (Figure 3.6). Between 1985 and 2005, U.S. residential per capita water use remained steady at about 0.38 m³ day. In most parts of the United States, household per capita water use declined due to efficiency improvements; however, these efficiency improvements were offset by population growth in the hottest, driest parts of the United States, where per capita water use is relatively high. Then, between 2005 and 2010, residential water use declined by 7 percent, or 2.64 km³—despite continued population growth—reducing water use to 0.33 m³ per capita per day in 2010. Household per capita water use declined in most U.S. states between 2005 and 2010, with the largest overall reductions occurring in Nevada, Texas, and Nebraska. Nationwide, household water use per capita per day in 2010 ranged from a low of 0.19 m³ in

Wisconsin to a high of 0.64 m³ in Idaho (Figure 3.7). As a region, water use was lowest on average in the Midwest, and highest in the Southwest and Northwest.

Water Use for Irrigation

Water withdrawals for agricultural irrigation have followed a history similar to other water use categories. Total water use for irrigation increased through much of the twentieth century, as did the extent of irrigated areas (Figure 3.8). Water use for irrigation peaked in 1980 at 210 km³ and has declined in nearly every period since.[3] By 2010, water use for irrigation was 160 km³, its lowest level in more than 40 years. Yet, irrigated areas have continued to expand, with 25 million hectares irrigated in 2010—the most land irrigated at any time in U.S. history.

As a result, the water intensity of U.S. agriculture, as measured by irrigation depth, has declined markedly over the past 60 years (Figure 3.9). In 1950, an average of 12,000 m³ per hectare of water was applied to U.S. farmland. By 2010, irrigation depth declined to 6,300 m³ per hectare. Reductions in water intensity could be due to several factors, including shifting to less water-intensive crops as well as improvements in irrigation technologies and practices. For example, since 1985, the area irrigated by surface flooding—the least efficient irrigation method—has declined, while the area irrigated by sprinkler and micro-irrigation methods has increased (Figure 3.10).

Conclusions

National water use has shown marked reductions in recent years. Total water withdrawals in the United States in 2010 were lower than they were in 1970, despite continued economic and population growth. This is evident in continued reductions in per capita water use, which was lower in 2010 than it was in 1945. Likewise, the economic productivity of water (dollars of gross domestic product per unit of water used) is higher than it has ever been, nearly tripling over the past three decades, from only $1.50 in 1980 (in 2009 dollars) to more than $4.30 (in 2009 dollars) of GDP per m³ used. These results show that the United States now produces far more wealth with far less water than at any time in the past.

Thermoelectric power plants represent the single largest use of water—both fresh and saline—in the United States. Thermoelectric power plants, which can be powered by fossil, geothermal, nuclear, and biomass fuels or the sun, use water for cooling purposes and for makeup water that replenishes boiler water lost through evaporation. However, water use for thermoelectric power plants is less than it was in 1970, an important reversal of a 25-year trend of increasing water use for producing energy. Continued water use reductions are possible by expanding energy-efficiency efforts, installing more dry cooling systems, and relying more heavily on renewable energy, such as wind and solar photovoltaics.

---

3. Irrigation water use includes water applied by an irrigation system to sustain plant growth in all agricultural and horticultural practices, as well as water that is used for pre-irrigation, frost protection, application of chemicals, weed control, field preparation, crop cooling, harvesting, dust suppression, and leaching salts from the root zone.

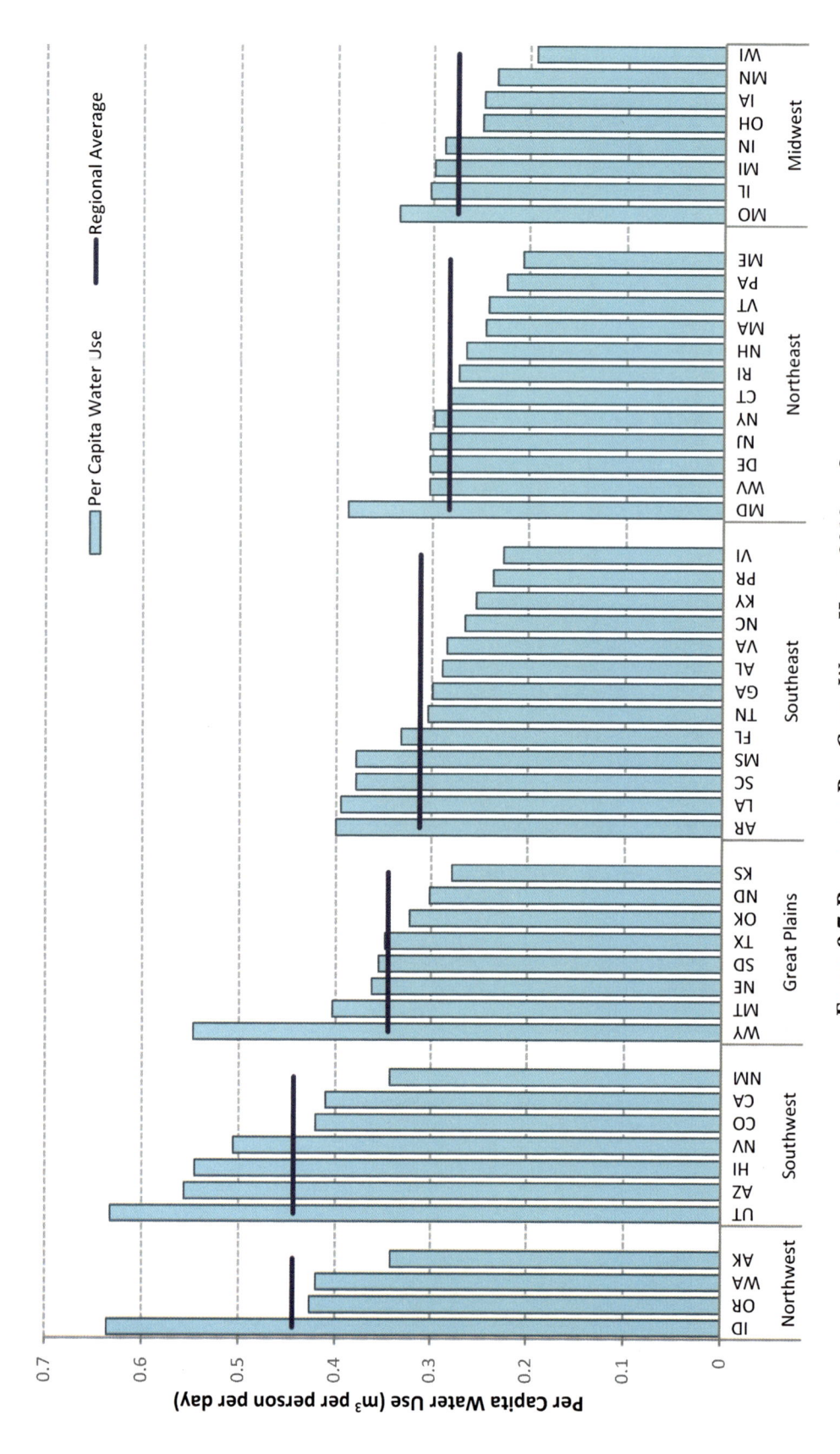

Figure 3.7 Residential Per-Capita Water Use in 2010, by State.
Note: Regional average is weighted by population.
Source: USGS 2014b.

Water Use Trends in the U.S. 51

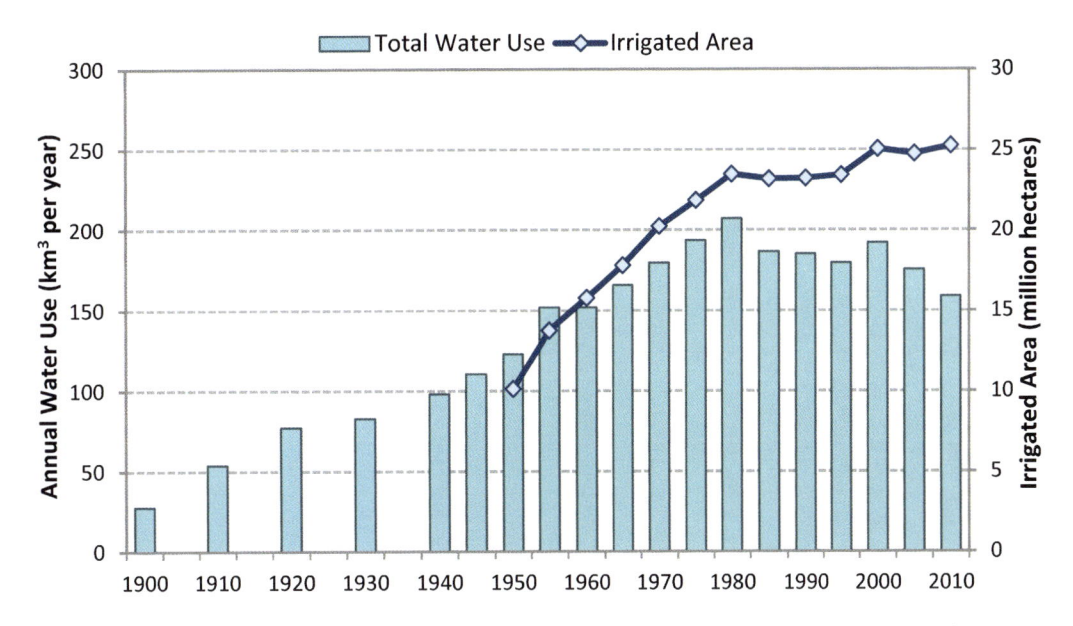

FIGURE 3.8 ANNUAL FRESHWATER USE FOR IRRIGATION (1900–2010) AND IRRIGATED AREA (1950–2010).

Sources: Data for 1900–1945 from CEQ 1991. Data on irrigated areas for 1950–1980 from the USGS water use data companion publications, *Estimated Use of Water in the United States*, which are published along with each data release: MacKichan 1951 and 1957; MacKichan and Kammerer 1961; Murray 1968; Murray and Reeves 1972 and 1977; Solley et al. 1983. Data on irrigated areas for 1985–2010 from USGS 2014b. Water use data for 1950–2010 from USGS 2014a..

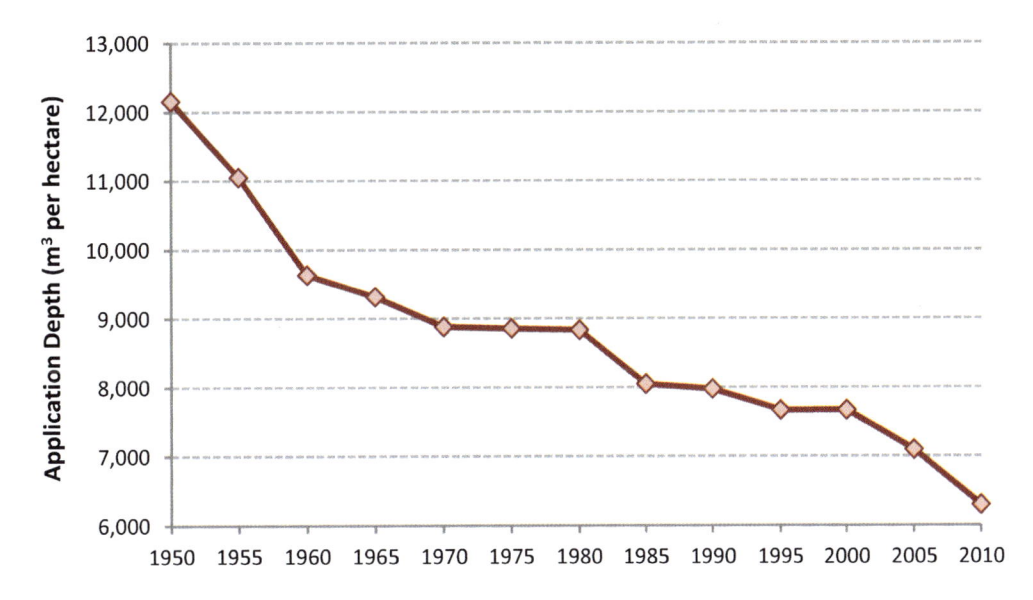

FIGURE 3.9 AVERAGE APPLICATION DEPTH, 1950–2010.

Sources: Data on irrigated areas for 1950–1980 from the USGS water use data companion publications, *Estimated Use of Water in the United States*, which are published along with each data release: MacKichan 1951 and 1957; MacKichan and Kammerer 1961; Murray 1968; Murray and Reeves 1972 and 1977; Solley et al. 1983. Data on irrigated areas for 1985–2010 from USGS 2014b. Water use data for 1950–2010 from USGS 2014a.

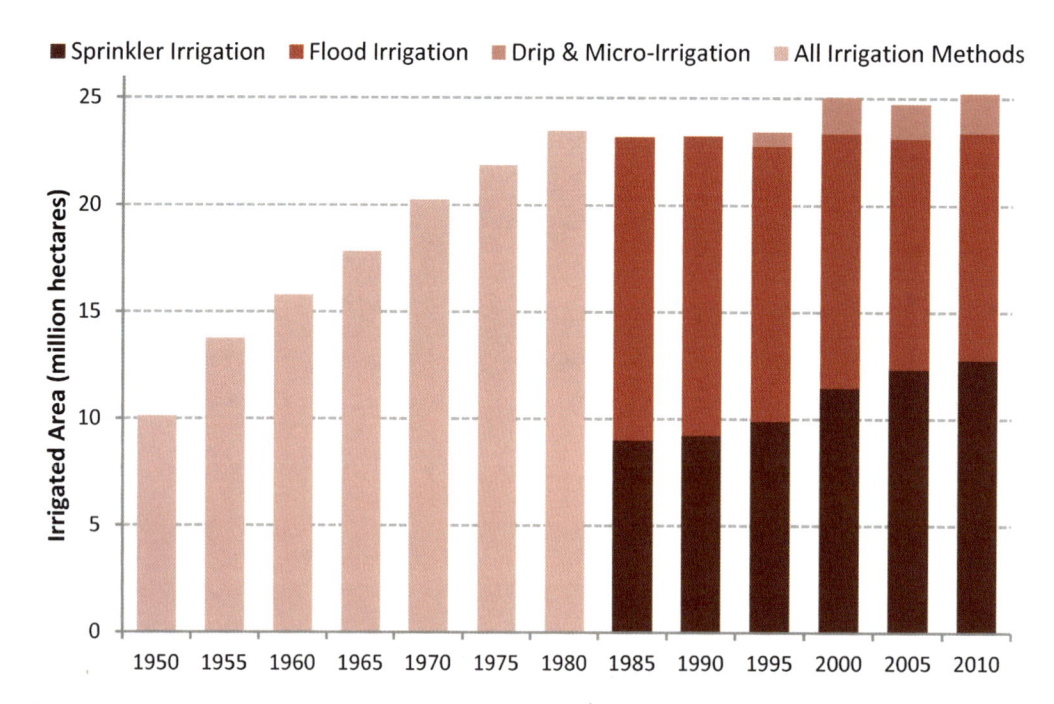

FIGURE 3.10 IRRIGATED AREA IN MILLION HECTARES, 1950–2010, BY IRRIGATION METHOD.
Sources: Data for 1985–2010 from USGS 2014b. Data for 1950–1980 from the USGS water use data companion publications, *Estimated Use of Water in the United States*, which are published along with each data release: MacKichan 1951 and 1957; MacKichan and Kammerer 1961; Murray 1968; Murray and Reeves 1972 and 1977; Solley et al. 1983. Irrigated areas by type were not available before 1985. Areas employing drip and micro-irrigation were included in sprinkler irrigation for 1985 and 1990.

Municipal and industrial water use represents the amount of water withdrawn to meet the needs of cities, towns, and small communities, including household uses; as well as commercial, industrial, institutional, and mining uses to produce the goods and services society desires. M&I water use peaked in 1980 and has been steadily declining since. By 2010, M&I water use was less than it was in 1965. Household water use, by contrast, has been steadily increasing since the 1950s but, for the first time ever, decreased between 2005 and 2010. Indeed, household per capita water use declined in 38 U.S. states and territories between 2005 and 2010, with the largest reductions in Nevada, Texas, and Nebraska.

Water used for agricultural irrigation also continued a declining trend in 2010, while irrigated areas continued to increase. Water use for agricultural irrigation has followed a pattern similar to other sectors. Total water use for irrigation increased through much of the twentieth century (along with irrigated areas), peaked in 1980, and has declined in nearly every period since. By 2010, water use for irrigation was at its lowest level in more than 40 years, despite continued growth in the number of hectares irrigated.

Considerable progress has been made in managing the nation's water and using it more effectively. In addition, USGS and other entities have greatly improved the process used to collect and evaluate the data. However, national water use remains high, and many freshwater systems are under stress from overuse. Continued improvements in water use and management will likely be hindered by continued population growth, economic expansion, and climate change, contributing to increasing tensions over scarce water resources. But this is not a foregone conclusion. In order to ensure that water use efficiency and productivity continue to improve, we must expand efforts to develop and deploy the technologies and policies that contribute to the effective use of our limited water resources in our homes, businesses, and on our farms.

References

California Energy Commission (CEC). 2016. California Energy Commission—Tracking Progress: Once-Through Cooling Phase-Out. http://www.energy.ca.gov/renewables/tracking_progress/documents/once_through_cooling.pdf.

Committee on USGS Water Resources Research. 2002. Estimating Water Use in the United States: A New Paradigm for the National Water-Use Information Program. Washington, DC: Water Science and Technology Board, Division on Earth and Life Studies, National Research Council, The National Academies Press. http://www.nap.edu/download/10484.

Cooley, H., J. Fulton, and P. H. Gleick. 2011. Water for Energy: Future Water Needs for Electricity in the Intermountain West. Oakland, CA: Pacific Institute.

Council on Environmental Quality (CEQ). 1991. 22nd Annual Report of the Council on Environmental Quality together with the President's Message to Congress. Washington, DC: U.S. Government Printing Office.

Gleick, P. H. 2003. Water Use. In *Annual Review of Environment and Resources* 28, pp. 275–314. doi: 10.1146/annurev.energy.28.040202.122849

Gleick, P. H. 2012. The Water of the United States: Freshwater Availability and Use. In *A Twenty-First Century U.S. Water Policy*, edited by J. Christian-Smith, P. H. Gleick, H. Cooley, L. Allen, A. Vanderwarker, and K. A. Berry, pp. 3–22. New York: Oxford University Press.

Hutson, S. S., N. L. Barber, J. F. Kenny, K. S. Linsey, D. S. Lumia, and M. A. Maupin. 2004. *Estimated Use of Water in the United States in 2000*. USGS Circular 1268. USGS. http://pubs.usgs.gov/circ/2004/circ1268/.

MacKichan, K. A. 1951. *Estimated Use of Water in the United States, 1950*. USGS Circular 115.

———. 1957. *Estimated Use of Water in the United States, 1955*. USGS Circular 398.

MacKichan, K. A., and J. C. Kammerer. 1961. *Estimated Use of Water in the United States, 1960*. USGS Circular 456.

Maupin, M. A., J. F. Kenny, S. S. Hutson, J. K. Lovelace, N. L. Barber, and K. S. Linsey. 2014. *Estimated Use of Water in the United States in 2010*. USGS Circular 1405.

Murray, C. R. 1968. *Estimated Use of Water in the United States in 1965*. USGS Circular 556.

Murray, C. R., and E. B. Reeves. 1972. *Estimated Use of Water in the United States in 1970*. USGS Circular 676.

———. 1977. *Estimated Use of Water in the United States in 1975*. USGS Circular 765.

Solley, W. B., E. B. Chase, and W. B. Mann IV. 1983. *Estimated Use of Water in the United States in 1980*. USGS Circular 1001.

Solley, W. B., C. F. Merk, and R. R. Pierce. 1988. *Estimated Use of Water in the United States in 1985*. USGS Circular 1004.

United States Bureau of the Census. 1975. Bicentennial Edition: Historical Statistics of the United States, Colonial Times to 1970. Washington, DC: Department of Commerce. https://www.census.gov/library/publications/1975/compendia/hist_stats_colonial-1970.html.

United States Environmental Protection Agency (U.S. EPA). 2014. 40 CFR Parts 122 and 125: National Pollutant Discharge Elimination System—Final Regulations to Establish Requirements for Cooling Water Intake Structures at Existing Facilities and Amend Requirements at Phase I Facilities; Final Rule. *Federal Registry*, 79(158): 48312.

United States Geological Survey (USGS). 2002. Concepts for National Assessment of Water Availability and Use. Report to Congress Circular 1223. Reston, VA. http://pubs.usgs.gov/circ/circ1223/pdf/C1223.pdf.

———. 2016. Changes in Water Use Categories. In *Water Use in the United States*. May. http://water.usgs.gov/watuse/WU-Category-Changes.html.

———. 2014a. United States, State-Level Data File, Table 14. In *Trends in Estimated Water Use in the United States*, 1950–2010. http://water.usgs.gov/watuse/data/2010/index.html.

———. 2014b. United States, County-Level Data File. http://water.usgs.gov/watuse/data/2010/index.html.

———. 2014c. National Water Census—Water Use. February 24. http://water.usgs.gov/watercensus/water-use.html.

Walsh, J., D. Wuebbles, K. Hayhoe, J. Kossin, K. Kunkel, G. Stephens, P. Thorne, R. Vose, M. Wehner, J. Willis, D. Anderson, S. Doney, R. Feely, P. Hennon, V. Kharin, T. Knutson, F. Landerer, T. Lenton, J. Kennedy, and R. Somerville. 2014. Our Changing Climate: Climate Change Impacts in the United States. Chapter 2 in *The Third National Climate Assessment*, edited by J. M. Melillo, T. C. Richmond, and G. W. Yohe. U.S. Global Change Research Program, 19–67. doi:10.7930/J0KW5CXT

Williamson, S. H. 2015. What Was the U.S. GDP Then? http://www.measuringworth.com/usgdp/.

CHAPTER 4

The Water Footprint of California's Energy System, 1990-2012

Julian Fulton and Heather Cooley

Introduction

Water and energy are interlinked and interconnected in a wide variety of ways. Water and sewerage systems use energy to pump, store, treat, and heat water. Energy systems use and pollute water for hydropower generation, extraction and processing of fuels, energy transformation, and end uses. A substantial amount of energy used in homes is used to heat water. This relationship—often referred to as the water–energy nexus—has received substantial attention in recent years. Indeed, in 2014 it was the theme of World Water Day and the focus of the United Nations World Water Development Report.

This chapter examines the impacts of energy systems on water resources. Energy policies are increasingly driven by the need to curtail greenhouse gas emissions in light of climate change. Despite growing recognition of the global water crisis and the potential for climate change to exacerbate these concerns (Gleick 2010; Vörösmarty et al. 2010; Oki 2006), policy makers have often failed to consider the implications of energy policies on water resources. We use the case of California's energy system from 1990 to 2012 to examine how energy policies have affected demands on water resources and provide insight into potential climate mitigation policies. We use a water footprint approach to highlight three features of California's energy-related water footprint (EWF), including (1) the *intensity*, or volume of water consumed for the state's energy system; (2) the *type* of water consumed in the form of "blue" or "green" water; and (3) the *location* where the water consumption occurred—that is, inside or outside of California.

Background

Water availability has posed real and perceived constraints on California's energy system. Most directly, seasonal precipitation and snowpack in the Sierra Nevada mountain range determine the state's hydropower generation, which provides an average of about 15 percent of in-state electricity. During drought years, hydropower generation is cur-

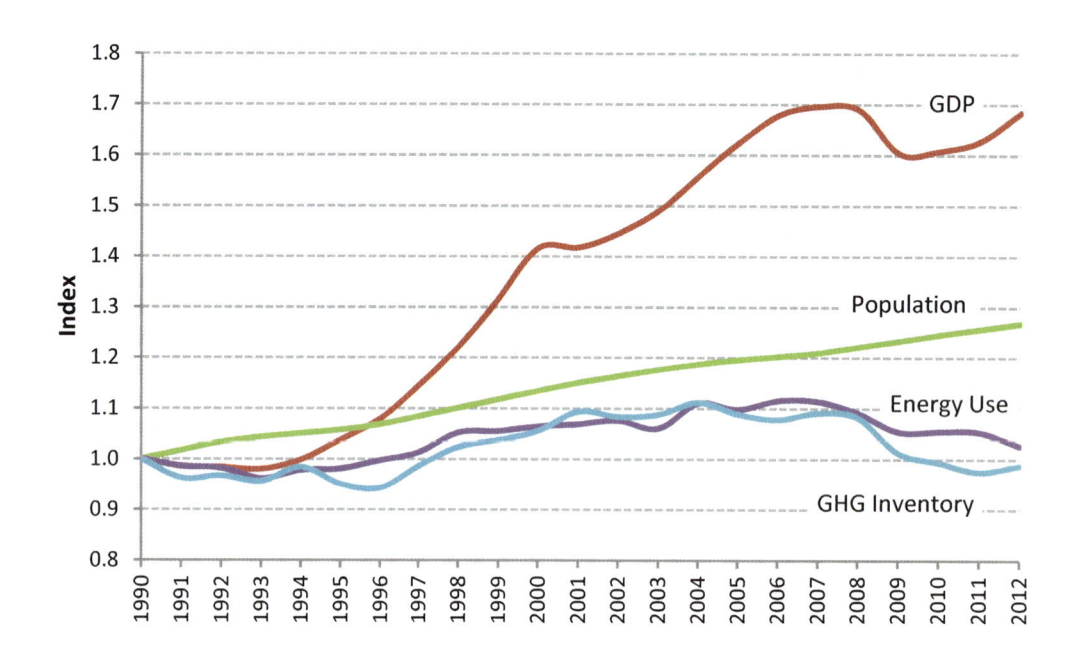

Figure 4.1 Changes in California GDP, Population, Energy Use, and Energy Greenhouse Gas Inventory, 1990–2012 (Index: 1990=1.0).
Source: U.S. EIA 2017c; CARB 2007; CARB 2014; and CDOF n.d.

tailed, forcing the state's grid operator to source electricity from other in-state generators or import more electricity from other states to meet demand. This trend was apparent most recently in the four-year period from October 2011 through September 2015, when California's hydropower generation was 57,000 GWh below average, costing ratepayers approximately $2.0 billion (Gleick 2016) in the form of more expensive makeup power. Other phases of energy production have also faced constraints: some groups have called for a ban on further development of California's shale oil resources using hydraulic fracturing and other well-stimulation techniques due to the drought and other water-supply constraints (Onishi 2014) and regulators have required some central solar power developments to use more expensive low-water-use cooling systems because of limits on water availability (CEC 2013).

Over the past several decades, California has emerged as a leader in energy efficiency, renewable energy generation, and greenhouse gas (GHG) management. In 2012, California's total energy use was only 2.6 percent higher than in 1990 (U.S. EIA 2017c), and GHG emissions from the energy sector were below 1990 levels (CARB 2007; CARB 2014). Meanwhile, during the same period the state's population increased by 27 percent and gross domestic product grew by 68 percent (Figure 4.1) (CDOF n.d.). These energy achievements were made through aggressive greenhouse gas management policies, including a low-carbon fuel standard, a renewables portfolio standard for electric utilities, and a cap-and-trade program, combined with energy-efficiency programs, demographic changes, rising energy prices, and changing consumer preferences (McCollum et al. 2012; Sudarshan 2013). Each of these changes has resulted in shifts in the amount and type of fuel use as well as in production technologies and locations.

Evaluating Water-Energy Links

In this analysis, we define California's energy system as the full range of energy consumed within the state's borders, including electricity and direct use of fuels for the household, industrial, commercial, and transportation sectors. California's energy system underwent significant changes between 1990 and 2012, making it an important time period to study, but also complicating data collection efforts. To account for complex and dynamic energy patterns, we utilized the framework of the California Energy Balance (CALEB) database, maintained by Lawrence Berkeley National Laboratories (CEC 2012). CALEB contains highly disaggregated data on annual energy supply, transformation, and end-use consumption for 30 distinct energy products, from 1990 to 2008. We used data in physical units (barrels of oil, million cubic feet of natural gas, etc.) from CALEB to quantify energy product flows over time. Following methods in de la Rue du Can (2013), we updated physical unit statistics for the years 2009–2012. While some energy products consumed in California are from in-state sources, others are imported from neighbors and distant trading partners. To identify the origin and type of imported energy products, we used data from the California Energy Commission on electricity (CEC 2017a), and from the Energy Information Administration on natural gas (U.S. EIA 2017a) and oil and ethanol (U.S. EIA 2017b).

Nearly every stage in the production of energy products consumes water, whether through evaporation, contamination, or other ways in which water is unavailable for reuse in the same river basin (Gleick et al. 2011). We characterize the EWF of an energy product by its "blue" and "green" components (Falkenmark and Rockström 2006): the blue water footprint (blue EWF) of an energy product refers to the consumption of surface or ground water, such as evaporation of water for power plant cooling; the green water footprint (green EWF) refers to the consumption of precipitation and in-situ soil moisture, such as through transpiration from the production of bioenergy feedstocks (Gerbens-Leenes et al. 2009). The related "gray" water footprint—that is, the volume of water to assimilate pollutants into water bodies at levels that meet governing standards—is not addressed explicitly in this analysis due to lack of data, although we address such water-quality concerns qualitatively.

Blue EWF factors for energy extraction, processing, and electricity generation were derived from several sources and are shown in Table 4.1. Meldrum et al. (2013) recently completed a review and harmonization of life-cycle water-use factors on various electricity fuel cycle and generation technologies. We used reported median consumptive use factors for natural gas, coal, nuclear, solar, wind, and geothermal power. We used a related study from the National Renewable Energy Laboratory (NREL) for consumptive use factors for biomass and hydropower (Macknick et al. 2011). All these factors were further weighted for the composition of California's electricity consumption when different types of fuel cycle, generation, and cooling technologies could be identified by location and year. Table 4.2 shows blue and green EWF factors used for extraction, processing, and refining of liquid fuels. Consumptive water-use factors for oil products were taken from Wu and Chiu (2011). For bioethanol production, we used country-level weighted average factors from Mekonnen and Hoekstra (2010), including refining and on-farm green and blue water requirements of bioethanol feedstocks. Further details on calculation steps for EWF factors can be found in Fulton and Cooley (2015).

TABLE 4.1 Median Consumptive Water-Use Factors for Electricity Production Technologies Used to Calculate California's Blue EWF

Fuel	Location	Fuel Cycle (l water per MWh)	Generation (l water per MWh)	Source
Coal	All	96	1,900	Meldrum et al. 2013
Natural Gas	All	24*	740	Meldrum et al. 2013
Nuclear	All	210	1,800	Meldrum et al. 2013
Conventional Hydropower	All	17,000†	-	Macknick et al. 2011
Geothermal	All	-	2,300	Meldrum et al. 2013
Biomass	All	-	2,100	Macknick et al. 2011
Solar PV	All	-	330	Meldrum et al. 2013
Solar Thermal	All	-	4,000	Meldrum et al. 2013
Wind	All	-	4	Meldrum et al. 2013
Unspecified Imported Electricity	All	1,300	1,400	Meldrum et al. 2013

Notes: EWF factors are weighted by extraction, processing, and electricity generation technologies pertaining to California's energy system. See Fulton and Cooley (2015) for further details. Numbers rounded to two significant figures.

*The equivalent factor for direct use of natural gas is 0.13 l water/m³ gas.

†This quantity refers to evaporative losses from reservoirs, which often serve other uses such as storage for flood control, urban and agricultural water supply, and recreation. However, as no methodology exists to accurately allocate consumption among the various uses, we used existing assumptions in the literature that all evaporative losses are attributable to electricity production (Macknick et al. 2011).

Blue and green EWF factors (e.g., liters of water per liter of ethanol) were multiplied by physical units of energy consumed in California (e.g., liters of ethanol) for each year between 1990 and 2012. This method assumed that blue and green EWF factors did not change over the 23-year time frame. In reality, we expect that many of these factors likely have decreased due to efficiency improvements, weather, etc. Many of these factors were derived using data from around the middle of our time series (2000), but we lack data with which to model changes before and after these points. Thus, results are indicative of how California's EWF has changed with respect to changes in its energy system, but exclude ongoing technical changes. Further research into how consumptive water-use factors have changed in the energy sector could enrich this approach and subsequent findings.

Water for California's Total Energy System

The amount of water required to support California's total energy system has changed significantly over the period examined (Figure 4.2). In 1990, the state's total EWF was about 2.1 cubic kilometers (km³), increasing by a factor of three to 7.7 km³ in 2012. The

TABLE 4.2 Median Consumptive Water-Use Factors for Liquid Fuel Production Used to Calculate California's Blue and Green EWF

Fuel	Location	Extraction Farming (l water per l fuel) Green Water	Extraction Farming (l water per l fuel) Blue Water	Refining (l water per l fuel)	Source
Crude Oil	Alaska & California	n/a	5.4	1.5	Wu and Chiu 2011
Crude Oil	Foreign Countries	n/a	3.0	1.5	Wu and Chiu 2011
Ethanol	California	n/a	n/a	3	Wu and Chiu 2011
Ethanol	USA (Corn)	1,200	150	3	Mekonnen and Hoekstra 2010
Ethanol	Brazil (Sugar)	1,200	54	3	Mekonnen and Hoekstra 2010
Ethanol	Canada (Corn)	1,100	13	3	Mekonnen and Hoekstra 2010
Ethanol	China (Corn)	1,800	170	3	Mekonnen and Hoekstra 2010
Ethanol	Costa Rica (Sugar)	1,400	250	3	Mekonnen and Hoekstra 2010
Ethanol	El Salvador (Sugar)	1,500	54	3	Mekonnen and Hoekstra 2010
Ethanol	Guatemala (Sugar)	1,300	130	3	Mekonnen and Hoekstra 2010
Ethanol	Jamaica (Sugar)	2,100	270	3	Mekonnen and Hoekstra 2010
Ethanol	Nicaragua (Sugar)	1,500	160	3	Mekonnen and Hoekstra 2010
Ethanol	Trinidad & Tobago (Sugar)	2,200	78	3	Mekonnen and Hoekstra 2010
Ethanol	Other (Sugar)	1,400	580	3	Mekonnen and Hoekstra 2010

bulk of the change is attributable to water consumed for ethanol production, which increased from 0.2 km³ in 1990 to 6.3 km³ in 2012. Indeed, California's EWF is highly sensitive to the role of ethanol, and we discuss this role at greater length below, after examining the EWF of other energy sources.

The EWF of California's natural gas consumption for the residential, commercial, industrial, and electric power sectors increased from 0.005 km³ in 1990 to 0.013 km³ in 2012, representing a 150 percent increase over this period. The consumption of natural gas, however, increased by only 24 percent during this period. This disparity resulted from the growing application of hydraulic fracturing techniques around the U.S. to extract unconventional natural gas resources, which doubled the technology-weighted water intensity of California's natural gas consumption between 1990 and 2012, from 0.1 to 0.2 liters per cubic meters. Despite this growth, natural gas remained a relatively small

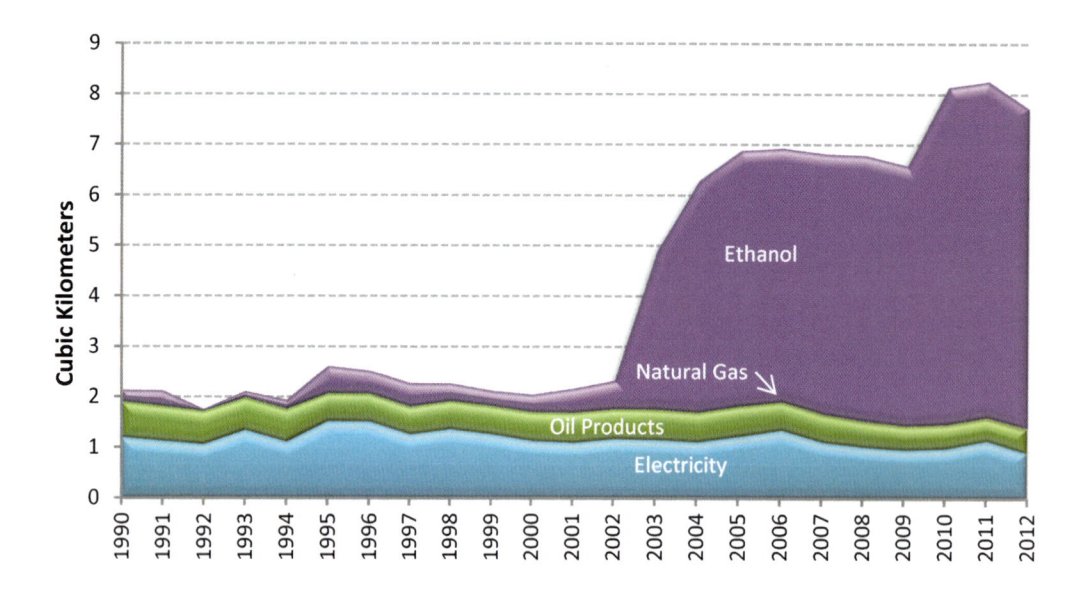

Figure 4.2 CALIFORNIA'S ENERGY-WATER FOOTPRINT, 1990–2012, BY ENERGY TYPE.
Source: Fulton and Cooley 2015.

component of the state's total EWF. However, there is regional variation in the water intensity and impacts in shale gas exploitation, making natural gas an important energy product to monitor and manage in California's future energy–water portfolio.

The EWF of oil products consumed in California declined from 0.7 km³ in 1990 to less than 0.5 km³ in 2012, representing a 30 percent decrease. During this period, however, the quantity of oil products consumed in California declined by only 2 percent. Therefore, the drop in oil's EWF was due primarily to shifting from more water-intensive oil production in California to less water-intensive production locations. In 1990, California produced around half of its domestic demand; however, by 2010 that number had dropped to 37 percent (CEC 2017b).

The EWF of California's electricity consumption first increased from 1.2 km³ in 1990 to 1.5 km³ in 1995 and then dropped substantially to 0.9 km³ in 2012. The relatively high degree of variability compared to other energy products is due to the complexity of California's portfolio of generation sources and the wide range in water requirements for those different generation technologies. While total electricity consumption increased over this period, most of this electricity was produced by relatively less water-intensive generation technologies, such as gas turbine or combined-cycle natural gas power plants, wind turbines, and solar photovoltaics. Hydroelectric generation, an extremely water-intensive form of electricity generation due to high evaporative losses from reservoirs, also decreased as a share of California's total electricity portfolio, in part due to changes in the state's electricity mix and in part due to droughts during this period.

Since 1990, there have been dramatic changes in the "type" of water consumed—green vs. blue water (Figure 4.3). In 1990, only 10 percent of California's EWF was green water and the remaining 90 percent was blue water, of which 63 percent was attributable to the electricity sector and 35 percent to oil products. Since 2003, however, green water has dominated California's EWF, and in 2012, blue water made up only 27 percent of the state's EWF. Plant-based ethanol accounts for all of this green water and 33 percent of the

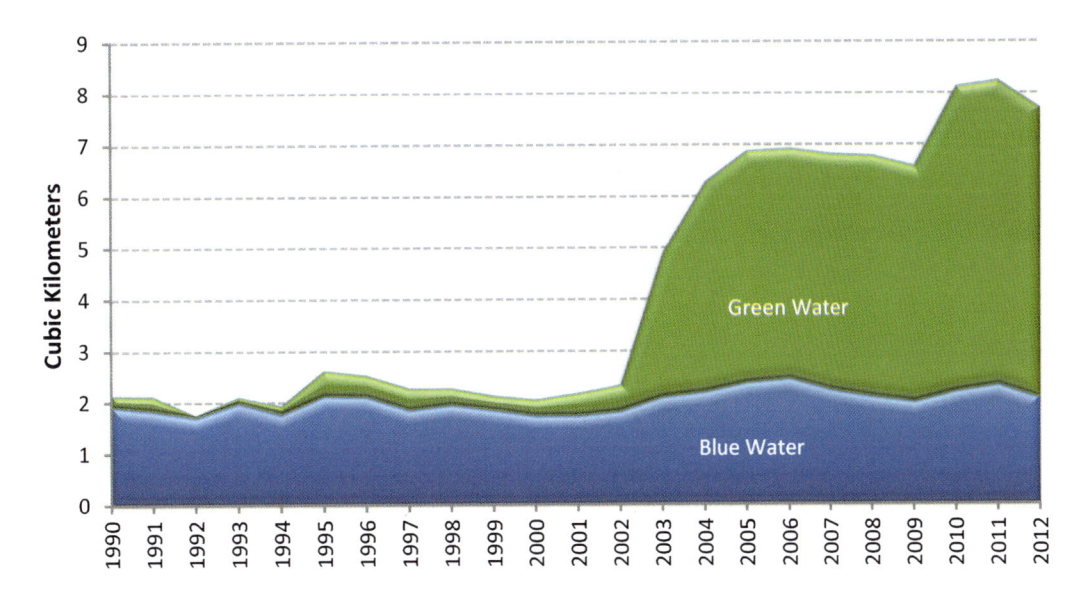

FIGURE 4.3 CALIFORNIA'S ENERGY–WATER FOOTPRINT, 1990–2012, BY TYPE OF WATER.
Source: Fulton and Cooley 2015.

blue water; while electricity, oil products, and natural gas make up the remainder of the blue EWF.

The location of blue and green water use is relevant to local water resource concerns. Figure 4.4 shows California's EWF by internal and external sources, including the U.S. and foreign countries. In 1990, 1.0 km^3 (or about half) of California's total EWF was internal to the state; that is, using California's water resources (for comparison, this represented about 3 percent of total in-state consumptive use for all purposes (Solley et al 1993)). By 2012, the volume of California's internal EWF was slightly smaller (0.9 km^3), but it made up just 11 percent of the state's total EWF. This means that all the increase in California's EWF occurred outside of the state's borders. Indeed, much of this growth occurred in ethanol-growing regions of the U.S. Midwest, but also substantially in other countries where ethanol and oil extraction have increased.

Summary

An examination of the water footprint of California's energy system sheds light on the amount, type, and location of water consumed to produce the state's energy products. Understanding these linkages is of growing importance as the impacts of climate change on water and energy resources intensifies and as efforts to adapt to and mitigate these impacts are implemented. Our assessment highlights the need for more careful, integrated consideration of the implications of the water–energy nexus for water resource and energy system planning.

California's EWF has substantially increased over recent decades without utilizing more of the state's water resources, but rather relying more heavily on external sources of water. The increase in the EWF has been primarily associated with a large increase in the use of biofuels in the form of ethanol for the transportation sector. Biofuels depend heavily on green water—precipitation used directly by biofuel crops in the field.

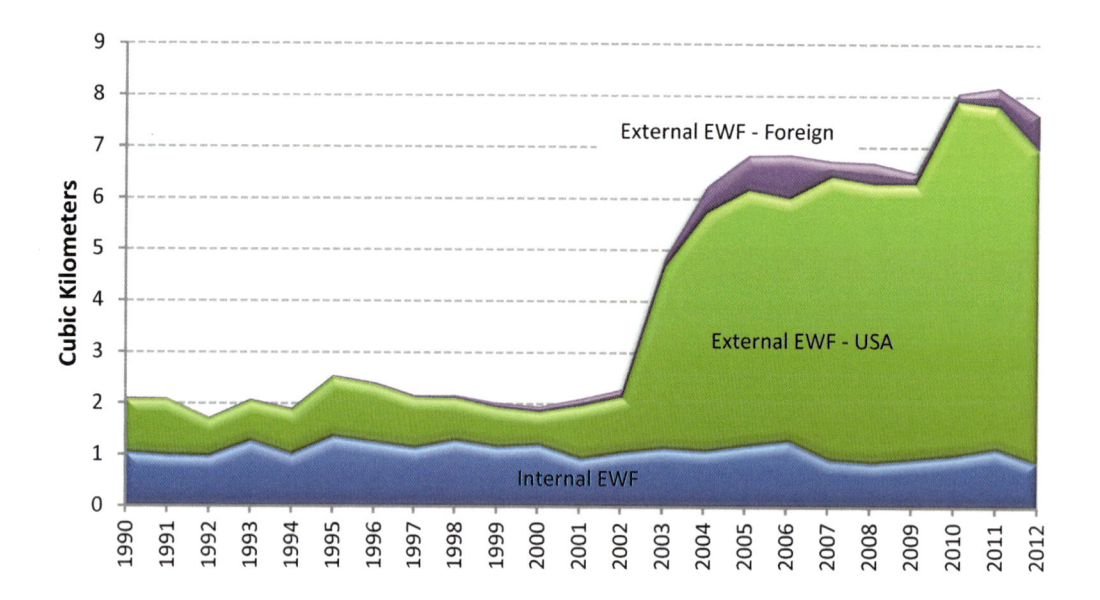

Figure 4.4 California's Energy–Water Footprint, 1990–2012, by Location.
Source: Fulton and Cooley 2015.

While green water utilization may have added benefits in that it does not require pumping or associated infrastructure, it also links California's energy future directly to future precipitation and crop choices in biofuel-growing regions. To the extent that California's increased ethanol demand has relied on blue water, its energy system has also become linked to surface and groundwater management issues in those regions, such as the overpumping of the Ogallala aquifer. The Midwest drought of 2011–2012 highlights one risk of these linkages, as this drought constrained the ethanol supply and resulted in higher ethanol prices in California markets (U.S. EIA 2012; Langholtz et al. 2014). Moreover, foreign sources of ethanol, which have constituted up to 12 percent of California's supply, may face similar climate-related challenges in the future (Haberl et al. 2011; De Lucena et al. 2009).

Although we do not present the gray water (pollution-related) footprint of ethanol here, factors provided from Mekonnen & Hoekstra (2010) indicate that California's gray EWF associated with ethanol consumption ranged from one to two cubic kilometers per year. This gray water footprint is associated with the runoff of excess fertilizers and pesticides from croplands into regional water bodies. As most of California's gray EWF is related to biomass production within the Mississippi River Basin, California's energy system requires an additional 0.2 percent to 0.4 percent of the average annual discharge of the Mississippi River to bring pollutants to acceptable levels. We note that the initial use of ethanol as a substitute for methyl tert-butyl ether (MTBE) was brought about by water-quality concerns in the state's urban groundwater basins; however, this effort may have shifted water-quality burdens outside the state rather than mitigated them altogether. This initial finding could be refined with further analysis of the pollutant persistence and relative impacts of these burdens. Nevertheless, these burdens may yet pose supply risks to California's energy system, as producing regions grapple with trade-offs between high agricultural yields and low water quality from runoff (Dominguez-Faus et al. 2009).

Water-quality concerns exist with other bioenergy sources as well, as with the extraction and processing of other fuels and electricity generation.

Many of these observed trends in California's EWF are linked to state energy policies, highlighting one consequence of failing to consider energy and water objectives together. Increased reliance on bioethanol was initially driven by the need for an alternative gasoline oxygenate following an executive order banning MTBE in 2003. More recent energy policies have encouraged additional ethanol blending in gasoline to meet state greenhouse gas targets. California's Low Carbon Fuel Standard (LCFS) of 2007, pursuant to its landmark Global Warming Solutions Act of 2006, has reinforced demand for bioethanol as a means to reduce the greenhouse gas intensity of transportation fuels. Although early LCFS policy assessments raised the issue of water demands and impacts from increased biofuel production (Farrell and Sperling 2007), any subsequent efforts to track or address water-related impacts of these energy policies have been lacking (CARB 2011).

Expected trends in California's biofuel demand pose deeper consideration for integrated research and policy. Since 2009, bioethanol has been blended into California reformulated gasoline to 10 percent by volume, and an emerging market for E85 (85 percent ethanol fuel) is likely to increase the state's demand for bioethanol. These developments have been further abetted by a broader policy environment, including the federal Renewable Fuel Standard (RFS), which since 2007 has mandated an increasing share of biofuels in U.S. transportation energy. A recent study assessed the regional water impacts of various potential RFS-technology policy scenarios, highlighting the need for attention to local effects and integrated approaches to federal policy (Jordaan et al. 2013). Still, California holds a unique position in the national biofuels landscape, as the state with the largest demand yet little economically viable in-state production capacity (U.S. EIA 2015). State-level energy policies have played, and will continue to play, a strong role in determining California's biofuel demand. Our research suggests that expected trends would substantially increase and further externalize the state's EWF in the future and that a closer examination of associated trade-offs and climate risks is needed.

Shifts in other energy products have also driven the externalization of California's EWF. In-state crude oil extraction has declined since the mid-1980s, the demand having been made up by Alaskan oil initially, then imports from foreign sources. In this case, the blue water footprint of most sources of foreign oil is lower than that of California or Alaska, so California's blue EWF declined by 31 percent as a result of this shift (despite near-constant overall supply). While this effect was unlikely intentional, it is not surprising that current efforts to "re-shore" energy production face increasing opposition, partly on the grounds of impacts to local water resources (Jordaan et al. 2013). Still, if California's consumption of oil products does not drop, water impacts may continue to accrue inside and outside the state's borders.

Electricity is another sector where consideration of water resources inside and outside of California is important (Sattler et al. 2012; Sathaye et al. 2013). Imported electricity has long been an important component of the state's energy portfolio, providing a flexible supply when hydropower potential is low or other factors restrict in-state generation. Yet, when California's grid operator outsources electricity, the state's EWF goes up because out-of-state thermoelectric sources, especially older coal plants, tend to be more water intensive than newer in-state plants and more likely to use fresh water (instead of saline water) for cooling (Ruddell and Adams 2014). Because out-of-state electricity also tends to be more greenhouse-gas intensive, we see greenhouse gas-driven energy policies having a synergistic effect in reducing California's EWF. Such synergy is not necessarily the case in other contexts. For example, in China, where electricity production in the arid north uses dry cooling and is therefore less water-intensive, energy efficiency goes down in such systems, resulting in higher greenhouse gas-per-kilowatt hour produced (Zhang et al. 2014).

As California's energy policies have sought to mitigate climate change, water systems and resources have received little attention. When energy policies have considered impacts to water, such as the MTBE ban, policy outcomes may have simply shifted water-related burdens rather than alleviated them. Given the exigencies of both climate change *and* the global water crisis, the interconnectedness of energy and water systems deserves closer attention in both academic and policy arenas. Climate and water goals are not mutually exclusive in energy policy; rather, to the extent that existing energy sources are fungible, climate and water goals can be achieved simultaneously. Additionally, many renewable sources of energy already have few water impacts. Policy makers should seek to ask questions about unforeseen or unintended water-related consequences of proposed energy policies and pathways. Analytical tools, such as the water footprint used here, provide a starting place and a framework to answer such questions.

Further research should focus more precisely on characterizing the relative impacts and risks of water footprint assessments such as California's EWF. Weighting green, blue, and gray water footprint values by their relevant water stress, opportunity costs, and water-quality impacts can lead to better decision making by energy supply-chain managers and energy-policy designers. Interconnected water and energy systems need not be a source of risk for California or other entities; rather, integrated analysis and deeper understanding of these essentially linked resources can increase productivity at the water-energy nexus and simultaneously support climate-change mitigation and adaptation strategies.

Acknowledgements

We wish to thank our collaborators and colleagues at the California Department of Water Resources and the U.S. Environmental Protection Agency for their support and feedback.

References

California Air Resources Board (CARB). 2007. Staff Report: California 1990 Greenhouse Gas Emissions Level and 2020 Emissions Limit. Sacramento, CA: CARB.
———. 2011. Staff Report: Initial Statement of Reasons for Proposed Rulemaking: Proposed Amendments to the Low Carbon Fuel Standard. http://www.arb.ca.gov/regact/2011/lcfs2011/lcfsisor.pdf.
———. 2014. 2014 Edition California Greenhouse Gas Emission Inventory. Sacramento, CA: CARB.
California Department of Finance (CDOF). Forecasting Reports for Demographics and Economics. http://www.dof.ca.gov/Forecasting/ (accessed in 2017).
California Energy Commission (CEC). 2012. California Energy Balance Database (CALEB). http://www.energy.ca.gov/2012publications/CEC-500-2012-FS/CEC-500-2012-FS-006.pdf.
———. 2013. Ivanpah Solar Electric Generating System: Sacramento, California. http://www.energy.ca.gov/sitingcases/ivanpah/.
———. 2017a. Energy Almanac: Total Electricity System Power. http://www.energy.ca.gov/almanac/electricity_data/total_system_power.html.
———. 2017b. Energy Almanac: Oil Supply Sources to California Refineries. http://www.energy.

ca.gov/almanac/petroleum_data/statistics/crude_oil_receipts.html.

De la Rue du Can, S., A. Hasanbeigi, and J. Sathaye. 2013. California Energy Balance Update and Decomposition Analysis for the Industry and Building Sectors: Final Project Report, Sacramento, CA: Lawrence Berkeley National Laboratory and California Energy Commission.

De Lucena, A. F. P., A. S. Szklo, R. Schaeffer, R. R. de Souza, B. S. M. C. Borba, I. V. L. da Costa, A. O. P. Júnior, and S. H. F. da Cunha. 2009. The Vulnerability of Renewable Energy to Climate Change in Brazil. *Energy Policy* 37, 879–889.

Dominguez-Faus, R., S. E. Powers, J. G. Burken, and P. J. Alvarez. 2009. The Water Footprint of Biofuels: A Drink or Drive Issue? *Environmental Science & Technology* 43, 3005–3010.

Falkenmark, M., and J. Rockström. 2006. The New Blue and Green Water Paradigm: Breaking New Ground for Water Resources Planning and Management. *Journal of Water Resources Planning and Management* 132 (June), 129–132.

Farrell, A. E., and D. Sperling. 2007. A Low-Carbon Fuel Standard for California. http://www.arb.ca.gov/fuels/lcfs/lcfs_uc_p1.pdf.

Fulton, J., and H. Cooley. 2015. The Water Footprint of California's Energy System, 1990–2012. *Environmental Science & Technology* 49(6), 3314–3321.

Gerbens-Leenes, W., A. Y. Hoekstra, and T. H. van der Meer. 2009. The Water Footprint of Bioenergy. *Proceedings of the National Academy of Sciences (PNAS)* 106, 10219–10223.

Gleick, P. H. 2010. Climate Change, Exponential Curves, Water Resources, and Unprecedented Threats to Humanity. *Climate Change* 100, 125–129.

———. 2016. Impacts of California's Ongoing Drought: Hydroelectricity Generation. Oakland, CA: Pacific Institute.

Gleick, P. H., J. Christian-Smith, and H. Cooley. 2011. Water-Use Efficiency and Productivity: Rethinking the Basin Approach. *Water International* 36(7), 784–798.

Haberl, H., K.-H. Erb, F. Krausmann, A. Bondeau, C. Lauk, C. Müller, C. Plutzar, and J. K. Steinberger. 2011. Global Bioenergy Potentials from Agricultural Land in 2050: Sensitivity to Climate Change, Diets and Yields. *Biomass Bioenergy* 35, 4753–4769.

Jordaan, S. M., L. Diaz Anadon, E. Mielke, and D. P. Schrag. 2013. Regional Water Implications of Reducing Oil Imports with Liquid Transportation Fuel Alternatives in the United States. *Environmental Science & Technology* 47, 11976–11984.

Langholtz, M., E. Webb, B. L. Preston, A. Turhollow, N. Breuer, L. Eaton, A. W. King, S. Sokhansanj, S. S. Nair, and M. Downing. 2014. Climate Risk Management for the U.S. Cellulosic Biofuels Supply Chain. *Climate Risk Management* 3, 96–115.

Macknick, J., R. Newmark, G. Heath, and K. C. Hallett. 2011. A Review of Operational Water Consumption and Withdrawal Factors for Electricity Generating Technologies. Golden, CO: National Renewable Energy Laboratory.

McCollum, D., C. Yang, S. Yeh, and J. Ogden. 2012. Deep Greenhouse Gas Reduction Scenarios for California—Strategic Implications from the CA-TIMES Energy-Economic Systems Model. *Energy Strategy Reviews* 1, 19–32.

Mekonnen, M. M, and A. Y. Hoekstra. 2010. The Green, Blue and Grey Water Footprint of Crops and Derived Crop Products. Delft, The Netherlands: UNESCO-IHE.

Meldrum, J., S. Nettles-Anderson, G. Heath, and J. Macknick. 2013. Life Cycle Water Use for Electricity Generation: A Review and Harmonization of Literature Estimates. *Environmental Research Letters* 8, 015031.

Oki, T., and S. Kanae. 2006. Global Hydrological Cycles and World Water Resources. *Science* 313, 1068–1072.

Onishi, N. 2014 (May 5). California's Thirst Shapes Debate Over Fracking. *New York Times*. https://www.nytimes.com/2014/05/15/us/californias-thirst-shapes-debate-over-fracking.html.

Ruddell, B., and E. Adams. 2014. Embedded Resource Accounting for Coupled Natural Human Systems: An Application to Water Resource Impacts of the Western US Electrical Energy Trade. *Water Resources Research* 50, 1–16.

Sathaye, J. A., L. L. Dale, P. H. Larsen, G. A. Fitts, K. Koy, S. M. Lewis, and A. F. P. de Lucena. 2013. Estimating Impacts of Warming Temperatures on California's Electricity System. *Global Environmental Change* 23, 499–511.

Sattler, S., J. Macknick, D. Yates, F. Flores-Lopez, A. Lopez, and J. Rogers. 2012. Linking Electricity and Water Models to Assess Electricity Choices at Water-Relevant Scales. *Environmental Research Letters* 7, 045804.

Solley, W. B., R. R. Pierce, and H. A. Perlman. 1993. Estimated Use of Water in the United States in 1990. U.S. Geological Survey Circular 1081.

Sudarshan, A. 2013. Deconstructing the Rosenfeld Curve: Making Sense of California's Low Electricity Intensity. *Energy Economics* 39, 197–207.

U.S. Energy Information Administration (U.S. EIA). 2012. Today in Energy: Drought Has Significant Effect on Corn Crop Condition, Projected Ethanol Production. https://www.eia.gov/todayinenergy/detail.php?id=7770.

———. 2015. State Energy Data System: Table F4: Fuel Ethanol Consumption Estimates. http://www.eia.gov/state/seds/sep_fuel/html/pdf/fuel_use_en.pdf.

———. 2017a. California International and Interstate Movements of Natural Gas by State. http://www.eia.gov/dnav/ng/ng_move_ist_a2dcu_sca_a.htm.

———. 2017b. Petroleum and Other Liquids: Company Level Imports. https://www.eia.gov/petroleum/imports/companylevel/archive/.

———. 2017c. State Energy Data System (SEDS) 1960–2015. http://www.eia.gov/state/seds/seds-data-complete.cfm?sid=CA.

Vörösmarty, C. J., P. B. McIntyre, M. O. Gessner, D. Dudgeon, A. Prusevich, P. Green, S. Glidden, S. E. Bunn, C. A. Sullivan, C. R. Liermann, et al. 2010. Global Threats to Human Water Security and River Biodiversity. *Nature* 467, 555–561.

Wu, M., and Y. Chiu. 2011. Consumptive Water Use in the Production of Ethanol and Petroleum Gasoline—2011 Update. Argonne, IL: Argonne National Laboratory.

Zhang, C., L. D. Anadon, H. Mo, Z. Zhao, and Z. Liu. 2014. Water–Carbon Trade-off in China's Coal Power Industry. *Environmental Science & Technology* 48, 11082–11089.

CHAPTER 5

The Nature and Impact of the 2012-2016 California Drought

Peter H. Gleick

Introduction

California experienced a severe drought beginning in 2012 and extending for five years through the end of the 2016 "water year".[1] The 2017 water year, in contrast, was extraordinarily wet, putting an end—at least for now—to the hydrological drought and reducing drought concerns and restrictions. California's five-year drought was felt as a significant shortfall in the amount of water available in the form of rain, snow, runoff, and groundwater compared to that demanded by all the different economic and ecological sectors of California. This summary offers an overview of the hydrologic conditions behind the drought and some insights into its impacts on agriculture, ecosystems, hydropower production, and urban centers. Because of the length and severity of the drought, and because all the consequences have yet to be catalogued or analyzed, this summary offers only a partial overview.

The Hydrological and Climatological Conditions behind the California Drought

Drought can be defined and measured in many ways—from meteorological drought to hydrological drought, to soil-moisture deficits, to a shortage of water for some defined economic or environmental demand (Wilhite and Glantz 1985). Box 5.1 describes various drought terms and definitions. No single metric or indicator is sufficient to characterize drought. In California, the recent drought includes characteristics of all these variables: reduced precipitation, increased water loss due to higher temperatures, below-average snowfall and earlier snowmelt, low streamflow, depleted soil moisture, reduced storage in reservoirs, and shortages in water deliveries to users. For the purposes of this

---

1. Each water year begins on October 1 of the prior calendar year and ends on the following September 30; water year 2016 refers to the period October 1, 2015, through September 30, 2016.

> **BOX 5.1** Definitions of Drought
>
> The term "drought" has many definitions. What is considered a drought in a wet region differs from that in a dry region. At its simplest, drought is a shortfall in precipitation over an extended period of time, which leads to a shortage of water for specific human or ecological needs. This definition includes both the effects of natural hydrologic variability and the demands placed on water resources by humans and ecosystems.
>
> Operational definitions of drought typically include data and information on changes in precipitation rates or soil moisture compared to historical averages, but the National Drought Mitigation Center (NDMC), which defines and tracks U.S. drought data, recommends that different regions and water users develop indices and metrics, and drought response and mitigation strategies most appropriate to local needs.
>
> **Meteorological Drought**
> Meteorological drought is defined based on a measure of "dryness"—usually quantified as precipitation shortfall—compared to a long-term average and the duration of the dry period.
>
> **Hydrological Drought**
> Hydrological drought is usually a consequence of meteorological drought and measured by the degree of impact on a hydrological variable (such as snowpack, streamflow, soil moisture, reservoir or lake levels, groundwater), with resulting social and economic impacts.
>
> **Agricultural Drought**
> Agricultural drought looks at how characteristics of meteorological and hydrological drought affect agricultural production and water availability for irrigation and the production of food and fiber.
>
> *Sources*: NWS 2012; NDMC 2016.

assessment, we define drought here in a straightforward manner: not having sufficient water to do what society wants; that is, a mismatch between the amounts of water nature provides and the amounts of water that humans and the environment demand. This is consistent with the definition used by the National Drought Mitigation Center (2016):

> In the most general sense, drought originates from a deficiency of precipitation over an extended period of time—usually a season or more—resulting in a water shortage for some activity, group, or environmental sector. Its impacts result from the interplay between the natural event (less precipitation than expected) and the demand people place on water supply, and human activities can exacerbate the impacts of drought.

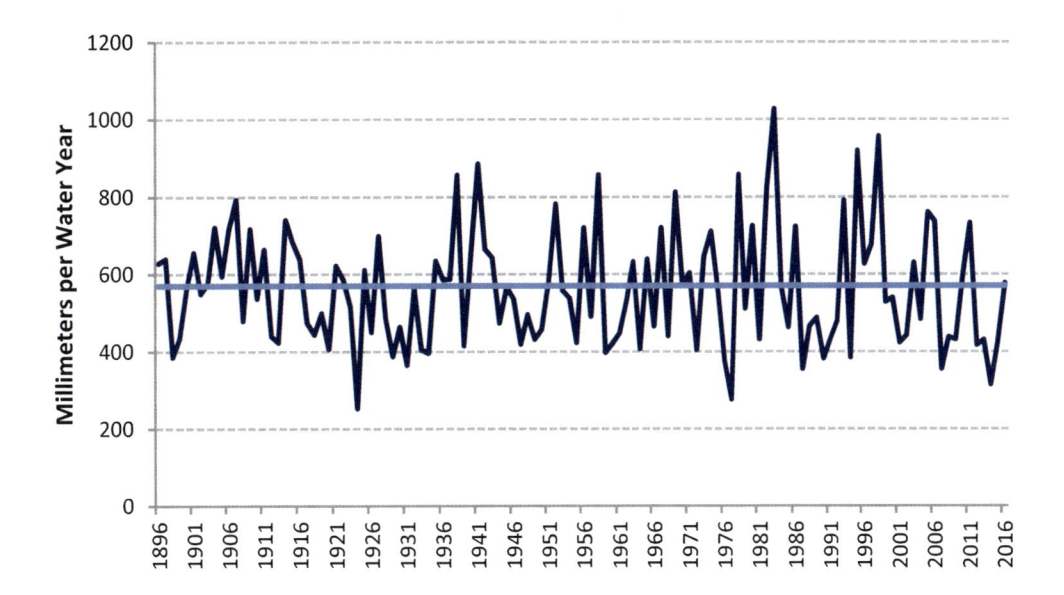

FIGURE 5.1 CALIFORNIA WATER YEAR PRECIPITATION, 1896–2016 (IN MILLIMETERS PER WATER YEAR).
Note: The graph also plots the 20th century average of 570 mm per year.
Source: NOAA (2015–2017).

Precipitation

A key driver in the California drought was a reduction in precipitation (rain and snow) between 2012 and 2016. Figure 5.1 shows the annual "water year" precipitation for California from 1895 through 2016. In an average year over the 20th century (1901 to 2000), California received 22.5 inches (570 mm) of precipitation. Precipitation over the five years of drought was 24% below normal, averaging 17.1 inches (434 mm), although no individual years were, by themselves, record low years. The deviation from average over the years from 2012 to 2016 was nearly 27 inches (678 mm)—in other words, five average years would have totaled 112 inches of precipitation, but the state only received 85 inches over this period. In addition, while most analysts have been describing the drought as a "five-year drought," seven of the past 10 years have been drier than average; and between 2000 and 2016, there have only been very short periods of time when no part of the state was in drought as measured by the National Drought Mitigation Center index (Figure 5.2).

Temperature

California's drought wasn't only influenced by low precipitation; the state also experienced far higher than normal temperatures, which worsened the water deficit by increasing evaporation and transpiration rates. Figure 5.3 shows average annual California temperatures from 1896 to 2016, plotted with a second-order polynomial trend line. During the five years of drought from 2012 to the end of the 2016 water year, average temperatures reached more than 2.8 degrees F (1.5 degrees C) above normal (the 1901 to 2000 average was 14.1 degrees C)—a dramatic departure. While the effect of rising temperatures on the hydrologic balance has not yet been assessed quantitatively, the net effect was a

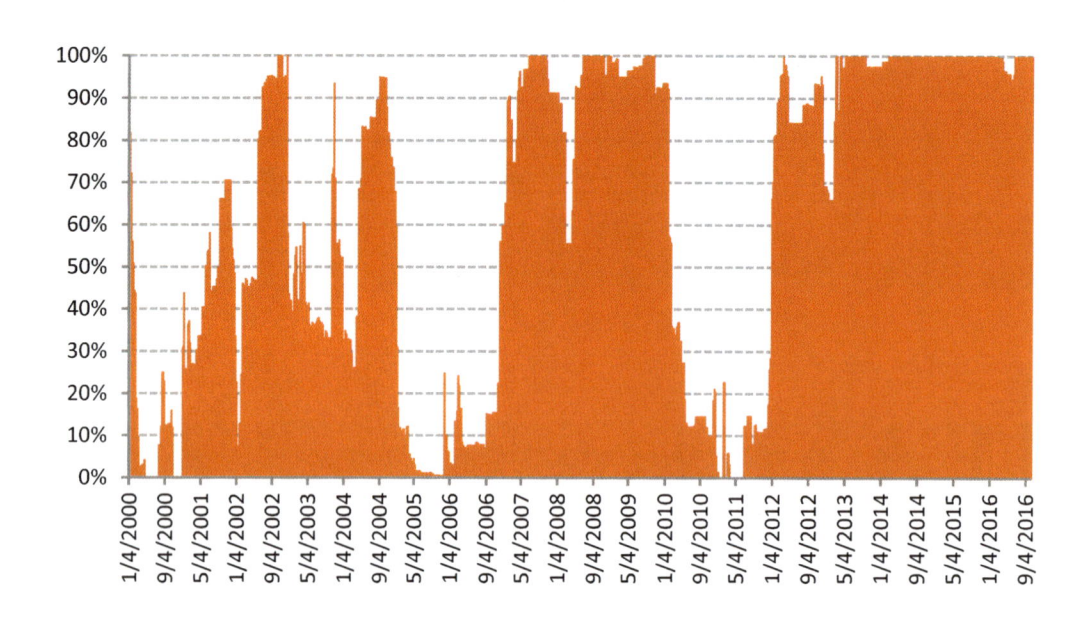

FIGURE 5.2 PERCENTAGE OF CALIFORNIA'S AREA IN DROUGHT FROM JANUARY 2000 TO OCTOBER 2016.
Source: NDMC 2016.

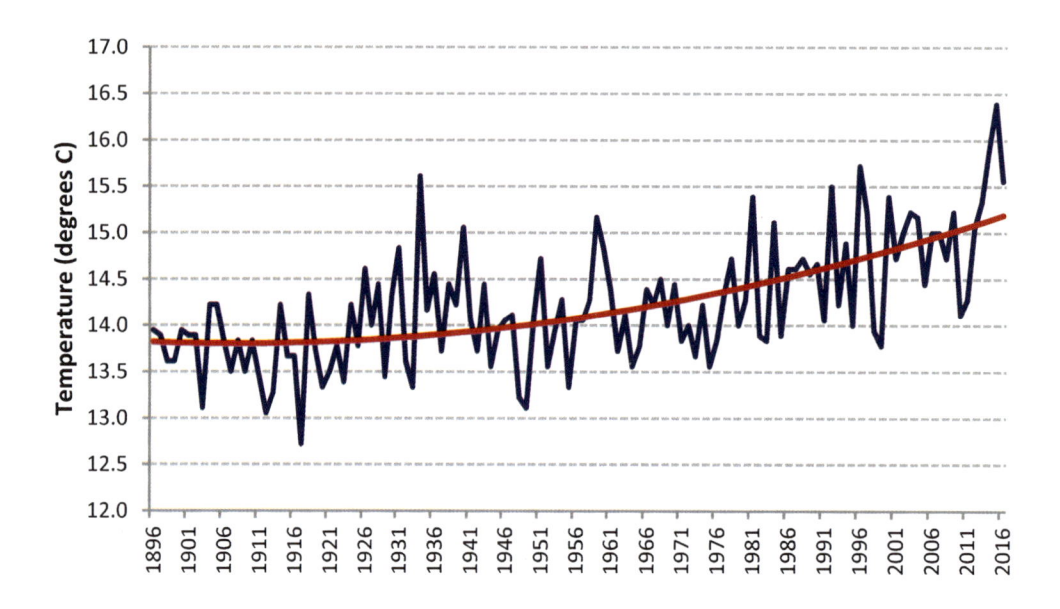

FIGURE 5.3 CALIFORNIA WATER YEAR TEMPERATURE, 1896–2016 (IN DEGREES CELSIUS).
Notes: The graph also plots the polynomial trend line (second order), showing a significant increase over the past 120 years. The average for 1901–2000 was 14.1 degrees C.
Source: NOAA 2017.

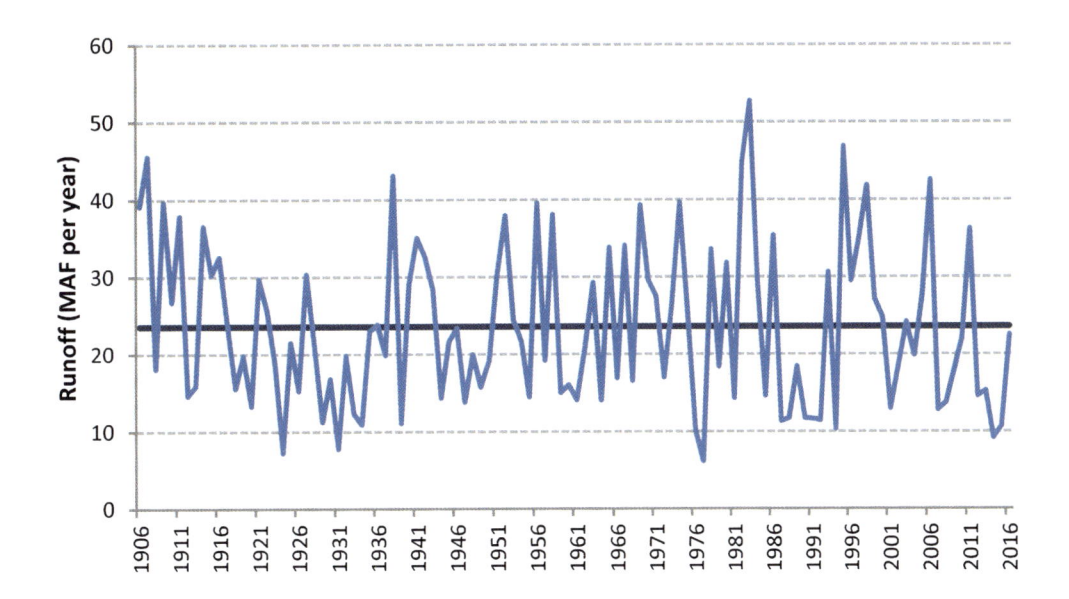

FIGURE 5.4 UNIMPAIRED RUNOFF FOR THE SACRAMENTO–SAN JOAQUIN INDEX, WATER YEARS 1906–2016.

Note: Also shown is the average runoff for the same period.
Source: CDWR 2017 (data accessed October 2017).

reduction in water supply and an increase in water demand through higher evaporation from reservoirs and soil surfaces, and greater evapotranspiration from natural vegetation and irrigated crops. Some climate scientists have noted that the rise in temperature experienced in California can be attributed, in part, to human-induced climate change (Diffenbaugh et al. 2015; Mann and Gleick 2015).

River Runoff

Decreased precipitation and increased temperature lead directly to a reduction in streamflow in California rivers. California's river system is complex, with a series of large streams and rivers draining the Sierra Nevada Mountains and coastal ranges. While some significant rivers run off directly into the Pacific Ocean, most drain into the Sacramento and San Joaquin watersheds and ultimately into the Sacramento–San Joaquin Delta.

Most water used for the urban and agricultural sectors in California comes from the Sacramento and San Joaquin river systems, and the California Department of Water Resources prepares a regular assessment of the "unimpaired flow" in these watersheds; that is, the amount of water that would have flowed in the absence of dams and human withdrawals. These indices (Figure 5.4) show how the drought affected overall river flow. A standard metric used to evaluate these river flows is whether they represent a "wet," "above normal," below normal," "dry," or "critical" year type. Table 5.1 shows these metrics for the past 17 years (from 2000 to 2016) for the Sacramento and San Joaquin river systems. Over this period, the total runoff "deficit," measured as the difference between the long-term average runoff of these basins and the amount of runoff that actually occurred, was over 55 million acre-feet (over 65 billion cubic meters) (Figure 5.5).

TABLE 5.1 Water-Year Index for the Sacramento and San Joaquin Valleys, 2000–2016

Year	Sacramento Valley	San Joaquin Valley
2000	AN	AN
2001	D	D
2002	D	D
2003	AN	BN
2004	BN	D
2005	AN	W
2006	W	W
2007	D	C
2008	C	C
2009	D	BN
2010	BN	AN
2011	W	W
2012	BN	D
2013	D	C
2014	C	C
2015	C	C
2016	BN	D

Note: Years shown in red had both river systems below normal, dry, or critically dry. Years shown in blue had both river systems wet or above normal.
Source: CDWR 2017.

		Year Type			
W	Wet	AN	Above normal	BN	Below normal
	D	Dry		C	Critical

Water in Reservoir Storage

California has long experienced natural extreme hydrologic events. As shown in Figure 5.4, periods of both wet and dry years are evident in the long-term record of runoff, including the severe Dust Bowl drought in the late 1920s and early 1930s, the 1987–1992 drought, as well as the wetter periods in the early 1940s and mid- to late-1990s. To compensate for these extremes, California has built an extensive system of storage dams that hold water in wet years for use in dry years and to balance intra-annual variability. Over the past century, more than 42 million acre-feet of storage has been built in dams paid for by federal, state, and local sources. Figure 5.6 shows the cumulative available reservoir storage volume in California since the mid-1800s.

During droughts, water deliveries to users are maintained by drawing down water stored in California reservoirs. Over time, if dry conditions persist, storage volumes may fall to low levels, and historical levels of water deliveries cannot be maintained. In wet periods, these reservoirs may fill to capacity; in dry periods—such as the droughts of 1976–1977, 1987–1992, 2007–2010, and 2012–2016—storage volumes decline.

The recent drought in California reduced water levels in all major reservoirs. Reservoir

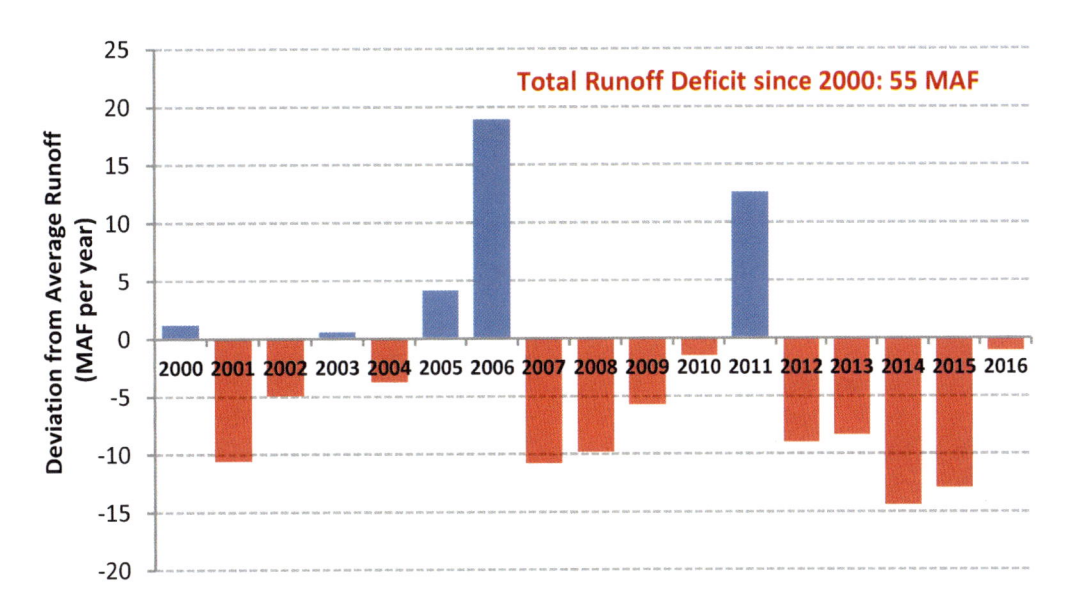

FIGURE 5.5 DEVIATION FROM 1906–2016 AVERAGE RUNOFF, BY WATER YEAR, FOR THE SACRAMENTO–SAN JOAQUIN RIVER SYSTEMS (IN MILLION ACRE-FEET PER YEAR).
Note: 1 AF = 1233 cubic meters.
Source: CDWR 2017.

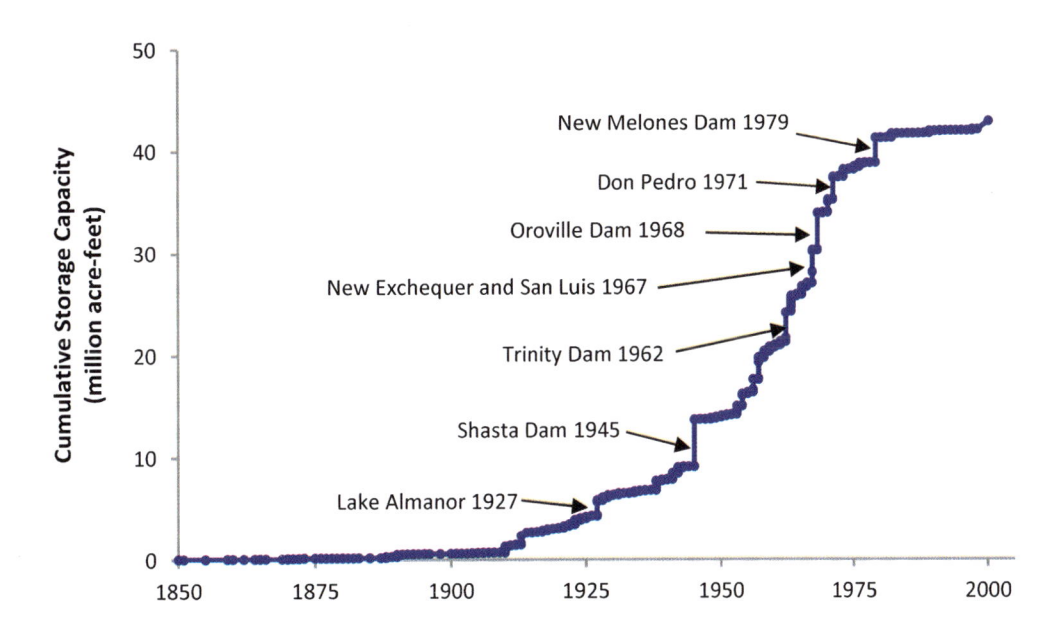

FIGURE 5.6 CUMULATIVE STORAGE CAPACITY BEHIND CALIFORNIA DAMS (IN MILLION ACRE-FEET).
Source: CDWR/DSD 2017.

storage at the end of the dry season in 2015 was less than half of the volume of water typically available, and a quarter of total storage volume. Modest rains during the winter of 2015–2016 refilled part of California's reservoir storage, including the two largest reservoirs (Shasta and Oroville), but failed—statewide—to bring reservoir storage levels up to normal. By the end of the 2016 water year, reservoir storage was around 70 percent of average for the date and less than half of total storage volume.

Groundwater

Groundwater is the third critical component of California's water supply system. Groundwater is a "stock"—a reserve drawn down to make up for shortfalls of more renewable flows of precipitation and runoff. Groundwater can be a renewable resource if withdrawals and recharge balance each other over time, but California's groundwater resources have long been grossly out of balance, even in normal or wet years. During droughts, groundwater overdrafts become even larger. The California Department of Water Resources estimates that the long-term average overdraft of groundwater, statewide, is on the order of 1 to 2 million acre-feet per year (1.2 to 2.5 billion cubic meters per year), largely focused in the southern San Joaquin Valley region (CDWR 2013).

During the recent drought, however, when surface water deliveries to users were severely limited, groundwater overdraft expanded dramatically. Some estimates put the overdraft at 5 to 7.4 km^3 per year (4 to 6 million acre-feet per year) during the 2012–2016 period. For example, estimates from the GRACE satellite missions put groundwater losses during the drought periods of 2007–2010 and 2012–2015 at around 6.9 km^3 per year (5.6 million acre-feet/year) (Famiglietti 2014; Scanlon et al. 2012; Wang et al. 2016).

Because groundwater use is largely to completely unregulated in California, depending on the watershed, there seem to be no short-term constraints on the continued overpumping of groundwater. While this water use permits continued agricultural production at higher levels (see below) than would otherwise be possible during drought, it comes with some severe negative consequences—including the drying up of shallower groundwater wells in many communities, the dewatering of streams and rivers normally fed during dry periods by groundwater flows, and land compaction in geologies vulnerable to subsidence.

Impacts of the California Drought

The hydrological and climatological data roughly indicate the major water conditions facing the state during the drought. The actual impacts of drought are many and varied, depending on the nature and severity of the drought, local economic and environmental conditions, the kinds of infrastructure in a region, how that infrastructure is operated, and responses of local governments and water institutions. Among the impacts are changes in agricultural production, urban water deliveries, ecological health, and energy production.

In California, agriculture accounts for around 80 percent of human uses of water; the rest goes to "urban" uses, which include satisfying residential, commercial, industrial, and institutional water demands. During the drought, deliveries of surface water from state and federal water projects to some agricultural users were substantially reduced.

These reductions were largely made up by increases in groundwater extraction and reductions in groundwater storage. In 2014, the third year of the drought, Governor Jerry Brown called for a voluntary 20 percent reduction in urban water use from 2013 levels (State of California 2014); one year later, with the persistence of the drought, the Governor called for a 25 percent mandatory reduction in statewide urban potable water use, with variations for different climates, prior conservation efforts, and other factors (State of California 2015a). Combined with low runoff, depleted reservoirs, and the related cutbacks in surface water deliveries for agriculture, water shortfalls could be expected to lead to ecological damages, reductions in agricultural production, lost hydropower, and economic impacts to urban users.

Measuring and quantifying drought impacts is difficult. Not all impacts have an easily quantifiable economic measure, such as dollars or jobs lost. For example, damages to native fish populations or forests, loss of habitat for migrating waterfowl, or health impacts from wildfires are hard to measure in economic terms. Even for more traditional economic sectors, like agriculture, data are often not collected or distributed in a timely manner, making it difficult to evaluate the full costs of drought. Some data are not collected at all. Moreover, when data are available, it is often difficult to separate the economic impacts of water cutbacks from other factors that affect economic productivity, output, and value—such as international crop prices, the effects of crop insurance programs, and larger factors influencing California's overall economy and employment levels.

Below, some drought impacts are described from recent analyses. These should not be considered comprehensive or complete—they represent snapshots of some of the consequences of the drought at the time of writing this overview, and as additional data are made available, these impacts must be updated.

Agricultural Revenue

California is one of the world's most productive agricultural regions and the United States' largest agricultural producer, supplying both U.S. and international markets with more than 400 different farm products. In 2013, total California farm output was valued at $50.2 billion, or about one-tenth of the total for the entire nation (in 2015 dollars). Two-thirds of this amount, $33.5 billion, was from crops; about 26 percent ($13 billion) from livestock, poultry, and livestock products; and the rest ($2.4 billion) from nursery, greenhouse, and floriculture (NASS 2014; NASS 2015). California is also the nation's largest agricultural exporter, with annual exports reaching a record $21.5 billion in 2013 (CDFA 2015). Here, we evaluate those impacts by examining some key indicators of the sector's overall health through the first several years of drought: income and employment (see Cooley et al. 2015 for a detailed assessment).

Gross farm income in California has been increasing since 2000, even during the most recent drought. Adjusted for inflation, farm income from 2000 to 2011—up to the beginning of the drought—increased from $38.6 billion to $49.5 billion (Figure 5.7).[2] Then, during the first three years of the drought (2012–2014), income continued to grow, reaching a record high of $56.9 billion in 2014. Much of the increase was due to strong crop prices

---

2. Agricultural income includes all crop and livestock receipts, the value of home consumption, inventory adjustments, other farm-related income, and direct government payments.

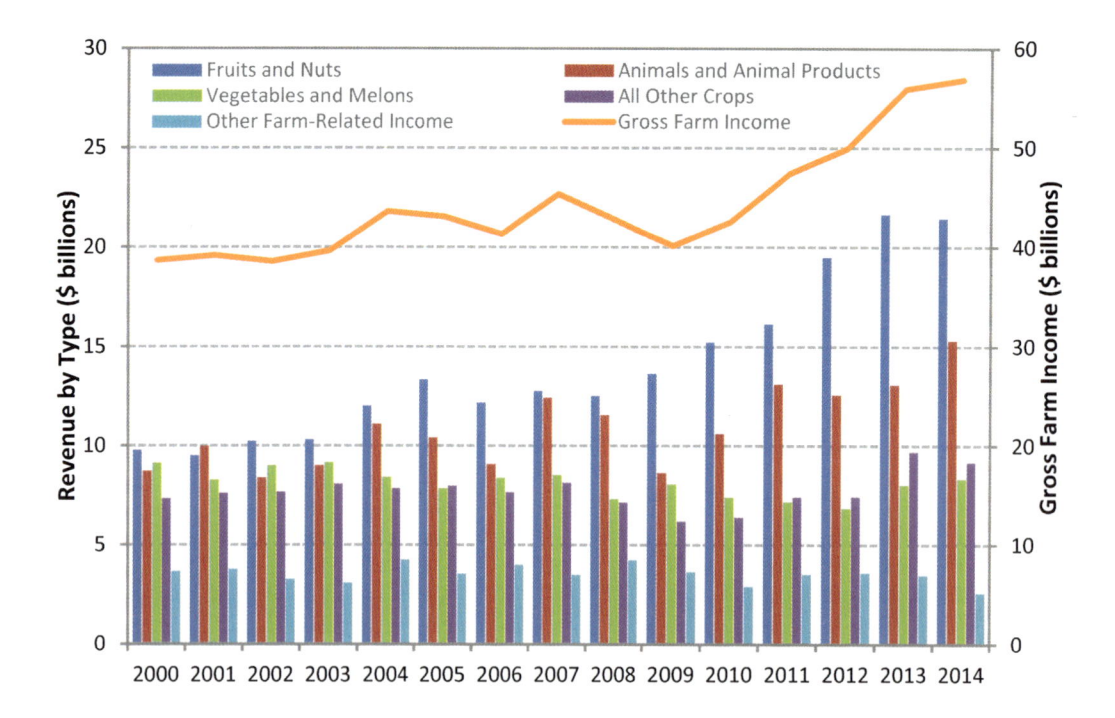

FIGURE 5.7 CALIFORNIA CROP REVENUE BY CROP TYPE (IN BILLIONS OF 2015 DOLLARS).
Notes: "All Other Crops" includes cotton, feed crops, food grains, oil crops, and all other crops. "Other Farm-Related Income" includes home consumption, inventory adjustment, direct government payments, forest products sold, gross imputed rental value of farm dwellings, machine hire, total commodity insurance indemnities, and net cash rent received by operator landlords.
Source: CEO Water Mandate 2014a.

and rising income from animal products and fruits and nuts. By 2014, these products accounted for nearly two-thirds of gross farm income. In 2014, although gross income from several crop types, including fruits and nuts, declined from 2013 levels, these reductions were mostly offset by increases in income from animal products (USDA/ERS 2017).

Production expenses also increased during the drought (Figure 5.8). Since 2000, production expenses increased from $30.9 billion to a record high of $41.2 billion in 2014; with large, sustained increases taking place during the latest drought years. Although net farm income was lower in 2014 than 2013, the most recent years have still been record-setting, thanks to large increases in gross income and despite the continuing drought; for the period 2000–2010, average net farm income was $10.4 billion, compared to an average of $16.7 billion for 2011–2014.

The long-term increase in farm income is attributable, in large part, to increases in crop-related income. Although animals and related products generate a lot of revenue, it is typically only about one-quarter of gross farm revenue. Crop-related income has been driven by three key factors:

> *First*, there has been a shift from lower- to higher-value crops, as evidenced by a reduction in the acreage planted in field crops and the expansion of acreage planted in fruit and nut crops. In 2014, for example, field crops generated on average $1,300 per acre, while vegetables gener-

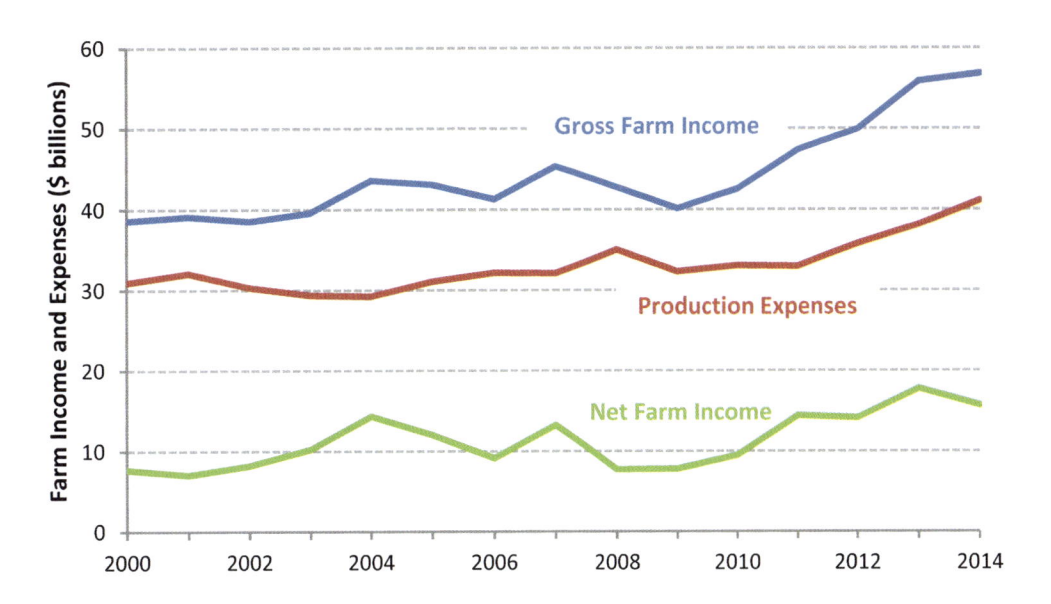

FIGURE 5.8 CALIFORNIA GROSS AND NET FARM INCOME, AND PRODUCTION EXPENSES (IN BILLIONS OF 2015 DOLLARS).
Source: USDA/ERS 2017.

ated $7,600 per acre and fruits and nuts generated $7,300 per acre. For the period 2000–2014, the amount of acreage harvested for fruits and nuts increased from 2.4 to 2.9 million acres, while the amount of acreage harvested for field crops decreased from 4.4 to 2.9 million acres. Total acreage irrigated, accounting for land left fallow during the drought, dropped by less than 10 percent.

Second, crop productivity, as measured by the tonnage of a given crop produced per acre, increased for key crops—including almonds, rice, strawberries, tomatoes, and walnuts. Tomato productivity, for example, was 35 tons per acre in 2000 but increased to 45 tons per acre in 2014.

Third, crop prices have increased for most crops grown in California. For example, almonds brought $2,600 per ton in 2000 but $6,400 per ton in 2014 (Cooley et al. 2015).

The impacts of the drought on California's agricultural sector through 2014 were less than expected. The resilience of the agricultural sector during the drought was due to several factors, including the sector's strong financial position before the drought began and the variety of response strategies employed. Perhaps most importantly, farmers massively increased groundwater pumping to make up for shortages of surface water. While actual groundwater use data are not available—a fundamental flaw in California water data—recent estimates are for massive groundwater depletion in large parts of the Central Valley agricultural region (Figure 5.9). Continued groundwater overdraft, while reducing the economic impacts of the drought for the agricultural sector now, has shifted the burden to others, including current and future generations forced to dig deeper wells, find alternative drinking water sources, and repair infrastructure damaged by subsidence. Water transfers have also played a role; however, the broader social and environmental impacts of these transfers are not well understood. Finally, short- and long-

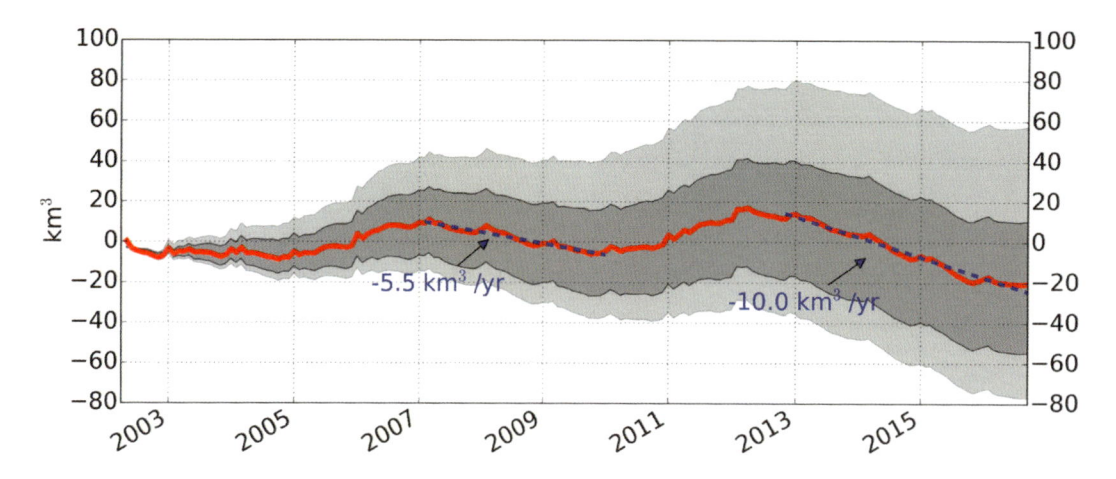

FIGURE 5.9 CENTRAL VALLEY GROUNDWATER CHANGE OVER TIME.
Notes: Monthly groundwater storage California's Central Valley from April 2002 to September 2016. Light gray shading shows the range of all ensembles and dark gray shows the inner quartiles. The red line is the individual ensemble member closest to the ensemble mean. The two blue lines are linear regressions for the drought periods in January 2007 to December 2009 and October 2012 to September 2016.
Source: Xiao et al. 2017.

term shifts in the types of crops grown and improvements in irrigation technologies and practices have also improved the resilience of the state's agricultural sector to extreme weather events.

Agricultural Employment

Agricultural employment data from 2014 suggest that the actual impact of the drought on farm jobs was much less than a loss of around 17,000 jobs initially projected by Howitt et al. (2014). In 2014, California agricultural employment reached a record high of 417,000 people (CEDD 2017a). According to the California Employment Development Department (2017b), agricultural employment in the third quarter of 2014—the period of peak farm employment—increased by 3,100 jobs from the same quarter in 2013. Agricultural employment would likely have been higher if there had been less land fallowed due to water shortages, but water availability is only one factor affecting it. The total number of jobs also depends on the types of crops grown, the irrigation method used, the use of new planting and harvesting machinery, and other details (Cooley et al. 2015).

Hydroelectric Power Generation

The State of California benefits from a diverse electricity generation system (Figure 5.10). More than 60 percent of in-state electricity in 2013 was generated by fossil fuels, almost entirely natural gas. Other sources—such as solar, wind, biomass, geothermal, and nuclear—made up 26 percent of the state's electricity, and renewable generation is growing rapidly. Hydropower systems generated approximately 12 percent of in-state electricity that year (Gleick 2015; Gleick 2016).

California Drought 2012–2016   79

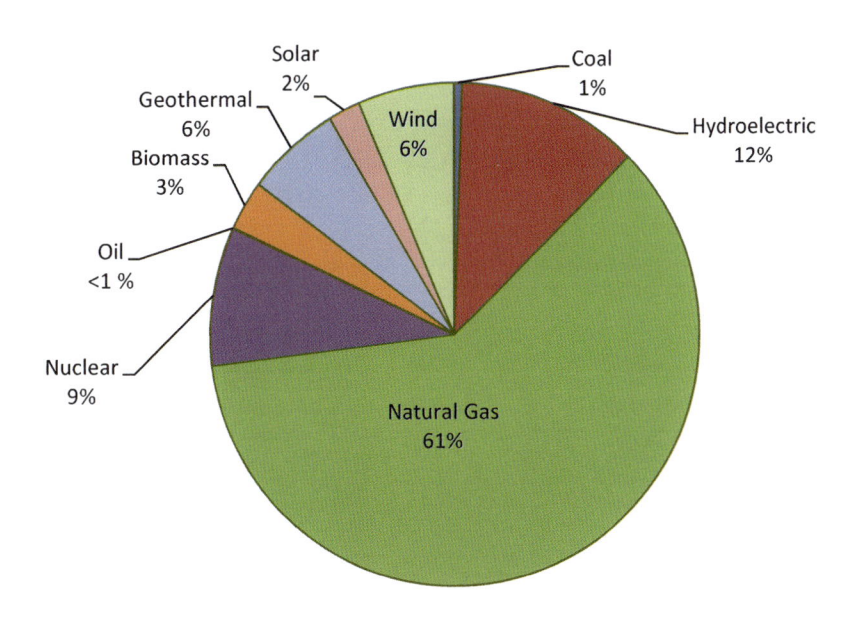

FIGURE 5.10 CALIFORNIA IN-STATE ELECTRICITY GENERATION BY SOURCE, 2013.

Notes: Additional electricity is generated in other states and sent to California, but details on the sources and variations due to drought are not available. This graph shows in-state generation by source.

Source: Gleick 2015.

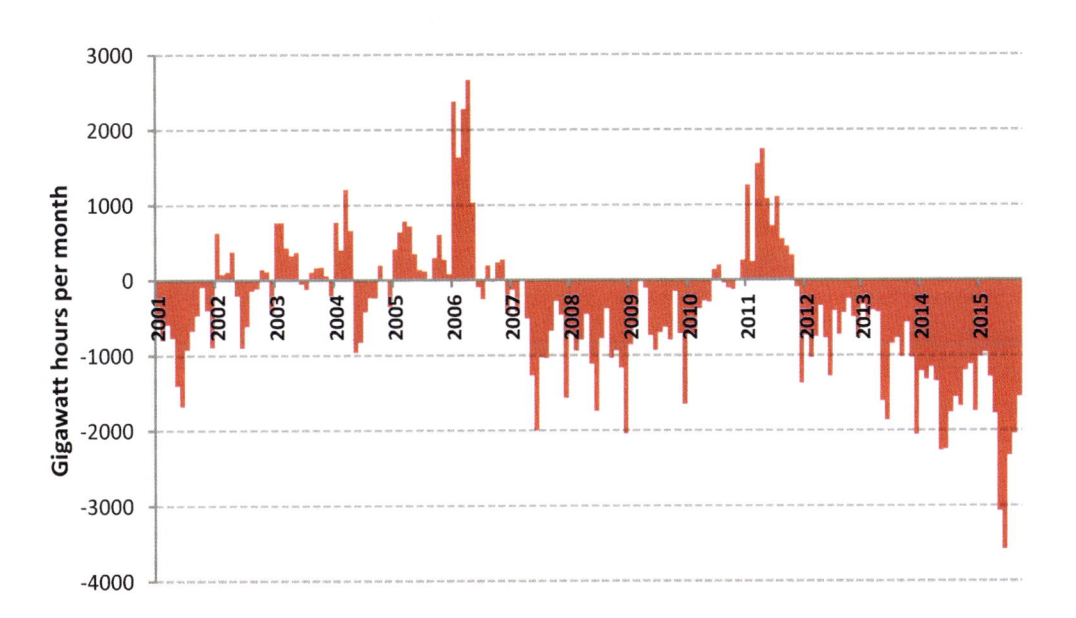

FIGURE 5.11 REDUCTIONS IN HYDROELECTRICITY GENERATION DUE TO CALIFORNIA DROUGHT, 2001 THROUGH SEPTEMBER 2015 (IN GIGAWATT-HOURS PER MONTH).

Source: Gleick 2016.

The amount of electricity generated from each source varies with availability, cost, the form and location of consumer demand, and other factors. Hydroelectricity production increases in winter and spring with increased runoff and decreases during late summer, fall, and early winter when natural runoff is low. During droughts, less water is available in rivers or stored in reservoirs and overall hydropower production drops. Figure 5.11 shows the dramatic monthly reductions in hydroelectricity generation due to drought, compared to the long-term average, for the first four years of the drought.

In California, reductions in hydropower production are made up primarily by burning more natural gas—the marginal energy supply—increasing purchases from out-of-state sources, and expanding wind and solar generation. Because hydropower is considerably less expensive (in both fixed and variable costs) than other forms of electricity, the drought led to a direct increase in electricity costs and the prices paid by California energy consumers. Gleick (2015; 2016) estimated that the total reductions in hydropower generation during the 2012–2016 drought increased statewide electricity costs by over $2 billion. On top of the direct economic costs of replacing lost hydroelectricity generation, the additional combustion of natural gas led to increased air pollution in the form of nitrous oxides (NO_x), volatile organic compounds (VOCs), sulfur oxides (SO_x), particulates (PM), carbon monoxide (CO), and carbon dioxide (CO_2)—the principal greenhouse gas responsible for climatic change. Overall, during the 2012–2016 period between 25 and 30 million tons of additional carbon dioxide, or around a 10 percent increase in CO_2-equivalent emissions from California power plants over the same five-year period, were emitted because of the drought, along with substantial quantities of NO_x, VOCs, PM, and other pollutants (Gleick 2016). Many of these pollutants are known contributors to the formation of smog and as triggers for asthma, and the economic costs of these health impacts have not been calculated.

Ecosystem Impacts

Severe impacts of the California drought have been felt by the state's freshwater fisheries, migratory bird habitat, and forests due to both changes in water availability and high temperatures, as noted above. A lack of detailed ecological data hinders producing an overall assessment of these impacts, but I note here some of the effects already observed.

Nearly 130 freshwater fish species are found in California, and two-thirds of them are endemic. Past water policies—including dam construction, water withdrawals, and water-quality threats—have caused some species to become extinct and others to be listed as threatened or endangered under federal and state law. One hundred species are already listed for protection under these laws or are expected to be listed in the future because of declining populations (Hanak et al. 2015). Populations of key species such as smelt (delta and longfin), salmon, steelhead, and others are at record low levels (CWIN 2016). For two years in a row (2015 and 2016), the annual cohort of winter-run Chinook salmon young was almost completely killed off by high temperatures in the Sacramento River (Associated Press 2016). Overall, the drought is worsening the risk of extinction for a large number of native fish species, including most remaining populations of salmon and steelhead, while also increasing conditions that favor invasive species.

Sudden and severe impacts also occur. In September 2015, 155,000 trout died in a fish hatchery on the American River when an algal bloom depleted oxygen levels in their water in a matter of minutes (Sabalow 2015). In addition, despite some legal protections for ecosystems, a series of emergency orders by the State Water Resources Control Board (SWRCB) have allowed several hundred thousand acre-feet of water to be taken from fish protection during the drought and given to agricultural users of the Central Valley Project (CVP) and State Water Project (SWP) (Kasler and Reese 2015).

Bird populations are also at risk. California's wetlands are key stopping points for migratory birds along the so-called "Pacific Flyway." These wetlands, greatly reduced in area from their historical extent, provide winter habitat for literally millions of aquatic birds, supplemented by flooded agricultural lands (primarily rice fields in the northern Sacramento Valley) in the winter. During droughts, the area of California wetlands decreases substantially and deliveries of water to remaining wildlife refuges also drop. Over the drought period, these deliveries were cut by 25 percent or more (Hanak et al. 2015).

A study conducted by the California Department of Fish and Wildlife in spring 2015, the "2015 Waterfowl Breeding Population Survey," showed declines in the total number of breeding duck populations to an estimated 315,580, compared to 448,750 in 2014 (CDFW 2015; Terrill 2015). Impacts during 2015 were also worsened due to cuts in acreage of temporary wetlands in rice fields.

The drought also worsened the risk, and reality, of fires and tree mortality in California's forests. During the first four years of drought from 2012 to 2015, tree mortality increased by an order of magnitude, with up to hundreds of dead trees per square kilometer in the Sierra Nevada and rapidly rising mortality rates as the drought continued (Young et al. 2017). Overall, more than a hundred million trees were estimated to have been killed by the drought, through a combination of water stress and high temperatures, worsened by secondary impacts of severe insect infestation (Asner et al. 2016; State of California 2015b). In October 2015, Governor Brown declared a state of emergency to help mobilize state and federal resources to remove dead and dying trees (State of California 2015b).

The large-scale tree die-off also worsened wildfire risk over the long term by adding to the fuel load in California forests. Fires in 2015 destroyed nearly 880,000 acres across all jurisdictions, and foresters fear the coming years will continue to see higher than normal fire rates (Table 5.2).

Overall Economic Well-Being

The overall impacts of the drought on the state's economy are difficult to quantify, in part because some of the most severe impacts, such as ecological damages (as noted above), cannot be easily measured in traditional economic terms. However, a qualitative assessment would suggest that the state has weathered the drought with little economic damage. Figure 5.12 shows the overall inflation-corrected gross state product (GSP) for all 50 U.S. states (in 2009 chained dollars). As seen in this figure, California's economy is expanding at a rate comparable to or exceeding that of other states. No specific drought signal can be seen.

There are three primary reasons for this:

> *First*, an increasing part of California's economy is not directly dependent on water-intensive activities. Only the agricultural sector relies on water as a primary input and this sector makes up only about 2.5 percent

TABLE 5.2 Number and Area of California Wildfires by Agency, 2015

Agency	Number of Fires	Acres Burned
CAL FIRE—State Responsibility	3,231	291,282
CAL FIRE—Local Government Contracts	2,556	6,137
Contract Counties	312	6,365
United States Forest Service	1,656	537,446
Bureau of Land Management	97	18,058
National Park Service	126	9,834
Bureau of Indian Affairs	178	360
United States Fish and Wildlife Service	12	23
Military	115	11,394
2015 Total	8,283	880,899
5-Year Average CAL FIRE (2011–2015)—Includes Local Government Contracts	5,431	156,406
5-Year Average (2011–2015)—All Agencies	7,836	633,180

Source: CDFFP 2016.

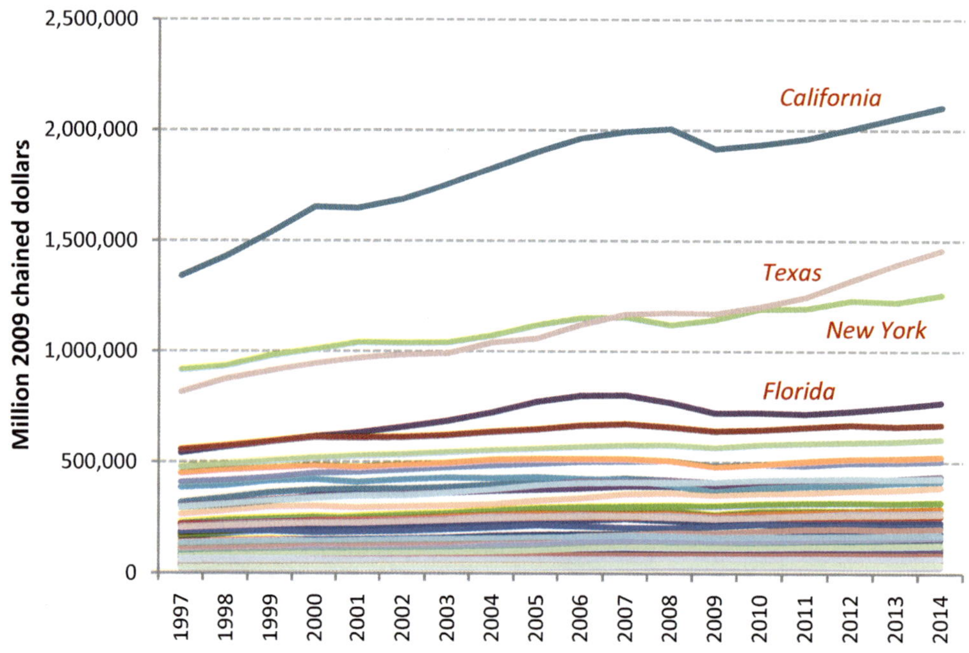

FIGURE 5.12 GROSS STATE PRODUCT (GSP) FOR ALL 50 U.S. STATES, 1997–2014 (IN MILLIONS OF 2009 CHAINED DOLLARS).
Note: The top four economies are labeled.
Source: U.S. Department of Commerce 2016.

of overall California gross state product (GSP). Growth in other sectors, especially the service sector that produces more revenue per unit of water used, has had a much greater impact on the economy. Professional and business services and the information industry together represent about 22 percent of the state's GSP and have grown more than 4 percent each year, on average, since 2011—faster than nearly all other sectors of the economy.

Second, groundwater overdraft has compensated for drought-induced water shortages, providing at least a short-term buffer from the economic impacts of the drought on this sector.

Third, many of the impacts of the drought are not quantified in traditional economic terms, and are not shown in time-series reporting of traditional economic indicators, such as GSP. Even though economic impacts are difficult to quantify, and quantifiable economic indicators suggest that there have been few impacts so far, continuing drought as well as the lingering effects of the current drought, may nonetheless have long-term economic repercussions.

The impacts of the drought should continue to be monitored, and efforts made to adapt to and mitigate the consequences of these kinds of extreme hydrologic events.

References

Asner, G. P., P. G. Brodrick, C. B. Anderson, N. Vaughn, D. E. Knapp, and R. E. Martin. 2016. Progressive Forest Canopy Water Loss during the 2012–2015 California Drought. *Proceedings of the National Academy of Sciences (PNAS)* 113(2): E249–E255. doi:10.1073/pnas.1523397113

Associated Press. 2016. California's Native Salmon Struggling after 5 Years of Drought. http://www.cbsnews.com/news/californias-native-salmon-struggling-after-5-years-of-drought/.

California Department of Fish and Wildlife (CDFW). 2015. 2015 California Waterfowl Breeding Population Survey. D. Skalos and M. Weaver. https://nrm.dfg.ca.gov/FileHandler.ashx?DocumentID=102218&inline.

California Department of Food and Agriculture (CDFA). 2015. California Agricultural Production Statistics. http://www.cdfa.ca.gov/statistics/.

California Department of Forestry and Fire Protection (CDFFP). 2016. California Wildfire Activity Statistics, Complete Redbook for 2015. http://www.fire.ca.gov/downloads/redbooks/2015_Redbook/2015_Redbook_FINAL.PDF.

California Department of Water Resources (CDWR). 2013. California Water Plan 2013 Update. http://www.water.ca.gov/waterplan/cwpu2013/final/.

———. 2017. Chronological Reconstructed Sacramento and San Joaquin Valley Water Year Hydrologic Classification Indices. http://cdec.water.ca.gov/cgi-progs/iodir/WSIHIST.

California Department of Water Resources, Division of Safety of Dams (CDWR/DSD). 2017 (September). Dams within Jurisdiction of the State of California. http://www.water.ca.gov/damsafety/docs/Dams%20by%20County_Sept%202017.pdf.

California Employment Development Department (CEDD). 2017a. Statewide Historical Annual Average Employment by Industry Data, 1990–2014. http://www.labormarketinfo.edd.ca.gov/LMID/Employment_by_Industry_Data.html.

———. 2017b. Agricultural Employment in California. http://www.labormarketinfo.edd.ca.gov/data/ca-agriculture.html.

California Water Impacts Network (CWIN). 2016. Bay-Delta Fish Profiles. https://c-win.org/bay-delta-fish-profiles/.

Cooley, H., K. Donnelly, R. Phurisamban, and M. Subramanian. 2015 (August). Impacts of California's Ongoing Drought: Agriculture. Oakland, CA: Pacific Institute. http://pacinst.org/wp-content/uploads/2015/08/ImpactsOnCaliforniaDrought-Ag.pdf.

Diffenbaugh, N. S., D. L. Swain, and D. Touma. 2015. Anthropogenic Warming has Increased Drought Risk in California. *Proceedings of the National Academy of Sciences (PNAS)* 112(13): 3931–3936. doi:10.1073/pnas.1422385112

Famiglietti, J. S. 2014. The Global Groundwater Crisis. *Nature Climate Change* 4(11): 945–948. doi:10.1038/nclimate2425

Gleick, P. H. 2015. Impacts of California's Ongoing Drought: Hydroelectricity Generation. Oakland, CA: Pacific Institute. March 17. http://pacinst.org/publication/impacts-of-californias-ongoing-drought-hydroelectricity-generation/.

———. 2016. Impacts of California's Ongoing Drought: Hydroelectricity Generation 2015 Update. Oakland, CA: Pacific Institute. February. http://pacinst.org/publication/impacts-of-californias-ongoing-drought-hydroelectricity-generation-2015-update/.

Hanak, E., J. Mount, C. Chappelle, J. Lund, J. Medellín-Azuara, P. Moyle, and N. Seavy. 2015. What If California's Drought Continues? San Francisco, CA: Public Policy Institute of California. http://www.ppic.org/content/pubs/report/R_815EHR.pdf.

Howitt, R. E., J. Medellin-Azuara, D. MacEwan, J. Lund, and D. Sumner. 2014. Economic Analysis of the 2014 Drought for California Agriculture. Davis, CA: UC-Davis Center for Watershed Sciences. https://watershed.ucdavis.edu/files/content/news/Economic_Impact_of_the_2014_California_Water_Drought.pdf.

Kasler, D., and P. Reese. 2015. Imperiled Fish Add to California's Drought Stress. June 6. *Sacramento Bee*. http://www.sacbee.com/news/state/california/water-and-drought/article23294532.html.

Mann, M. E., and P. H. Gleick. 2015. Climate Change and California Drought in the 21st Century. *Proceedings of the National Academy of Sciences (PNAS)* 112(13): 3858–3859. doi:10.1073/pnas.1503667112

National Agricultural Statistics Service (NASS). 2014. Crop Values Annual Summary. http://usda.mannlib.cornell.edu/usda/nass/CropValuSu//2010s/2014/CropValuSu-02-14-2014.pdf.

National Agricultural Statistics Service (NASS). 2015b. California Agricultural Statistics, 2015 Crop Year. http://usda.mannlib.cornell.edu/usda/nass/CropValuSu/2010s/2015/CropValuSu-02-24-2015_correction.pdf.

National Drought Mitigation Center (NDMC). 2016. What is Drought? University of Nebraska, Lincoln, Nebraska. http://drought.unl.edu/DroughtBasics/WhatisDrought.aspx.

National Oceanic and Atmospheric Administration (NOAA). 2015–2017. Climate at a Glance. http://www.ncdc.noaa.gov/cag/.

National Weather Service (NWS). 2012 (October). Drought Fact Sheet. National Oceanic and Atmospheric Administration (NOAA). http://www.nws.noaa.gov/om/csd/graphics/content/outreach/brochures/FactSheet_Drought.pdf.

Sabalow, R. 2015. More than 155,000 Trout Die at American River Hatchery. *Sacramento Bee*. September 8. http://www.sacbee.com/news/local/environment/article34416483.html.

Scanlon, B. R., L. Longuevergne, and D. Long. 2012. Ground Referencing GRACE Satellite Estimates of Groundwater Storage Changes in the California Central Valley, USA. *Water Resources Research* 48(4): 1–9. doi:10.1029/2011WR011312

State of California. 2014. Governor Brown Declares Drought State of Emergency. January 17. https://www.gov.ca.gov/news.php?id=18368.

State of California. 2015a. California Drought Executive Order B-29-15. April 1. https://www.gov.ca.gov/docs/4.1.15_Executive_Order.pdf.

State of California. 2015b. Tree Mortality State of Emergency. https://www.gov.ca.gov/docs/10.30.15_Tree_Mortality_State_of_Emergency.pdf.

Terrill, T. 2015. Waterfowl Population Declining in California. *Daily Democrat*. July 14. http://www.dailydemocrat.com/article/NI/20150714/NEWS/150719954.

United States Department of Agriculture, Economic Research Service (USDA/ERS). 2017. Farm Income and Wealth Statistics. http://www.ers.usda.gov/data-products/farm-income-and-wealth-statistics/value-added-years-by-state.aspx.

United States Department of Commerce, Bureau of Economic Analysis. 2016. http://www.bea.gov/regional/downloadzip.cfm.

Wilhite D. A., and M. H. Glantz. 1985. Understanding the Drought Phenomenon: The Role of Definitions. *Water International* 10(3): 111–120.

Wang, S.-Y. S., Y.-H. Lin, R. R. Gillies, and K. Hakala. 2016. Indications for Protracted Groundwater Depletion after Drought over the Central Valley of California. *Journal of Hydrometeorology* 17(3). doi:10.1175/JHM-D-15-0105.1.

Xiao, M., A. Koppa, Z. Mekonnen, B. R. Pagán, S. Zhan, Q. Cao, A. Aierken, H. Lee, and D. P. Lettenmaier. 2017. How Much Groundwater Did California's Central Valley Lose during the 2012–2016 Drought? *Geophysical Research Letters* 44(10): 4872–4879. doi:10.1002/2017GL073333.

Young, D. J. N., J. T. Stevens, J. M. Earles, J. Moore, A. Ellis, A. L. Jirka, and A. M. Latimer, 2017. Long-Term Climate and Competition Explain Forest Mortality Patterns under Extreme Drought. *Ecology Letters* 20(1): 78–86. doi:10.1111/ele.12711.

CHAPTER 6

Water Trading in Theory and Practice

Michael Cohen

The contents of this chapter appeared in slightly different form in Cooley et al. 2015, Incentive-Based Instruments for Water Management

Growing pressure on the world's limited freshwater resources adversely affects our social, economic, and environmental well-being. Devastating droughts have destabilized parts of the Middle East, dried up streams and lakes in Australia, California, and the Colorado River Basin; decreased agricultural productivity, and threatened community access to water. Existing water supplies are already overallocated in many areas—an imbalance expected to worsen in coming years due to population growth, climate change, global shifts toward more water-intensive meat-based diets, and other factors. For example, the Colorado River Basin, whose waters support some 40 million people in two countries, currently suffers from a structural deficit—where diversions and losses exceed average annual runoff by 1.4 km^3 per year, projected to increase to 3.8 km^3 per year by 2060 (USBR 2012).

Water allocations, controlled by custom or law and rooted in historic practice and development patterns, often do not reflect current or expected future demands. Globally, agriculture consumes 70 to 80 percent of developed freshwater resources, producing tension with rising demand in the municipal and industrial sectors. This tension between limited resources and historic allocation patterns has prompted a variety of approaches to bridge the apparent gap between demand and supply. These include true-cost water pricing, social norming, and other public campaigns to increase public recognition of water's value and importance; supply augmentation schemes such as desalination and wastewater reuse (see *World's Water* Vol. 8, Ch. 6 and Vol. 5, Ch. 3); increasing urban and agricultural water-use efficiency and conservation (see *World's Water* Vol. 6, Ch. 6 and Vol. 4, Ch. 5 and Ch. 6); and market-based mechanisms to reallocate, or shift, water from one user to another.

Market-based water reallocation mechanisms, often known as water trading or water markets, have received increased attention and support in recent years because of their perceived adaptability and ability to meet changing water needs. This chapter discusses water trading in theory and practice—including its environmental, economic, and social performance, and the conditions needed for implementation, to assess its potential to address the water challenges described above.

Water Trading, Transfers, Markets, and Banks

A variety of terms and phrases describe water reallocation mechanisms, including water transfers, water trading, water markets, and water banks; as well as formal and informal water transfers and trades. At times, these terms are used interchangeably. Box 6.1 offers definitions of these terms.

Water Trading in Theory

Water trading is perhaps the best known and most widely used method of reallocating water. In some cases, purchasing or leasing water from existing users has proven to be less expensive, more flexible, and less time-consuming than developing new water supplies—such as constructing new diversion structures or desalination plants. This is especially true in regions where total renewable water supplies are heavily or overallocated. Similarly, water trading is generally a more accepted method for reallocating water than state appropriation or condemnation of existing water rights (Culp et al. 2014; NRC 1992). Successful examples of water trading in Australia and other locations—combined with neoclassic economic theory suggesting that market mechanisms can optimize resource allocation—have focused attention on this mechanism in both academic literature and popular media, as well as among those working to improve water supply reliability.

An extensive body of literature argues that water trading improves the economic efficiency of water through reallocation from lower- to higher-value uses (Glennon 2005; Dellapenna 2000; Bjornlund and McKay 2002). The germinal study entitled *Water and Choice in the Colorado Basin* (NRC 1968) recommended that water in the western United States be transferred from irrigation, which generates relatively low returns per unit of water, to high-value non-agricultural uses. More recent research has continued to emphasize the potential value created by water trading. For example, models used to project California's economic costs under a dry climate change projection (Medellín-Azuara et al. 2008), found significantly increased benefits with market-based reallocations. Newlin et al. (2002) and Jenkins et al. (2004) asserted that water trading could dramatically reduce Southern California's water scarcity costs. Water trading is attractive because it tends to minimize the impact on existing rights holders by providing compensation and, in many cases, additional security for existing water rights, while providing opportunities to those with new or increasing demands (NRC 1992).

Water can be made available for trading from a variety of activities—including fallowing fields, crop shifting, and, in some cases, by shifting from surface water diversions to groundwater extraction where groundwater is not included in water rights constraints. Trades can also be linked to water conservation and efficiency efforts, including increasing irrigation efficiency and decreasing system losses, such as by lining canals or constructing operating reservoirs that generate surplus water. Water trades can also refer to conditional arrangements, such as options or dry-year leasing arrangements where an urban water agency provides a farmer an annual fee to reserve a right to call for water under certain conditions, such as drought or interruption of other urban supplies.

Institutional and physical water banks facilitate water trading, at a variety of geographic and temporal scales. Local water users can store water in an underlying aquifer

BOX 6.1 A Note on Terminology Used in This Chapter

Water trading. The temporary or permanent transfer of the right to use water in exchange for some form of compensation.

Informal water trading. The sale of a specified volume of water for a limited period of time, which does not involve actual contracts or occurs outside of a recognized legal or administrative framework.

Water transfers. The National Research Council (NRC) of the United States National Academies defines water transfers as changes in the point of diversion, type of use, or location of water use (NRC et al. 1992). The term "water transfers" encompasses a broad range of market-based and non-market water reallocation mechanisms of varying periods, geographic scales, and arrangements. Water transfers can range from time-limited leases or conditional arrangements to the permanent transfer (i.e., sale) of a water right. They can range in scale from a change in type of use on an existing parcel of land—such as when a water right shifts from irrigation to municipal use when agricultural land is purchased and converted to housing—to inter-basin transfers, such as when a city purchases or leases water from a different watershed. "Water transfer" is also used to refer to non-market redistribution, rather than reallocation, of water. Zhao et al. (2015) note that China transfers some 26 km^3 of water annually, roughly 4.5% of total water use, but these actions do not refer to market-based transfers.

Water banks. A water bank is a mechanism for changing the time or location of water use. Water banking, as with water transfers, can refer to market-based or non-market activities. The term "water bank" can refer to an actual institution or to the physical storage of water. Water banks as institutions may function as (i) brokers that connect buyers and sellers of water rights or leases, providing an important communication function; (ii) clearinghouses that directly purchase or lease water from willing sellers and aggregate supplies for subsequent sale to others; (iii) facilitators that expedite water transfers using existing storage or conveyance facilities (Culp et al. 2014); or (iv) trusts that hold or otherwise manage water rights or entitlements for a specific purpose, such as streamflow augmentation (O'Donnell and Colby 2010). When serving as facilitators, water banks may perform various administrative and technical functions, including the confirmation of water rights and screening of potential buyers (Clifford 2012). Water banks may also refer to physical storage, either in surface reservoirs or in aquifers; which, in turn, may be a component of a larger water transfer or simply a mechanism enabling an entitlement holder to store water for future use.

Water markets. The term "water market" also has a variety of meanings. It generally refers to the system under which market-based water transfers can occur, especially where such transactions include multiple buyers and sellers. A water market can also refer to informal transactions involving the direct sale of water that does not involve the lease or sale of water rights. Informal water transactions can include purchasing bottled water or water from a tanker truck, a common practice in many parts of the developing world that lack a reliable piped water supply.

for future use; a group of rights-holders in one or more irrigation districts can pool a portion of their water for lease to a distant urban area via a broker, such as Colorado's Super Ditch (McMahon and Smith 2013); and a water bank may exchange and store water for a different state, such as the Arizona Water Bank's storage agreement with Nevada (Megdal et al. 2014). Some of these banking agreements may terminate after a single irrigation season, while others may persist for decades. Some water banks, such as California's Drought Water Bank, may function for limited periods in response to specific conditions (Clifford 2012).

The concept and the practice of water trading have critics. Questions of externalities, commodification, and the special nature of water itself highlight the challenges faced by implementing or expanding water trading. Freyfogle (1996) asserted that externalities (third-party impacts), intrinsic to the very nature of water itself, pose such an insurmountable obstacle that water trading does not and cannot work. Many of these externalities arise from the physical properties of water: it's heavy, unwieldy, and easily contaminated; it sometimes has dramatic seasonal and year-to-year variability; and it can be easily lost through evaporation, seepage, or runoff (Salzman 2006). Further, these externalities may be borne by disparate parties, such as the environment or future generations, challenging efforts to compensate those injured by trading (Freyfogle 1996). Salzman (2006) argued that custom, history, and religion in many parts of the world treat drinking water as a common property resource, rather than a tradable commodity. Similarly, Zellmer and Harder (2007) asserted that water differs fundamentally from other resources treated as property, due to its public attributes.

Legal and institutional challenges also impede the implementation and performance of water trading. For example, irrigation districts and water courts often do not recognize a legal property right to water saved by conservation or efficiency (Hundley 2001), precluding efforts of irrigators to lease or sell water conserved by investing in efficiency improvements. Additionally, existing institutions often impose significant costs on those attempting to dedicate water to non-traditional uses such as instream flows (Getches 1985).[1] These problems have tested the resilience of water trading regimes, which have shown some flexibility in adapting to new values and goals but often impose high transaction costs (Colby et al. 1991).

Water Trading in Practice

Despite these difficulties, water trading exists, to varying degrees and forms, in countries around the world. The most active water trading markets have been developed in Australia and the western United States. Australian experience includes both short-term trades (referred to as "allocation trading") and long-term trades (referred to as "entitlement trading"). The total value of water trading in Australia in fiscal year 2012–2013 exceeded $1.4 billion (NWC 2013). The Murray-Darling Basin, Australia's largest river system, has an active and well-documented water market first established more than 30 years ago (Grafton et al. 2012). That market accounts for 98 percent of all allocation trades and 78

---

1. "Instream flows" refers, at the most basic level, to water flowing within a stream channel. Many jurisdictions now permit property owners, be they the state itself or private individuals, to dedicate water rights to augment instream flows, affording legal protection to a specific quantity or rate of flow.

percent of all entitlement trades within Australia, by volume. The Murray-Darling Basin figures prominently in discussions about water trading, as an example of an incentive-based system that successfully transitioned from a non-market system (Grafton et al. 2012). In fiscal year 2012–2013, the total volume of short-term (allocation) trading within the Murray-Darling Basin increased 44 percent from the previous year, from almost 4.3 km^3 to 6.0 km^3. This represents about 50 percent of total surface water use in the basin. The total volume of long-term trades, however, decreased by about 14 percent over that period, to about 1 km^3. A national study found that these permanent entitlement trades often offset the temporary allocation trades, as irrigators planting perennial crops—such as grapes or almonds—purchased entitlements to meet expected future demand, but then sold a portion of the temporary allocations associated with these entitlements to generate revenue (Frontier Economics and Australia National Water Commission 2007).

In the western United States, the scale of water trading is considerably smaller. A database compiled by the University of California at Santa Barbara (UCSB) shows notifications for more than 4,000 water trades in 12 states in the western United States from the years 1987–2009.[2] In 2009, the database reports almost 640,000 acre-feet[3] (0.79 km^3) of water traded in California, through 36 trades with a total value of about $234 million (all values adjusted to 2014 dollars). More than 80 percent of this water was leased rather than sold. According to the database, 15 of these trades, accounting for about 88,000 acre-feet (0.11 km^3) of total volume, occurred within one agricultural district. However, the UCSB database only records the initial year a water trade is reported, and thus does not reflect the volume of multi-year trading agreements. That means that a review of 2009 trading activity does not reflect previous multi-year trades that may still have been active in 2009, so the values reported above understate trading activity in 2009.

A comprehensive review of water trading in California reports about 1.5 MAF (1.8 km^3) of water were traded in 2009, a dry year (Hanak and Stryjewski 2012). Volumes reported for 2011, a wet year and the most recent year for which data were available, were about 5 percent lower. In 2011, 42 percent of the water traded went to municipal and industrial users, 37 percent to agricultural users, 17 percent for environmental purposes, and the remainder to mixed uses. Because of limited data, the review does not include trading activity within irrigation districts or similar user associations, although some estimates suggest that such intra-district activity accounted for several hundred thousand acre-feet of water, a third of total water supplies within some of the larger irrigation districts. Hanak and Stryjewski (2012) did not provide total dollar values associated with the California water market, though they noted that prices of temporary water transfers had increased from an average of $30–$40 per acre-foot per year in one region in the mid-1990s to $180 per acre-foot per year in 2011; while prices in another basin rose to an average of $400 per acre-foot per year. The authors noted the shifting trend from short-term to longer-term leases and permanent trades.

---

2. The database summary notes that "The data are drawn from water transactions reported in the monthly trade journal the *Water Strategist* and its predecessor the *Water Intelligence Monthly* from 1987 through February 2010." These data reflect published reports that in some cases do not reflect final transfer agreements. For example, the database reports that the Imperial Irrigation District–San Diego County Water Authority water transfer began in 1997, although the final transfer agreement was not actually signed and the transfer did not begin (at different volumes than the database reports) until October 2003. The Bren School water transfer database is available at http://www.bren.ucsb.edu/news/water_transfers.htm.

3. An acre-foot is the conventional unit of water measurement in the U.S. West, equivalent to 325,851 gallons or 1,233 cubic meters.

California is also home to the largest United States water trade to date. The San Diego County Water Authority (SDCWA) entered into a 45-year contract in 2003, with an option for a 30-year extension, with the Imperial Irrigation District (IID), one of the largest irrigation districts in the country. Under the terms of the agreement, the SDCWA pays the IID to reduce its diversion of Colorado River water, while the Authority diverts a like amount farther upstream. After a 15-year period intended to create time to address ecological and public health impacts resulting from the trade, the IID will shift to efficiency-based methods (such as lining canals and constructing regulating reservoirs) to generate the water to be conserved. In essence, the Authority is paying the District to improve the efficiency of its operations and to retain the water conserved. The trade is ramping up to a maximum volume of 200,000 acre-feet per year by 2021, representing about 25 percent of the County's total water supply. In 2014, the price for the water was $594 per acre-foot, plus an additional $445 per acre-foot to a different agency to convey the water through its facilities. This total, which does not include additional payments to offset the environmental impacts of the trade, is about half of what the Authority has contracted to pay for water generated by a new desalination plant on the coast (Fikes 2015; Cooley and Phurisamban 2016).

As noted above, water trading occurs within sectors; such as from agricultural users to other agricultural users, between the agricultural, municipal, and industrial sectors; and, less frequently, from any of these to the environment (Brewer et al. 2007). Howe (1998) found that the source and destination of water trades varied dramatically by state in the U.S. West in the 1990s, with more than 75 percent of trades moving water out of agriculture in Colorado and Wyoming, but only accounting for about 20 percent of total trades in Arizona. Figure 6.1, from the California Department of Water Resources, shows the relative proportions of water trading within and between different sectors in one region of California in 2013. Although water trading is often promoted as a means to move water from agriculture to urban uses, nearly three-quarters of the 270,000 acre-feet (0.3 km^3) of water traded in California in 2013 occurred between agricultural users. Interestingly, nearly 25,000 acre-feet of water were traded from municipal and industrial (M&I) uses to agriculture, which was nearly half of the volume of water traded from agriculture to M&I uses.

There is considerable experience with water trading markets in countries outside of the United States. Chile's Limarí Basin enjoys water-rights trading and water transfers, enabled by three large state-built reservoirs and robust local water organizations. Older information indicated that the actual number of water trades in Chile's Limarí Basin averaged about 33 each year (Romano and Leporati 2002), although water trading has been more limited in the rest of the country (Bauer 1997). Mexico's National Water Law of 1992 established a formal water market with tradable concessions that formed the basis for active markets in several parts of the country (Thobani 1997), with nearly 3,700 registered water transfer requests in 2006 alone (Conagua 2012). Hearne (1998) reports very active water trading of both temporary and permanent water concessions in Mexico's Mexicali Valley in the mid-1990s, with a total annual trading volume of 0.86 km^3, almost 30 percent of total water use in the region.

In Spain, informal trades, sales, and short-term exchanges of water are common, while formal transfers of long-term water rights are generally limited to groundwater (Albiac et al. 2006). In Spain's Alicante Basin, several irrigation districts auction their annual water allocations to district farmers (Albiac et al. 2006), creating a strong incentive to

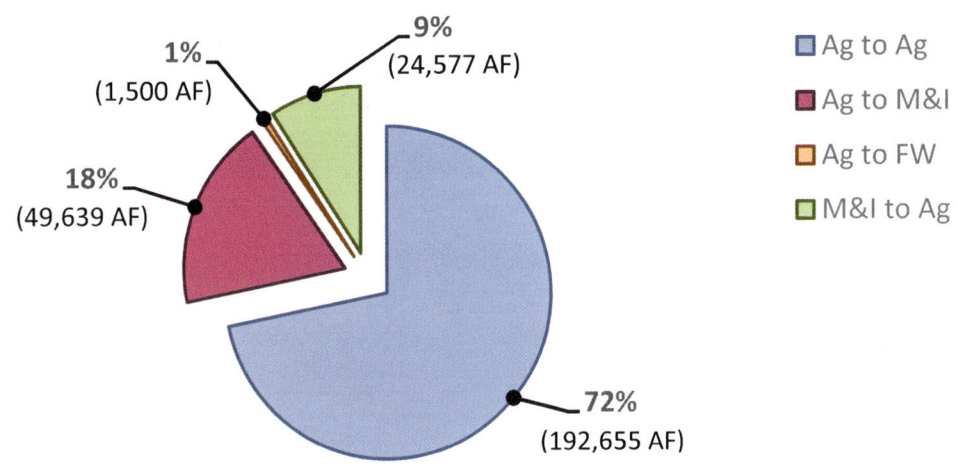

Water Trading 93

FIGURE 6.1 NON-PROJECT WATER TRANSFERS WITHIN THE SACRAMENTO-SAN JOAQUIN WATERSHEDS IN 2013.
Note: Ag – agriculture; FW – fish and wildlife; M&I – municipal and industrial; AF – acre-foot.
Source: California Department of Water Resources 2014.

improve water-use efficiency and shift toward higher value crops. England has encouraged water trading for more than a decade, although only about 60 trades have occurred to date (EPSRC 2013).

South Africa's Water Act of 1998 provided a framework for water trading. Historically, agricultural irrigators traded water rights within their sector, mediated by the national Department of Water Affairs and Forestry (Farolfi and Perret 2002). In 2001, mining companies seeking to expand operations in northern South Africa successfully negotiated a temporary trade of 13 million m³ from neighboring farmers—representing more than 70 percent of their annual allocation—in exchange for the current equivalent of about $1 million. These funds, used to help rehabilitate the local irrigation infrastructure, represented less than 0.1 percent of the mines' development costs, reflecting a significant economic disparity between the two interests (Farolfi and Perret 2002).

In Asia, India and Pakistan have informal water trading, in which well owners may sell some of the water they extract to neighboring farms or residents (Easter et al. 1999). Moench et al. (2003) described an active but largely unregulated water trading system in Chennai, India, where private companies meet as much as 35 percent of urban water demand by delivering raw or purified well water purchased from farmers in surrounding areas or extracted from the companies' wells, to urban consumers. This private sector engagement helps meet a demand for water that the intermittent municipal water supply does not satisfy, though the price is much higher. Moench et al. (2003) reported that the price of water for urban customers can be 1,000 times higher than the price paid to the peri-urban farmers supplying the water. Also in Asia, in a rare international water trade, the Bishkek Treaty of 1998 committed Kyrgyzstan to deliver water via the Syr Darya River to Uzbekistan and Kazakhstan in exchange for compensation (Ambec et al. 2013).

China reportedly has small, local water markets (Grafton et al. 2010). In Oman, the local *falaj* irrigation systems purchase short-term allocations of water based on units of time rather than volume (e.g., a certain duration of water delivery) in a village-based auction (Al-Marshudi 2007).

Water banks are less widespread than water trading because they require additional expertise, funding, and governance structures. Water banks appear to be most prevalent in the western United States, although there are examples in several other countries. In Australia, brokerage-type water banks are active in both the Murray-Darling Basin and in northern Victoria, where the banks post information about pricing and availability (O'Donnell and Colby 2010). Mexico's National Water Commission reported that the 13 state-based water banks in the country broker thousands of water trades annually (Conagua 2012). In three basins in Spain, water banks operated by local water agencies, known as "exchange centers," have successfully brokered water trades that have lessened groundwater overdraft (Garrido and Llamas 2009). The presence of three reservoirs in Chile's Limarí Basin facilitates the large number of water trades in the region (Bauer 1997): physical storage rather than an institutional bank facilitates the water trades.

In 2003, nine states in the western United States had functioning state-operated water banks, although their level of activity varied dramatically and several are no longer active. From 1995–2003, for example, Texas's water bank only reported one transaction (Clifford 2012). California's Drought Water Bank functioned for a limited period in the early 1990s, providing a mechanism to facilitate and expedite water trading between agriculture and cities during a multi-year drought, while also ensuring minimum instream flows and providing limited groundwater recharge. The Drought Water Bank purchased, held, and sold water, primarily from northern California agricultural users to southern municipal and industrial users, though about half of the more than 800,000 acre-feet purchased in 1991 was dedicated to instream flows (20 percent) and to recharge aquifers (32 percent) (Dinar et al. 1997). Idaho operates water banks to manage storage in reservoirs; and in Oregon, river conservancies operate as water trusts to purchase or lease water rights to supplement instream flows (Clifford 2012). The Northern Colorado Water Conservancy District maintains a webpage[4] that functions as an online bulletin board connecting those seeking to acquire water with those who have water to rent—an example of a brokerage-type water bank. The very active water trading within the Conservancy District is attributable to the equal volume and priority of each share available for trade, the absence of any requirement to preserve return flows or protect downstream or junior priority users, and the fact that trading only requires the approval of the district itself, not a water court, as is the case for most other trades within Colorado (Howe and Goemans 2003).

The Colorado River Basin, shown in Figure 6.2, boasts a large number of creative approaches to water banking. In 1999, the federal government adopted a new rule permitting interstate banking agreements within the basin (43 CFR 414). To date, Arizona has diverted and stored more than 600,000 acre-feet of Colorado River water for southern Nevada, and a southern California water agency has diverted and stored more than 161,000 acre-feet for southern Nevada. In 2007, the seven Colorado River Basin states in the United States adopted a new set of rules for managing the river that, among other

---

4. http://www.northernwater.org/AllotteeInformation/RentalWater.aspx.

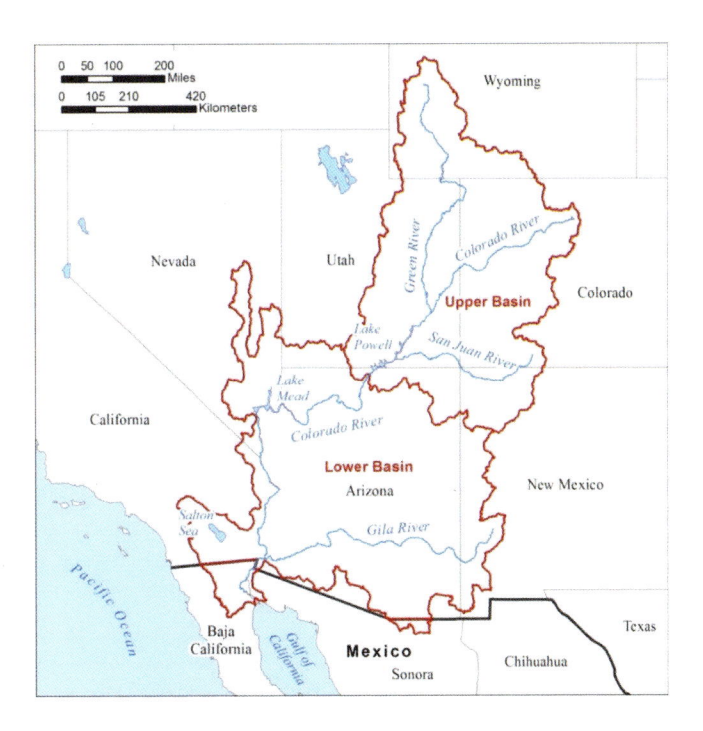

FIGURE 6.2 THE COLORADO RIVER BASIN.
Source: Cohen et al. 2013.

key developments, permitted entitlement holders in Arizona, California, and Nevada to invest in various water-efficiency projects within their own states and store a percentage of the conserved water in Lake Mead for later use (73 Fed. Reg. 19873). To date, more than 1.1 million acre-feet have been stored in Lake Mead under this new program. More recently, four large municipal water agencies in the basin, in cooperation with the United States Bureau of Reclamation (USBR), agreed to invest $11 million in fallowing and efficiency improvements and to, in effect, "return" the conserved water to Colorado River Basin system storage, rather than claiming it for themselves. In this instance, the USBR acts as a water bank by obtaining water through a reverse auction process, augmenting system storage for the benefit of the system as a whole.

Environmental, Economic, and Social Performance

The primary goal of water trading is to promote economic efficiency by reallocating water from lower- to higher-value uses. In some cases, water trading has been used for environmental or recreational purposes, reflecting the increasing societal value ascribed to instream flows. In this section, we evaluate the environmental, social, and economic performance of water trading.

Economic Performance

Although there are many articles and studies modeling the potential economic benefits of water trading, the number of detailed economic assessments of existing water trades is surprisingly limited. Some studies on local impacts suggest positive net economic performance, but these studies typically do not describe changes in the distribution of impacts, and they rarely describe broader economic impacts. Assessing the economic performance of water trading is frequently limited to documenting trading activity and quantifying the number, volume, and value of reported water trades. A more comprehensive analysis would require surveys to estimate the number and volume of additional water trades that users would like to make, as a means to assess the disparity between availability and demand. An even more robust analysis would compare the ability of different methods—such as water trading, demand-side management, and supply augmentation—to meet specific water demands, and the cost of those methods, as well as assess impacts to third parties who may be affected.

The previous section describes a range of countries where different forms of water trading occur. In most of these regions, limited data preclude detailed assessment of the number or volume of water trading activities. In several locations, such as the Murray-Darling Basin and the Northern Colorado Water Conservancy District, water trades occur frequently, often for small volumes, suggesting a robust and active market with low transaction costs (Howe and Goemans 2003). In other areas, there tend to be fewer but larger transactions.

The Imperial Irrigation District–San Diego County Water Authority agriculture-to-urban water trade, described earlier, currently provides about 15 percent of San Diego County's water supply. The long-term water trade is cost effective from San Diego's perspective but, due to significant externalities, may not be from the broader society's perspective. Total transaction costs for this water trade have exceeded $175 million in attorney fees, plus an additional $171 million in mitigation fees to date to offset public health and environmental impacts. In addition, in 2003 the State of California agreed to cover all direct mitigation costs in excess of a pre-determined financial cap for the water trade parties. The magnitude of these additional mitigation costs—primarily for managing dust emissions—will not be known for many years, but costs are expected to run into the hundreds of millions of dollars (Cohen 2014). As suggested by the Imperial Valley–San Diego example, a narrow focus on direct economic performance and specific water costs may ignore the broader economic impacts of water trading.

Despite its size and importance, there have not been any economic analyses of the Imperial Valley–San Diego County water trade that assess revenues, agricultural production lost due to fallowing, value of transfer payments, relative value of the water in San Diego, or employment impacts. There are limited regional or district-level assessments of water trading, as well as an extensive body of literature on macro-economic trends, and expected or modeled benefits of water trading. These assessments of "net" economic benefit at the state or regional level, expressed in terms of net increase in employment or revenue, can mask disparities between areas of origin and importing areas, and even within the areas of origin themselves.

In one study, the income and employment gains found in regions in California that imported water via trades exceeded the net losses (total compensation often failed to cover foregone crop revenue) in exporting areas (Howitt 1998). In 1991, trading activity generated an average net income loss in water-exporting areas equivalent to about 5 percent of net agricultural activity, though this varied within different parts of the state. However, agricultural areas importing water saw total gains greater than the losses in exporting areas: net agricultural water trading activity was positive, as water moved from lower-value crops to higher-value crops (Howitt 1998). In another example, an agricultural community in California exporting water to urban areas from 1987 to 1992 saw a 26 percent decrease in the number of farms overall, but this masked a 70 percent loss in the number of small farms and the loss of almost half of the number of produce-packing facilities in the area (Meinzen-Dick and Pradhan 2005).

The Northern Colorado Water Conservancy District, introduced previously, has a very active water market in part because of low transaction costs. Much of the trading activity in the district is short term and low volume, especially in comparison with trading activity in the same water basin but outside of the district. Municipal and industrial (M&I) users buy district water rights to meet expected future demand and then lease some of this water back to district irrigators. This rising M&I demand has increased the price of imported water rights (known as allotments) within the district (Howe 2011). Within the relatively prosperous district, this has improved economic performance. However, in other regions, particularly in economically depressed rural areas, selling water out of the area has exacerbated local economic decline, causing property values to fall and the tax base to shrink (Howe 2011).

In Australia, water trading has enabled the expansion of the wine industry and other high-value crops such as almonds. Over time, the dairy industry in one part of the Murray-Darling Basin transitioned from a small purchaser to a net seller of water entitlements, primarily to expanding wine and nut producers in other parts of the basin. These expanding producers have also exhibited a shift from the former model of shared irrigation infrastructure (such as canals) to direct extraction from the river by individual irrigators—in other words, from a communal to a more flexible individual approach to irrigation (Frontier Economics and Australia NWC 2007).

Water trading within the Murray-Darling Basin grew and matured within the context of the devastating drought from 2001 through 2009 that afflicted the region. The national water trading assessment noted the challenge of disentangling the economic impacts of the drought from those of water trading itself, generally concluding that trading offered irrigators an additional revenue stream, plus additional flexibility and resilience within the face of a severely limited water supply. Without water trading, some sectors, such as the dairy industry, would have seen even greater losses. Trading also offered a mechanism to adjust for historic water apportionments, facilitating the voluntary sale of water from less productive to more productive lands and uses (Frontier Economics and Australia NWC 2007; Heberger 2012).

The active participation of the Australian government in water trading increased prices and participation but may also have increased total water use within the basin. A large survey (n=520) of those selling entitlements or allocations to the Australian environmental water program found that sellers believed they received a higher price from the government than they would have from other private agents, or that the government was the only purchaser in the market. The survey also found that sellers reportedly used 69 to 77 percent of their water allocations prior to trading it to the government (Wheeler and Cheesman 2013). That is, survey respondents reported selling portions of their allocations that they were not otherwise using. The sale and subsequent activation or use of these "sleeper" or "dozer" rights is not a reallocation so much as an expansion of total water use.

Environmental Performance

Water trading has occasionally been used to obtain water for ecological purposes, to augment streamflows, and to address water-quality concerns (such as temperature) in threatened reaches. The environmental performance of water trading is highly variable, depending on the type of trade and site-specific conditions. The benefits of voluntary, incentive-based water acquisition include greater community support, especially relative to regulatory takings.[5] However, water trading can also generate large environmental externalities, adversely affecting either natural habitats or downstream users, or both (NRC 1992). For example, when water for trading is generated by improving efficiency or by fallowing land, the trade may reduce the amount of excess runoff supporting local habitat and may diminish instream flows. On the other hand, some water trades may improve local instream flows by decreasing diversions and contaminant loadings. Where water is traded to downstream users using the existing stream as a conveyance, trading could offer measurable environmental benefits. Where water is traded out of the basin or alters the timing and magnitude of flows, adverse impacts are likely to occur. Unfortunately, there do not appear to be published assessments of the relative impacts of water trading on streamflow. In the following, we discuss the environmental performance of several examples of water trading.

Water trading is used in some areas to return water to river channels to support protected species or threatened habitats, and for general ecosystem restoration (Tarlock 2014). In most areas, such activity still represents only a tiny fraction of total water use.[6] For example, the Colorado Water Trust (CWT) brokered a lease agreement between two state agencies, increasing low-season flows in the White River by 3,000 acre-feet of water three times over a 10-year period in order to lower the temperature of river flows to benefit fish (CWT 2015). The Columbia Basin Water Transactions Program (CBWTP), active for more than a decade, works with partner organizations in four Western states to acquire and dedicate water for instream flows within the basin. In 2013, 45 transactions led to the acquisition of more than 48,000 acre-feet (0.06 km^3) of water, costing about $13.9 million and benefiting some 276 miles (444 km) of streams, the fish and wildlife, and the communities that depend on them (National Fish and Wildlife Foundation 2014). Bonneville Power Administration (BPA), in cooperation with the Northwest Power and Conservation Council, provides some of the funding for the program due, in part, to concerns about endangered species. In California, environmental water purchases averaged 152,000 acre-feet (0.19 km^3) per year, accounting for about 14 percent of trading activity between 1982 and 2011, but less than 0.5 percent of total water use in the state (Figure 6.3).

The Australian government has invested more than $3 billion to date to purchase entitlements and allocations for environmental water, protecting ecological resources to enable and expedite water trading between non-governmental users. In 2008–2009, the federal government purchased nearly 1 km^3 of long-term water entitlements and 1.7 km3 of short-term allocations, at a total cost of about $2 billion (adjusted to 2014 dollars). The

---

5. A regulatory taking occurs when a government regulation limits or infringes upon a private property right to such an extent that it deprives the owner of some or all of the value of that property. Not all water rights are necessarily considered "property" rights.

6. Such instream flows typically require additional legal conditions, such as explicit recognition of instream flow rights, improved monitoring and measurement, and the acceptance of local entitlement holders.

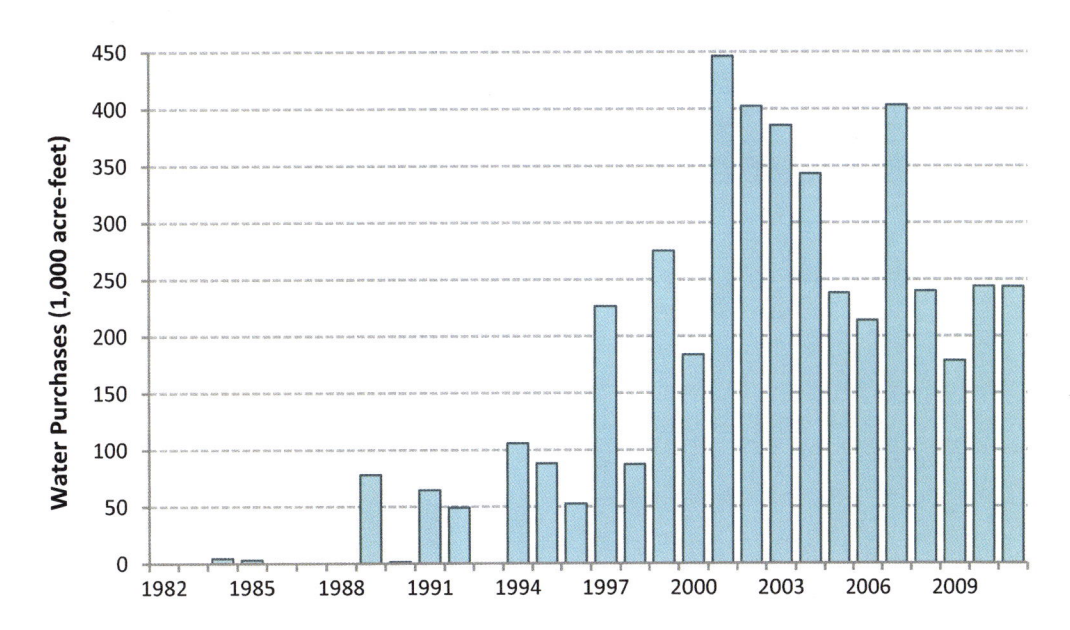

FIGURE 6.3 WATER PURCHASES FOR THE ENVIRONMENT IN CALIFORNIA, 1982–2011.
Source: Hanak and Stryjewski 2012.

price for this water ranged from about $218 million to $306 million per km^3 ($269 to $377 per acre-foot). Local interest in this environmental water buyback program has been strong, with the Australian government receiving nearly 7,600 applications to sell water from 2007 to early 2012. Water entitlement sales for the environment account for roughly 25 percent of total entitlement trading activity (Wheeler and Cheesman 2013). Some irrigators and state governments in Australia oppose the instream buyback program due to concerns about the loss of agricultural productivity (Grafton and Horne 2014), and it was cut dramatically when the Labor Party fell from power in September 2013.

Water trades that do not take account of environmental factors can inadvertently create a host of adverse environmental impacts by altering the timing, quantity, and quality of return flows and harming riparian and wetland habitats and the species that depend upon them. Some trades, such as the trade from California's Owens Valley to Los Angeles early in the 20th century, adversely affect public health by increasing the amount of dust emissions from exposed lakebed and fallowed land in the area (LA DWP 2013). Groundwater substitution, in which a user trades surface water and increases groundwater extraction, can lead to overextraction and land subsidence; depleting springs and seeps and harming future generations (Brown et al. 2015).

Water trading can also diminish groundwater recharge rates, whether the water is generated via fallowing or increased efficiency. In the southern Indian state of Tamil Nadu, farmers irrigating with groundwater have increased extraction rates and sold the excess to water tanker trucks serving urban populations. This increased groundwater extraction lowered the water table, increasing pumping costs for other irrigators or drying up their wells entirely (Meinzen-Dick and Pradhan 2005).

Efforts to mitigate the environmental impacts of water trading have had mixed success. In Spain, a proposal to add a small environmental mitigation fee to each unit of water traded was insufficient to overcome the strong opposition of environmental and social organizations to water trading (Albiac et al. 2006). In California, state commitments to mitigate the environmental and public health impacts of the Imperial–San Diego water trade have yet to materialize, potentially jeopardizing several listed species and likely resulting in the loss of open water and wetland habitats that support several hundred species of birds and a major bird migration stopover (Cohen and Hyun 2006).

Water trading occurs in regions of water scarcity, where water resources in particular have already undergone dramatic transformation. Dams, canals, and diversions have already altered the timing and magnitude of streamflows throughout many of the regions now turning to water trading (Worster 1985). Determining the additional impacts of water trading upon this altered landscape would be difficult. An alternative basis for comparison could be the marginal or cumulative environmental impacts of water trading relative to the new impacts of additional water development. Water trading may prove to be less environmentally harmful than the construction of new dams and diversion projects, or even the construction of new desalination plants. On the other hand, water trading creates more adverse impacts than demand-side management efforts that may leave more water in natural systems.

Social/Equity Performance

Water trading is usually characterized as a market-based mechanism that reduces economic inefficiencies by reallocating water from lower- to higher-value uses. Trading has been used to meet explicit environmental objectives, but it is rarely employed to address equity challenges. Indeed, recent experience indicates water trading can exacerbate social and economic inequalities.

Limarí Basin, Chile

Unequal access to water markets due to unequal access to information or credit can distort outcomes and reduce market efficiency. Romano and Leporati (2002) argued that the water trading in Chile's Limarí Basin suffers from several market distortions arising from disparities between the resources available to those trading water. Farmworkers fare poorly in trading activity because their water rights often are not fully recognized, they are not as well-organized as those purchasing the water, and they lack access to information on pricing (Romano and Leporati 2002). Dinar et al. (1997) noted that economic performance is affected by disparities in the value of water in different sectors and by the ability of those with limited means to participate in water trading.

Southern California

Water generated for trades by fallowing land can benefit water rights holders at the expense of farmworkers and equipment suppliers, potentially devastating rural communities (Loh and Gomez 1996; Gomez and Steding 1998). California's Owens Valley provides one of the early examples of the adverse impacts of trading water away from rural areas. In the early 1900s, agents secretly representing the City of Los Angeles (LA) covertly purchased land in the Owens Valley. In 1908, LA began a five-year construction project of a 419-mile pipeline to divert water from Owens Valley farmland to LA. Although Owens Valley irrigators had willingly sold their water through market transactions, they had not contemplated the plight of the valley as a whole. Over the next several years, agitators from the valley dynamited the pipeline several times in an unsuccessful attempt to protect their water supplies (Hundley 2001). In addition to the direct economic and social impacts on the Owens Valley, the water trade had desiccated Owens Lake by 1926, just 13 years after water first began flowing to LA, creating the single largest source of dust pollution in the United States. In the past decade, after years of litigation, LA has spent more than $1.4 billion on dust management efforts and has returned some of the water to Owens Lake.

San Luis Valley, Colorado

As demonstrated by efforts to destroy the infrastructure moving water out of the Owens Valley, local opposition to trading water can be strong. In the late 1980s, the Canadian owner of the 97,000-acre Baca Ranch in southern Colorado's San Luis Valley began buying water rights from other farms in San Luis Valley, allegedly to irrigate new crops. Local residents, who soon discovered that the true purpose of the purchases was to sell the water to Denver suburbs, 100 miles to the northeast, feared that their valley would experience the devastation seen in Owens Valley. They formed Citizens for San Luis Valley Water to fight the water trade, working with the local irrigation district to support a special ballot measure to raise local taxes to fund litigation against the proposed water sale. The ballot measure won with 92 percent of the vote. In 1991, the locals prevailed in court, stopping the proposed water trade. After Baca Ranch was subsequently sold, the new owner also attempted to sell the water out of the valley by sponsoring two statewide initiatives seen as efforts to support the water trade. In 1998, both initiatives failed, receiving less than 5 percent of the vote. With continued public pressure, the federal government purchased Baca Ranch in 2004 (Reimers 2013).

Imperial Valley, California

Water trading that promotes efficiency rather than fallowing of agricultural land can improve socioeconomic outcomes for both the area of origin and the destination. For example, a previous water trade from the Imperial Valley that began in 1989 relies on efficiency-based measures rather than fallowing to generate water for trade, creating additional employment while keeping land in production. The Imperial Irrigation District's *IID/MWD Water Conservation Program Final Construction Report* (2000) documented 24 separate system water conservation projects and programs implemented through 1999. The capital cost for these totaled $193 million (2014 dollars), with an additional $8.3 million in annual operations and maintenance costs. These improvements yield 0.13 km^3 (108,500 acre-feet) of conserved water per year. In addition to the jobs associated with the initial construction effort, the ongoing water trade supports about a dozen full-time positions for managing water deliveries and for annual operations and maintenance.

Water trading's social impacts vary based on several factors—including the relative economic health of the area of origin and the purchasing area, whether or not the water leaves the area of origin, the process used to trade the water, the relative economic and political power of the parties (Meinzen-Dick and Pradhan 2005), gender differences regarding access to and control of water (Zwarteveen 1997), the amount of trading activity in the area (Howe 2011), and the legitimacy of the water rights being traded (Meinzen-Dick and Pradhan 2005). Impacts often vary within the same community, as those with water rights or allocations to trade receive compensation, while third parties—such as irrigation equipment suppliers or farmworkers—may suffer a loss of revenue or income as a result of trading (Gomez and Steding 1998; Meinzen-Dick and Pradhan 2005).

Water trades within the same region typically have fewer or no adverse social or equity impacts. Howe (2011) noted the large number of small-volume, short-term water trades within an irrigation district as an example of positive economic and equity outcomes. Intersectoral trades, such as from agriculture to manufacturing or mining within the same region, may also generate positive economic and equity outcomes, as jobs shift from lower-income farm employment to higher-income industrial employment (Meinzen-Dick and Pradhan 2005). However, Zwarteveen (1997) noted that even such intraregional trades can generate differential impacts based on gender, requiring additional agricultural and domestic labor for women within households where men have left for new industrial jobs enabled by new water supplies. In places where rural agriculture provides subsistence and basic food security, reduced access to water can impose significant adverse impacts (Farolfi and Perret 2002).

Rural household access to water for domestic uses and for subsistence agriculture may have only informal community-level recognition that does not translate into tradable water rights. Water trading that does not recognize these informal or ad hoc water uses can adversely affect equity outcomes and prompt questions of legitimacy (Meinzen-Dick and Pradhan 2005). Formal, state-recognized water rights typically require the means and ability to register and defend them. In South Asia and other parts of the developing world, informal water-use arrangements that permit and enable water use and trading can be disrupted by formal rights-based trades and command-and-control reallocations (Meinzen-Dick and Pradhan 2005).

Zwarteveen (1997) noted that as men in Ecuador, Nepal, and Peru have migrated in search of employment, women have assumed a disproportionately large number of agricultural roles, even as formal and informal water rights continue to be held by the absent men. These geographic and gender disparities can generate adverse outcomes as water is traded by absentee owners. Conversely, trading within households—even in the form of recognition of joint ownership—can encourage investment in water-resource maintenance and productivity at the local level (Zwarteveen 1997). Similarly, water organizations in the developing world, where decisions may be made about trading water out of the community, tend to have limited female participation, potentially neglecting compensation for impacts that would have been identified if there were stronger female roles and participation (Zwarteveen 1997).

Water-trading mechanisms can privilege certain populations and marginalize others, especially when cultural practices differ. For example, New Mexico's cooperative irrigation systems, known as *acequias*, usually enjoy very senior water rights. However, they have fared poorly when defending their rights or seeking compensation for third-party impacts in state proceedings, where language and cultural practices favor English fluency and legal literacy (Meinzen-Dick and Pradhan 2005). Romano and Leporati (2002) found similar circumstances in Chile, where less-educated rural peasants fared poorly in trading water rights compared to more powerful non-agricultural interests.

Economic disparities also affect water-trading outcomes. As with the acequias, wealthy, powerful interests enjoy disproportionate advantages relative to many historic water rights holders. In South Africa in the late 1990s, mining interests sought to increase their production and activity in rural, water-scarce regions by purchasing water rights from small irrigators, at prices ten times higher than other irrigators were willing to offer.

Although the mines offered employment and generated greater returns per unit of water, they threatened to dewater local subsistence farms and adversely affect a broad swath of rural economies beyond the irrigators voluntarily selling their water (Farolfi and Perret 2002). A study of water trading in Chile's Limarí Valley found a similar impact, where increasing rural poverty was traced to water-rights sales from peasants to non-agricultural interests (Romano and Leporati 2002).

In regions with informal water trades that are functional at the community level, such as rural Nepal, demands from outlying urban areas for larger-scale trades can overwhelm local water management institutions. Trades from these rural areas might not reflect the true value of the many informal uses water has in the community (such as subsistence fishing or milling) or the full range of informal ownership and use rights within the community, meaning residents may be deprived of full compensation (Pant et al. 2008). Even within the community, the complex web of informal water-use arrangements can complicate informal trading agreements and, in turn, generate a range of economic impacts on those using the water who had not been consulted or had not participated in the trading arrangements (Pant et al. 2008).

As noted in the examples of the Owens and San Luis Valleys, those in areas of origin can strongly, sometimes violently, oppose the sale of water to outside interests. A national study of water trading in Australia found that this opposition can extend to local interests that trade their water rights to external interests (Frontier Economics and Australia NWC 2007). In addition to cultural and social bases for opposing such trades, trading can increase costs for those who do not sell, such as operations and maintenance costs associated with water storage and delivery structures. The economic and equity impacts of water traded from rural areas can accumulate with additional trading activity, reaching a tipping point where local demand for agricultural services falls below the level necessary to maintain operations, creating a cascading set of business failures and depressing the local tax base (Howe 2011). Agricultural areas importing traded water may also suffer from third-party impacts—in the form of increased competition, extended wait-times for water deliveries via shared infrastructure, and rising water tables that may threaten plant roots or require additional drainage (Frontier Economics and Australia NWC 2007).

Necessary, Enabling, and Limiting Conditions for Water Trades

Institutional arrangements determine the ultimate success or failure of formal water trading (Livingston 1998). Successful water trading requires secure and flexible water rights that recognize and protect users and others from externalities. Such institutional arrangements also need to be flexible enough to adapt to changing physical conditions as well as changing social norms, such as the growing interest in meeting environmental needs and protecting water quality (Livingston 1998). Recognizing and understanding these factors can help explain the varying successes and even the existence of water trading in different countries and regions within countries. Some factors, such as legal and transferable rights to use water, may be *necessary* for water trading to occur. Others, such as access to timely information about water available to trade, can *enable* water trading but may not be required for trading to occur. Still other factors, such as "no injury" regulations and "area of origin" protections, *limit* water trading or function as barriers or obstacles to trading.

Necessary conditions for successful water trades include:

- legal, transferable rights to use water;
- decoupling of water rights from land rights;
- contract adjudication and enforcement;
- means for buyers and sellers to communicate; and
- physical infrastructure to move water from point of sale to point of use.

Culp et al. (2014) asserted that water trading requires legally enforceable contracts that clearly and completely define the water right to be traded, an exclusive right to the water, and the recognized right to trade the water. Diversions or, better yet, consumptive-use water rights with clear title and quantified allocations that can be leased or sold can be described as marketable property rights, a necessary condition for water trading (Grafton et al. 2012). Government plays an important role in establishing these necessary conditions—documenting and, in some cases, allocating water rights themselves; establishing and maintaining the legal framework in which trading occurs; and, in many cases, financing the physical infrastructure to store and convey water and allow water trading to occur (Dinar et al. 1997). Strong and effective institutions that adjudicate and resolve disputes, enforce contracts, and monitor trading agreements are a necessary element in successful water markets (Zwarteveen 1997).

Typically, infrastructure is required to physically convey water from a seller to a buyer, or to store or otherwise manage water availability so that an agreed-upon volume can be conveyed to the buyer at the appropriate time. In some cases, creative agreements, sometimes known as in-lieu trades or exchange agreements, have enabled trades from unconnected or remote sources of water. For example, the Coachella Valley Water District (CVWD) is entitled to a share of California's State Water Project (SWP) water, but it lacks any means to access this water. Instead, it has executed an agreement with Metropolitan Water District (MWD), in which CVWD exchanges its share of SWP water with MWD for an equivalent volume of Colorado River water. While these trades can avoid requirements for connecting physical infrastructure, they do require sophisticated legal arrangements, management, and monitoring to ensure that the correct volumes of water move at the approved time.

Water trading can and does occur when necessary conditions are satisfied, but markets are much more robust and active when additional enabling conditions are met.

Enabling conditions for successful water trades include:

- water rights equivalency (as opposed to prioritized rights);
- water banks and contracts;
- relevant, available information;
- social cohesion;
- competitive markets with multiple participants of roughly equivalent economic power; and
- mechanisms to monitor and measure water flows and use.

Water Trading 105

One of the major factors contributing to Australia's successful adoption of water trading in the Murray-Darling Basin was the absence of prioritized water rights. This enabled water trading without concern for impacts on those holding less senior water rights. By contrast, in the western United States and other regions with prioritized water rights (also known as prior appropriation or seniority), an entitlement holder with a senior water right (determined by the date the right was first exercised or "perfected") could only sell or lease water after ensuring that more junior rights holders receive compensation or do not otherwise protest the transaction. This distinction helps explain the frequency of trades within irrigation districts where district members share a common priority right—such as the Northern Colorado Water Conservancy District—and the much lower number of transactions between those with different priorities. Common priority rights or water rights with equivalent seniority can be traded more readily than rights with different priority dates.

Water banks can enable water trading by connecting buyers and sellers, posting information on availability and transaction history and, in some cases, by physically storing water to match availability and demands. The existence of technically skilled staff and monitoring equipment increases the efficacy of water banks and can help resolve disputes. Where water banks do not exist or have limited capacity, water contracting can enable spot trading (Brown et al. 2015).

The availability of pertinent information can be considered both a necessary and an enabling condition, depending on the extent and type of information available. The availability of information on quantity, quality, location, and timing of water entitlements or allocations can enable trading by pairing sellers and buyers. Clear and timely information about prices also facilitates trading and decreases search costs (Levine et al. 2007).

Social cohesion can also enable water trading. Trading is more likely to occur where informal bonds exist, such as between neighbors or within an irrigation district or even between irrigators, relative to trading between parties with no common history. In some cases, irrigators will accept a lower bid from another irrigator than a higher bid from a municipal agency, particularly one from outside the basin or region. Water-rights holders may fear that indicating they have water to trade could be interpreted to mean that they do not need the water, jeopardizing the right or imposing political costs (Albiac et al. 2006).

Levine et al. (2007) argued that successful water trading requires the participation of multiple buyers and sellers, with roughly equivalent power. They contended that without these factors, market inefficiencies will result. In Australia's Murray-Darling Basin and within several U.S. irrigation districts, the satisfaction of these criteria has enabled active and successful water trading. In their absence, as seen in many agricultural-to-urban trades, a small number of economically powerful buyers has distorted markets and created significant externalities.

Measurement and monitoring increase transaction costs, but enable trading by providing verification of the timing and volume of water trades. Measurement and monitoring also increase confidence in water trading generally, assuaging concerns that water trading may simply increase total water use, rather than reallocate it. For example, more than 21,000 acre-feet of consumptive use rights had already been transferred from a large irrigation district along the Middle Rio Grande in central New Mexico to M&I use. Yet the irrigation district does not actually measure water deliveries or use (Oad and King

2005), challenging efforts to determine whether water trades actually result in a reallocation of water.

Limiting conditions that can hinder or reduce water trading include:

- no injury rule;
- anti-speculation doctrine;
- beneficial use doctrine;
- property rights/pre-conditions;
- high transaction costs; and
- spatial and temporal differences in supply and demand.

In many arid and semi-arid regions, water scarcity and variability dictate that upstream "return flows"—water diverted but not consumed that subsequently returns to the stream—are claimed and used by downstream users. To protect the rights of these downstream users, courts or regulators typically require that the quantity and timing of these return flows be maintained when upstream water is traded. These and similar protections, known as "no injury" rules, place the burden of proof that the trade will not harm or adversely affect other water rights on those wishing to sell or lease water. The "no injury" rule is the prevailing law in most of the western United States, intended to presumptively protect junior water rights holders from harm that may occur due to changes in the volume or timing of return flows from senior appropriators. Such rules dramatically increase transaction costs and reduce incentives for trading by requiring sellers to hire attorneys and hydrologists to prove no injury, or to otherwise compensate junior entitlement holders (Culp et al. 2014).

The anti-speculation doctrine requires buyers to describe the new location and use of the water, conditioning the trade on these terms and increasing transaction costs (Culp et al. 2014). The anti-speculation doctrine is intended to prevent hoarding and market distortion by those with the economic means to acquire large volumes of water (Grafton et al. 2010). In some areas, such as parts of Colorado, this doctrine is waived for municipal water agencies, enabling them to acquire water for unspecified future needs (Howe and Goemans 2003).

The beneficial use doctrine requires water rights be exercised, encouraging inefficient or unproductive uses as rights holders must "use it or lose it." Some jurisdictions have amended beneficial use requirements to enable rights holders to sell or lease the water they conserve or save by implementing efficiency measures; water they would otherwise simply lose to junior rights holders. Without explicit protection for such conservation measures, the beneficial use doctrine precludes water efficiency and hinders trading. For example, Colorado laws have historically explicitly prohibited users from selling or leasing water "salvaged" from conservation or efficiency measures (Culp et al. 2014).

Some kinds of water rights, such as non-consumptive, appurtenant water rights (common in wetter regions of the world) do not lend themselves to water trading.[7] Examples of such non-consumptive rights include rights to use or divert water to run mills or generate hydroelectric power.

Some markets limit participation to existing contractors or entitlement holders (Al-

---

7. An appurtenant water right is directly tied to the land itself, typically to lands adjacent to streams.

biac et al. 2006). A related barrier is a limitation on the purpose or use to which a buyer may apply water. For example, several states only allow state agencies, and not private individuals or non-profit organizations, to purchase or lease water for environmental purposes.

High transaction costs, driven by the various doctrines described above as well as by the need to overcome information constraints and related factors, hinder water trading. Similarly, the time required to complete a transaction may limit trading, particularly when buyers seek to meet a short-term demand such as an additional irrigation cycle or to offset a delivery disruption within an urban system; the ability to implement relatively fast trades will produce greater trading activity.

Finally, geographic and temporal mismatches between supply and demand can impose additional barriers to water trading, especially in the absence of appropriate physical infrastructure to bridge these gaps. Where storage and conveyances do not exist, those wishing to sell water may lack the means to physically deliver the water to a potential buyer, or be unable to deliver the water at the right time (Bauer 1997).

Conclusion

Overallocated rivers, projections that climate change will reduce runoff in many of the water-stressed regions of the world, and already degraded ecosystems and marginalized populations with limited access to water have all prompted an intense interest in water trading. This interest, buttressed by many thousands of articles extolling the theoretical ability of markets to allocate water rationally and the existing context in which agriculture consumes an overwhelming proportion of developed freshwater supplies, prompted this assessment of water trading in practice.

Water trading in the real world has generated decidedly mixed results, dependent primarily on the perspective of the analysis and the local legal, social, political, and economic conditions. Active water trading in several areas, such as Australia's Murray-Darling Basin and the Mexicali Valley, has led to temporary or permanent reallocation of more than a third of total annual water use in these areas. These examples, along with experience from places like specific irrigation districts in Colorado or California, markets in Chile, and more, indicate that several common factors are important in creating successful trades. These factors include water-rights equivalency (as opposed to a prioritized system of water rights), low transaction costs, limited or otherwise mitigated impacts to third parties, and credible and timely information about the price and availability of water for trading. In Australia, a $3 billion public subsidy (to date), in the form of federal purchases of water for environmental purposes, effectively removed a significant constraint on trading activity.

Replicating successes in other regions, particularly at scale, could be very difficult. For example, changing existing prioritized water rights to rights with equivalent priority in order to remove one of the larger obstacles to water trading in the western United States would require a massive regulatory taking likely to precipitate years of litigation, if it could be implemented at all. Building and instituting the necessary and enabling conditions for effective water trading, and removing limiting conditions, would require a significant, long-term investment of time, money, and institutional attention. As John Fleck wrote recently (Fleck 2016), "Ignoring the transaction costs of institutional change is the ag water economics equivalent of 'Imagine a frictionless plane.'"

Moreover, the volume or number of trades does not answer the question of whether such activity effectively reallocates water. Many water trades—the majority, in some locations—occur within the agricultural sector itself. While this may increase the economic productivity of the water, it does not address the broader objectives of water trading proponents who characterize trading as a mechanism to move water from the agricultural sector to meet growing demands in other sectors.

Water trading has been partly effective in helping to dedicate water for instream flows to meet environmental and recreational needs. In Australia, strong national support for ecological protection and a significant national investment have led to an impressive effort to identify and protect at-risk freshwater habitats. In the United States, several programs have produced more limited successes, with work continuing on these efforts in many states. On the other hand, water trades that are not made for environmental purposes can have significant adverse environmental impacts, worsening human and ecological health if these risks are not understood and addressed as part of the process.

Water trading can also lead to adverse socioeconomic impacts in the areas of origin. "Buy and dry" arrangements, where trading curtails agricultural productivity, have affected areas well beyond just the buyers and sellers; depressing tax bases, shuttering agricultural equipment suppliers, decreasing employment for farmworkers, and depopulating rural areas. Examples from around the world suggest that water trading can adversely and disproportionately affect poor and marginalized populations, including women, who may lack access to information or credit to negotiate with buyers on an equal footing, or who may be excluded from direct negotiations entirely.

The presence or absence of various necessary, enabling, and limiting conditions determines the success or failure of water trading in different areas. These existing conditions determine the magnitude of externalities and transactions costs. Any assessment of the potential for new or expanded water trading in a given area should start with a thorough appraisal of these existing conditions. Effective water markets have been developed and implemented in several areas, such as the Murray-Darling Basin, but proponents of new or expanded water trading should recognize the decades of effort and adaptive management associated with market development and implementation in these areas. Designing flexible, transparent, and effective water markets is neither fast nor easy.

In a limited number of areas with the necessary legal and technical conditions and with sufficient public investment, water trading has offered a timely, relatively inexpensive, and flexible mechanism to reallocate water between users. Achieving these successes has required determined effort involving accurate and transparent monitoring and measurement of water flows and use, significant and sometimes contentious legal changes to the nature of water rights themselves, the development and maintenance of publicly available databases reporting information on transactions, regional water planning, and construction and maintenance of appropriate infrastructure to convey water. Such significant institutional changes require broad public support and, importantly, a considerable amount of time to implement. Although water trading can reallocate water effectively, successful implementation requires a clear understanding of existing conditions and a determined, long-term effort to make the necessary changes and minimize externalities.

References

Albiac, J., M. Hanemann, J. Calatrava, J. Uche, and J. Tapia. 2006. The Rise and Fall of the Ebro Water Transfer. *Natural Resources Journal* 46: 729–757.

Al-Marshudi, A. S. 2007. The Falaj Irrigation System and Water Allocation Markets in Northern Oman. *Agricultural Water Management* 91(1–3): 71–77. doi:10.1016/j.agwat.2007.04.008.

Ambec, S., A. Dinar, and D. McKinney. 2013. Water Sharing Agreements Sustainable to Reduced Flows. *Journal of Environmental Economics and Management* 66(3): 639–655. doi:10.1016/j.jeem.2013.06.003

Bauer, C. J. 1997. Bringing Water Markets down to Earth: The Political Economy of Water Rights in Chile, 1976–1995. *World Development* 25(5): 639–656.

Bjornlund, H., and J. McKay. 2002. Aspects of Water Markets for Developing Countries: Experiences from Australia, Chile, and the US. *Environment and Development Economics* 7(04). doi:10.1017/S1355770X02000463.

Brewer, J., R. Glennon, A. Ker, and G. D. Libecap. 2007. Water Markets in the West: Prices, Trading, and Contractual Forms. Working Paper 13002. National Bureau of Economic Research. http://www.nber.org/papers/w13002.pdf.

Brown, C., U. Lall, and F. A. Souza Filho. 2015. Allocating Water in a Changing World: Incentive-Based Mechanisms for Climate Adaptation: Synthesis Review. https://assets.rockefellerfoundation.org/app/uploads/20160208145427/issuelab_23697.pdf.

California Department of Water Resources. 2014. Non-Project Water Transfers within the Sacramento/San Joaquin Watersheds. http://www.water.ca.gov/watertransfers/docs/2014/2012-2013_Transfers_Charts.pdf.

Clifford, P. 2012. Water Banking in Washington State: 2012 Report to the Legislature. Report to Legislature 12-11-055. Washington State Department of Ecology. https://fortress.wa.gov/ecy/publications/SummaryPages/1211055.html.

Cohen, M. J. 2014. Hazard's Toll: The Costs of Inaction at the Salton Sea. Oakland, CA: Pacific Institute. http://pacinst.org/publication/hazards-toll/.

Cohen, M. J., and K. H. Hyun. 2006. Hazard: The Future of the Salton Sea with No Restoration Plan. Oakland, CA: Pacific Institute. http://pacinst.org/publication/restoration-project-critical-to-salton-seas-future/.

Cohen, M. J., J. Christian-Smith, and J. Berggren. 2013. Water to Supply the Land: Irrigated Agriculture in the Colorado River Basin. Oakland, CA: Pacific Institute. http://pacinst.org/wp-content/uploads/2013/05/pacinst-crb-ag.pdf.

Colby, B. G., M. A. McGinnis, and K. A. Rait. 1991. Mitigating Environmental Externalities through Voluntary and Involuntary Water Reallocation: Nevada's Truckee-Carson River Basin. *Natural Resources Journal* 31: 757.

Colorado Water Trust (CWT). 2012. Colorado Water Trust Big Beaver Creek Reservoir/Big Beaver Creek & White River. http://www.coloradowatertrust.org/impact/projects/lake-avery-big-beaver-creek-white-river.

Cooley, H., M. Cohen, M. Heberger, and H. Rippman. 2015. Incentive-Based Instruments for Freshwater Management. Prepared for the Rockefeller Foundation and the Foundation Center. Oakland, CA: Pacific Institute.

Cooley, H., and R. Phurisamban. 2016. The Cost of Alternative Water Supply and Efficiency Options in California. Oakland, CA: Pacific Institute. http://pacinst.org/publication/cost-alternative-water-supply-efficiency-options-california/.

Conagua. 2012. Water Banks in Mexico. National Water Commission. https://www.gob.mx/cms/uploads/attachment/file/104934/Water_Banks_in_Mexico.pdf.

Culp, P., R. Glennon, and G. Libecap. 2014 (October). Shopping for Water: How the Market Can Mitigate Water Shortages in the American West. Discussion Paper 2014-05. Washington, DC: The Brookings Institute. http://www.hamiltonproject.org/files/downloads_and_links/market_mitigate_water_shortage_in_west_paper_glennon_final.pdf.

Dellapenna, J. 2000. The Importance of Getting Names Right: The Myth of Markets for Water. *William & Mary Environmental Law and Policy Review* 25(2): 317.

Dinar, A., M. W. Rosegrant, and R. S. Meinzen-Dick. 1997. Water Allocation Mechanisms: Principles and Examples. Policy Research Working Paper Series 1779. Washington, DC: The World Bank. http://econpapers.repec.org/paper/wbkwbrwps/1779.htm.

Easter, K. W., M. W. Rosegrant, and A. Dinar. 1999. Formal and Informal Markets for Water:

Institutions, Performance, and Constraints. *The World Bank Research Observer* 14(1): 99–116.

Engineering and Physical Sciences Research Council (EPSRC). 2013. Transforming Water Scarcity Through Trading. http://www.researchperspectives.org/rcuk/9B3A1B1A-616B-4D77-AFF1-8AB81AF87F72_Transforming-Water-Scarcity-Through-Trading.

Farolfi, S., and S. Perret. 2002. Inter-Sectoral Competition for Water Allocation in Rural South Africa: Analysing a Case Study through a Standard Environmental Economics Approach. Working Paper 2002-23. Department of Agricultural Economics, Extension and Rural Development. Pretoria, South Africa: University of Pretoria. http://agritrop.cirad.fr/511263/.

Freyfogle, E. T. 1996. Water Rights and the Commonwealth. *Environmental Law* 26: 27.

Fikes, B. J. 2015. State's Biggest Desal Plant to Open: What It Means. California, County Enter New Era with Opening of Poseidon Water Plant in Carlsbad. *San Diego Union-Tribune*. Dec. 13, updated Dec. 15. http://www.sandiegouniontribune.com/news/2015/dec/13/poseidon-water-desalination-carlsbad-opening/.

Fleck, J. 2016 (January). Pushback on the Export of Palo Verde Alfalfa. *JFleck at Inkstain*, http://wp.me/p4KlsT-3dd.

Frontier Economics, and Australia National Water Commission. 2007. The Economic and Social Impacts of Water Trading: Case Studies in the Victorian Murray Valley/Frontier Economics. PANDORA Electronic Collection. Canberra, ACT: National Water Commission. http://pandora.nla.gov.au/pan/79821/20071203-0953/svc044.wic032p.server-web.com/publications/docs/EconomicSocialWaterTrade.html.

Gomez, S., and A. Steding. 1998 (April). California Water Transfers: An Evaluation of the Economic Framework and a Spatial Analysis of the Potential Impacts. Oakland, CA: Pacific Institute. https://nrm.dfg.ca.gov/FileHandler.ashx?DocumentID=8756.

Garrido, A., and M. Ramón Llamas. 2009. Water Management in Spain: An Example of Changing Paradigms. *Policy and Strategic Behaviour in Water Resource Management* 125.

Getches, D. H. 1985. Competing Demands for the Colorado River. *University of Colorado Law Review* 56: 413.

Glennon, R. 2005. Water Scarcity, Marketing, and Privatization. *Texas Law Review* 83(7): 1873–1902.

Grafton, R. Q., and J. Horne. 2014. Water Markets in the Murray-Darling Basin. *Agricultural Water Management* 145: 61–71. doi:10.1016/j.agwat.2013.12.001.

Grafton, R. Q., C. Landry, G. D. Libecap, S. McGlennon, and B. O'Brien. 2010. An Integrated Assessment of Water Markets: Australia, Chile, China, South Africa and the USA. ICER Working Paper 32-2010. International Centre for Economic Research (ICER). https://ideas.repec.org/p/icr/wpicer/32-2010.html.

Grafton, R. Q., G. Libecap, E. Edwards, B. O'Brien, and C. Landry. 2012. Comparative Assessment of Water Markets: Insights from the Murray-Darling Basin of Australia and the Western USA. *Water Policy* 14: 175–193.

Hanak, E., and E. Stryjewski. 2012. California's Water Market, By the Numbers: Update 2012. Public Policy Institute of California. http://wee.ppic.org/content/pubs/report/R_1112EHR.pdf.

Hearne, R. R. 1998. Opportunities and Constraints to Improved Water Markets in Mexico. In *Markets for Water*, edited by K. W. Easter, M. W. Rosegrant, and A. Dinar, pp. 173-186. Natural Resource Management and Policy 15. Cambridge, MA: Springer.

Heberger, M. 2012. Australia's Millennium Drought: Impacts and Responses. In *The World's Water Volume 7: The Biennial Report on Freshwater*, edited by P. H. Gleick, pp. 97-125. Washington, DC: Island Press.

Howe, C. W. 1998. Water Markets in Colorado: Past Performance and Needed Changes. In *Markets for Water*, edited by K. W. Easter, M. W. Rosegrant, and A. Dinar, pp. 65-76. Natural Resource Management and Policy 15. Cambridge, MA: Springer.

———. 2011. The Efficient Water Market of the Northern Colorado Water Conservancy. District: Colorado, USA. IBE Review reports CS22. WP6 IBE EX-POST Case Studies. EPI Water. http://www.feem-project.net/epiwater/docs/d32-d6-1/CS22_Colorado.pdf.

Howe, C. W., and C. Goemans. 2003. Water Transfers and Their Impacts: Lessons from Three Colorado Water Markets. *Journal of the American Water Resources Association* 39(5): 1055–1065. doi:10.1111/j.1752-1688.2003.tb03692.x.

Howitt, R. E. 1998. Spot Prices, Option Prices, and Water Markets: An Analysis of Emerging Markets in California. In *Markets for Water*, edited by K. W. Easter, M. W. Rosegrant, and A. Dinar, pp. 119–140. Natural Resource Management and Policy 15. Cambridge, MA: Springer.

Hundley, N. Jr. 2001. The Great Thirst: Californians and Water—A History, Revised Edition. University of California Press. http://www.ucpress.edu/book.php?isbn=9780520224568.

Imperial Irrigation District and Metropolitan Water District of Southern California. 2000 (April).

Water Conservation Program: Final Construction Report. http://www.iid.com/home/showdocument?id=4060.

Jenkins, M. W., J. R. Lund, R. E. Howitt, A. J. Draper, S. M. Msangi, S. K. Tanaka, R. S. Ritzema, and G. F. Marques. 2004. Optimization of California's Water Supply System: Results and Insights. *JAWRA Journal of Water Resources Planning and Management* 130(4): 271–280. doi:10.1061/(ASCE)0733-9496(2004)130:4(271).

Levine, G., R. Barker, and C. C. Huang. 2007. Water Transfer from Agriculture to Urban Uses: Lessons Learned, with Policy Considerations. *Paddy and Water Environment* 5(4): 213–222. doi:10.1007/s10333-007-0092-8.

Livingston, M. L. 1998. Institutional Requisites for Efficient Water Markets. In *Markets for Water: Potential and Performance*, 19–33. Cambridge, MA: Springer. http://link.springer.com/chapter/10.1007/978-0-585-32088-5_2.

Loh, P., and S. V. Gomez. 1996. Water Transfers in California: A Framework for Sustainability and Justice. Oakland, CA: Pacific Institute. 23 pp.

Los Angeles Department of Water and Power (LA DWP). 2013. Owens Lake Master Project. https://www.ladwp.com/ladwp/faces/ladwp/aboutus/a-water/a-w-losangelesaqueduct/a-w-laa-owenslake.

McMahon, T. G., and M. G. Smith. 2013. The Arkansas Valley "Super Ditch"—An Analysis of Potential Economic Impacts. *JAWRA Journal of the American Water Resources Association* 49(1): 151–162. doi:10.1111/jawr.12005.

Medellín-Azuara, J., J. J. Harou, M. A. Olivares, K. Madani, J. R. Lund, R. E. Howitt, S. K. Tanaka, M. W. Jenkins, and T. Zhu. 2008. Adaptability and Adaptations of California's Water Supply System to Dry Climate Warming. *Climatic Change* 87(1): 75–90.

Megdal, S. B., P. Dillon, and K. Seasholes. 2014. Water Banks: Using Managed Aquifer Recharge to Meet Water Policy Objectives. *Water* 6(6): 1500–1514.

Meinzen-Dick, R. S., and R. Pradhan. 2005. Recognizing Multiple Water Uses in Intersectoral Water Transfers. In *Asian Irrigation in Transition: Responding to Challenges*, pp. 178–205. Bangkok, Thailand: Asian Institute of Technology.

Moench, M., A. Dixit, S. Janakarajan, M. S. Rathore, and S. Mudrakartha. 2003. The Fluid Mosaic: Water Governance in the Context of Variability, Uncertainty and Change: A Synthesis Paper. Kathmandu: Nepal Water Conservation Foundation; and Boulder, CO: Institute for Social and Environmental Transition.

National Fish and Wildlife Foundation. 2014. FY 2013 Annual Report: Columbia Basin Water Transactions Program. Portland, OR: National Fish and Wildlife Foundation. http://cbwtp.org/jsp/cbwtp/library/documents/NLB_CBWTP_Annual13_R8.pdf.

National Research Council (NRC). 1968. Water and Choice in the Colorado Basin: An Example of Alternatives in Water Management: A Report. Washington, DC: National Academies Press.

———. 1992. Water Transfers in the West: Efficiency, Equity, and the Environment. Washington, DC: National Academies Press.

National Water Commission (NWC). 2013. Australian Water Markets Report 2012–2013. Canberra, ACT: NWC. http://webarchive.nla.gov.au/gov/20160615064023/http://www.nwc.gov.au/publications/topic/water-industry/australian-water-markets-report-2012-13.

Newlin, B. D., M. W. Jenkins, J. R. Lund, and R. E. Howitt. 2002. Southern California Water Markets: Potential and Limitations. *Journal of Water Resources Planning and Management* 128(1): 21–32.

Oad, R., and J. P. King. 2005. Final Report: Irrigation Forbearance Feasibility Study in the Middle Rio Grande Conservancy District. Prepared for MRGCD by MBK International LLC. July 28.

O'Donnell, M., and B. Colby. 2010. Water Banks: A Tool for Enhancing Water Supply Reliability. Guidebook. Tucson: University of Arizona. http://www.climas.arizona.edu/sites/default/files/pdfewsr-banks-final-5-12-101.pdf.

Pant, D., M. Bhattarai, and G. Basnet. 2008. Implications of Bulk Water Transfer on Local Water Management Institutions: A Case Study of the Melamchi Water Supply Project in Nepal. Working Paper 78. Consultative Group on International Agricultural Research (CGIAR). http://indiaenvironmentportal.org.in/files/capriwp78.pdf.

Reimers, F. 2013 (May/June). Shifting Ground. *The Nature Conservancy*. https://www.nature.org/magazine/archives/shifting-ground-1.xml.

Romano, D., and M. Leporati. 2002. The Distributive Impact of the Water Market in Chile: A Case Study in Limari Province, 1981–1997. *Quarterly Journal of International Agriculture* 41(1): 41–58.

Salzman, J. 2006. Thirst: A Short History of Drinking Water. *Yale Journal of Law and the Humanities*

17(3). http://lsr.nellco.org/duke_fs/31.

Tarlock, A. D. 2014. Mexico and the United States Assume a Legal Duty to Provide Colorado River Delta Restoration Flows: An Important International Environmental and Water Law Precedent. *Review of European, Comparative & International Environmental Law* 23(1): 76–87. doi:10.1111/reel.12066.

Thobani, M. 1997. Formal Water Markets: Why, When, and How to Introduce Tradable Water Rights. *World Bank Research Observer* 12(2): 161–179.

U.S. Bureau of Reclamation (USBR). 2012. Colorado River Basin Water Supply and Demand Study. Washington, DC: Bureau of Reclamation, U.S. Department of the Interior. https://www.usbr.gov/lc/region/programs/crbstudy/finalreport/index.html.

Wheeler, S. A., and J. Cheesman. 2013. Key Findings from a Survey of Sellers to the Restoring the Balance Programme. *Economic Papers: A Journal of Applied Economics and Policy* 32(3): 340–352. doi:10.1111/1759-3441.12038.

Worster, D. 1985. *Rivers of Empire: Water, Aridity, and the Growth of the American West.* New York: Pantheon Books.

Zellmer, S. B., and J. Harder. 2007. Unbundling Property in Water. *Alabama Law Review* 59: 679.

Zhao, X., J. Liu, Q. Liu, M. R. Tillotson, D. Guan, and K. Hubacek. 2015. Physical and Virtual Water Transfers for Regional Water Stress Alleviation in China. *Proceedings of the National Academy of Sciences (PNAS)* 112(4): 1031-1035. doi:10.1073/pnas.1404130112.

Zwarteveen, M. Z. 1997. Water—From Basic Need to Commodity: A Discussion on Gender and Water Rights in the Context of Irrigation. *World Development* 25(8): 1335–1349.

CHAPTER 7

The Cost of Water Supply and Efficiency Options: A California Case

Heather Cooley and Rapichan Phurisamban

Introduction

Water is one of our most precious and valuable resources. Communities, farms, businesses, and natural ecosystems depend upon adequate and reliable supplies of clean water to satisfy a wide range of demands. In more and more regions of the world, pressures from economic and population growth, industrial pollution, and climate change have led to concerns over our ability to meet future water demands. In addition, some regions are making efforts to restore natural ecosystems by returning water previously used for human activities. As we approach the limits of traditional water supplies—a situation sometimes described as peak water (Gleick and Palaniappan 2010)—more effort is being made to improve the efficiency of water use and to develop alternative sources of water.

A key element in determining which water strategies to pursue is the relative cost of different alternatives. Limited data are available for some newer options, and there are methodological challenges in making appropriate comparisons. A new study from the Pacific Institute (Cooley and Phurisamban 2016) examined the cost of a range of efficiency and alternative supply options in urban areas for the state of California: storm water capture, water reuse, brackish and seawater desalination, and a selection of urban water conservation and efficiency measures. This assessment provides a best estimate for the cost of these options, expressed in dollars per unit water.[1] Some of these options also provide important co-benefits, such as reducing energy bills or reducing polluted runoff in coastal waterways. Where possible, these benefits are also integrated into the cost estimate; however, the economic value of most environmental costs and benefits is not well documented and not included in this analysis. There is a growing recognition that,

---

1. The original study used acre-feet as the water unit, reflecting the common unit used in the western United States. Here, we use cubic meters. One acre-foot of water is 1,233 cubic meters.

while difficult to quantify, these factors are economically relevant and further research and analysis on them is needed. It is appropriate and necessary that similar assessments be done in other water-stressed regions of the world.

Methods and Approach

This analysis uses methods developed in the field of energy economics to estimate the levelized cost of water in California. This method accounts for the full capital and operating costs of a project or device over its useful life and allows for a comparison of alternative projects with different scales of operations, investment and operating periods, or both (Short et al. 1995). For each alternative, a ratio of net costs (costs minus benefits) to the output achieved in physical terms is determined. For the purposes of this study, the output is a unit of water in the case of a new supply, or a unit of water savings in the case of an efficiency measure. Comprehensive summaries of the methodology for water supply and efficiency options are provided in Cooley and Phurisamban (2016).

Water Supply Projects

For water supply projects, the analysis considers the investment required to build new facilities and the associated operation and maintenance (O&M) costs over the lifetime of the facility. Key components include capital costs, O&M and replacement costs, discount rate, expected useful life, water production capacity, and average water yield. Capital cost represents a one-time expenditure over a fixed period to bring the project into operation and includes structures, land, equipment, labor, and allowances for unexpected costs or contingencies (generally assumed to be 20%–30% of construction costs). These costs are annualized over the life of the project and divided by the water production capacity. O&M and replacement costs are incurred during operation and typically vary with output levels. For projects that are currently in operation, we use average annual O&M costs whenever possible; otherwise, we use values from the most recent year available. The O&M costs are annualized over the life of the project and divided by the annual water yield. The annualized capital and variable costs are summed, resulting in an estimate of the cost of water expressed in 2015 dollars per cubic meter of water over the lifetime of the project. Because many project- and site-specific factors affect the cost of a project, we provide the 25th and 75th percentiles of the cost range for each water supply option, which are represented in this report as the low and high values, respectively.

Water Efficiency Measures

A water efficiency measure is an alternative to new or expanded physical supply and can also be evaluated using a levelized-cost approach. In this chapter, we use the term "conserved water" to refer to the water savings associated with an efficiency measure. The cost of conserved water from efficiency savings is based on the incremental cost of purchasing and installing a new, water-efficient device and any changes in operation and maintenance costs resulting from the investment (excluding water bill payments as they reflect the cost of water production). This cost is annualized over the life of the device and divided by the average annual volume of water conserved, resulting in an estimate of the cost of conserved water expressed in 2015 dollars per cubic meter of water.

For most efficiency measures, we assume that the customer is in the market for a new device because the old device has reached the end of its useful life (i.e., natural replacement). To estimate water savings and incremental cost under natural replacement, we develop two scenarios: a baseline and an efficient scenario. For the baseline scenario, we assume that the customer replaces the old device with a new device that uses the same amount of water. For our efficient scenario, we assume that the customer replaces the old device with a new, efficient model. Annual water savings are then calculated as the product of the difference in water use between the two models and the estimated average frequency of use. The incremental cost is the cost difference between a new efficient and a new inefficient device and is based on price surveys of available models. For some devices, such as faucet aerators and water brooms, we assume that the customer would not have made the investment otherwise, and thus the cost of the water-efficiency investment is the full cost of the device.

In this analysis, efficiency measures are evaluated from the perspective of the customer. This approach addresses costs and benefits to the water supplier—which are eventually passed on to the customer—as well as costs and benefits customers experience from implementing the efficiency measure. For example, a high-efficiency clothes washer uses less energy and produces less wastewater than inefficient models, thereby reducing the customer's energy and wastewater bills. When non-water benefits accrued over the lifetime of the device exceed the cost of the water conservation investment, the cost of conserved water may be negative; i.e., a positive return on investment.

Data Sources and Limitations

This analysis uses the best-available public information on the cost and yield of water supply projects and conservation and efficiency measures currently in operation or under consideration in California. Because costs vary widely around the world, care should be taken in making any assumptions for other regions, though the trends and methods may be similar elsewhere. Data sources include end-use and field studies, surveys, expert knowledge, and online resources. Data for actual and proposed water supply projects are drawn from state agencies, local water utilities, engineering estimates, and project documents. These costs can be affected by design errors, construction delays, changes in interest and financing options, and regulatory factors. For water reuse and desalination projects, water production volumes are based on plant capacity and average annual production, when available. For storm water capture projects, water yield is represented by groundwater recharge estimates. Operational decisions to produce less water would increase the levelized cost of a project.

Storm Water Capture

For more than a century, storm water has been viewed as a liability in California, and most urbanized areas were designed to remove this water as quickly as possible. Urban runoff washes pesticides, metals, and other pollutants into inland and coastal waters and can worsen erosion. Both the U.S. Environmental Protection Agency (EPA) and the State Water Resources Control Board (SWRCB) have determined that "stormwater and urban runoff are significant sources of water pollution that can threaten aquatic life and public health" (SWRCB 2014). Improving storm water management can improve water quality, while also reducing flood damage and boosting local water supplies. It also offers several non-water benefits, including enhancing wildlife habitat, reducing the urban heat island effect, improving community cohesion, and reducing greenhouse gas emissions (CNT 2010).

Increasingly, storm water is being viewed as a resource challenge and an asset in many water-scarce regions in California. In 2009, the SWRCB set a goal to increase the annual use of storm water over 2007 levels by at least 600 million cubic meters by 2020, and 1.2 billion cubic meters by 2030 (SWRCB 2013). They also developed, and are now implementing, a Storm Water Strategy to better manage this resource and optimize its use. In addition, a state law (the "Rainwater Capture Act," AB 275) passed in 2012 authorizes residential users and public and private utilities to install and operate rainwater capture systems that meet specified requirements for landscape use.

Local efforts to capture storm water are also expanding. For example, the Fresno-Clovis metropolitan area captures and recharges about 21 million cubic meters of storm water per year (DWR 2014b), while the Los Angeles Department of Water and Power and its partners actively capture about 36 million cubic meters of storm water annually and plan to recharge an additional 84 to 140 million cubic meters per year by 2035 (Geosyntec Consultants 2015). An analysis by Garrison et al. (2014) suggested that there is still significant potential for storm water capture in urbanized Southern California and the San Francisco Bay areas, which could contribute 520 to 780 million cubic meters per year to local water supplies.

Cost of Storm Water Capture

Measures to capture storm water were initially designed to improve water quality and provide flood relief. Increasingly, projects are also being designed to boost local water supplies at a variety of scales. For example, rain barrels or cisterns can be used at a residential or commercial building to capture and store rainwater onsite. Bioswales and spreading basins can capture storm water on a larger scale. The potential to capture and reuse storm water varies by soil properties, topography, and precipitation levels. Variability in the type of project and local conditions results in a wide range of costs for storm water capture projects. While storm water detention basins have been used for decades for flood control and/or groundwater recharge, data from older projects are incomplete and outdated. The Cooley and Phurisamban (2016) analysis includes 10 proposed storm water projects that were submitted for consideration to receive state funding.

Table 7.1 shows the cost estimates for centralized storm water capture projects, such as spreading basins. Estimates for distributed storm water capture systems, such as rain barrels or cisterns that may be installed at a household or building scale, are not included due to data limitations. Projects are grouped by size, with small projects defined as those with an annual yield of 0.35 to 1.9 million cubic meters and large projects as those with an annual yield of 8.0 to 9.9 million cubic meters.[2] The cost of small projects ranges from $0.48 to $1.04 per cubic meter, with a median cost of $0.95 per cubic meter. Larg-

---

2. Data for projects with expected annual yields between 1.9 and 8.0 million cubic meters were not available and thus are not included in this analysis.

TABLE 7.1 Storm Water Capture and Reuse Cost

	Sample Size	Storm Water Capture and Recharge ($ per m³) Low	Median	High	Groundwater Pumping and Treatment ($ per m³)	Total Cost ($ per m³) Low	Median	High
Small Project (≤1.85 million m³)	8	$0.48	$0.95	$1.04	$0.28	$0.76	$1.23	$1.32
Large Project (>8.0 million m³)	2	$0.19	$0.20	$0.21		$0.46	$0.48	$0.49

Notes: All cost estimates are rounded to the nearest cent and are shown in year 2015 dollars. Low and high costs represent the 25th and 75th percentile, respectively, of the estimated cost range. However, we report the full cost range for large storm water capture projects as only two projects are included in this analysis. Groundwater pumping and treatment costs are based on a median cost of $0.08 per cubic meter and $0.19 per cubic meter, respectively.

er projects exhibit significant economies of scale with a much lower levelized cost. The large projects, which employ a variety of techniques to capture storm water and recharge groundwater aquifers, cost $0.19 to $0.21 per cubic meter, with a median value of $0.20 per cubic meter. Costs at the higher end of the range reflect those that require additional infrastructure to convey storm water to recharge areas.

In addition to the cost to capture and store storm water, there is a cost to extract it from the aquifer and treat it to drinking water standards. These costs will vary based on groundwater quality and well depth. We estimate that groundwater pumping and treatment would cost an additional $0.28 per cubic meter.[3] Thus, the total cost of small projects ranges from $0.76 to $1.32 per cubic meter, with a median cost of $1.23. The total cost of large projects ranges from $0.46 to $0.49 per cubic meter, with a median cost of $0.48 per cubic meter.

Notably, these costs do not include some of the potential co-benefits of storm water capture projects, such as reducing pollution in nearby waterways, providing habitat, minimizing flooding, beautifying neighborhoods, and providing recreational opportunities, among others. Integrating these benefits into the economic analysis may significantly reduce the cost of water. Additional research is needed to quantify these benefits.

Water Recycling and Reuse

A variety of terms are used to describe water reuse. For the purposes of this chapter, the terms "water reuse" and "water recycling" are used to refer to wastewater that is intentionally captured, treated, and beneficially reused. "Municipal recycled water" refers to municipal wastewater that is collected from homes and businesses and conveyed to a nearby reclamation facility, where it undergoes treatment to meet standards suitable for reuse. Some wastewater can also be reused onsite with little or no treatment. For example, a home may have a gray water system that collects wastewater from a clothes washer and uses it to irrigate a garden, or an office building may be equipped with a wastewater treatment system to reuse a portion of the wastewater for flushing toilets and other non-

---

3. Groundwater pumping costs were calculated based on OCWD (2015), Upper Kings Basin IRWM Authority (2013), LACFCD (2013), City of Pasadena (2011), LADWP (2010), and MWDSC (2007). Treatment costs were based on MWDSC (2007).

potable applications. This analysis focuses solely on municipal recycled water because only limited data are available on the cost of onsite reuse systems. In coming years, as more onsite systems are put in place, additional information on their costs will become available.

Intentional reuse of treated wastewater has been practiced around the world for more than a century, and some regions are now heavily dependent on this water source, including Windhoek, Namibia, Singapore, and Israel. The earliest uses of recycled water were for agriculture, but today there is a broader set of recycled water applications, including for geothermal energy production, groundwater recharge, landscape irrigation, and industrial use; and in some regions, indirect or direct potable reuse. In California, between 1970 and 2009, the beneficial use of recycled water increased almost fourfold, mainly due to the growing cost and difficulty of finding new natural sources of water and changes in state law and policy to support water recycling infrastructure, production, and use. According to a 2009 statewide survey (the most recent available), California beneficially reuses about 860 million cubic meters of recycled water per year, or an estimated 13% of the wastewater generated (Newton et al. 2012). Tremendous additional opportunities exist to expand water reuse. An analysis by Cooley et al. (2014) estimated that the technical potential for water reuse in California was at least an additional 1.5 to 2.2 billion cubic meters per year.

Cost of Water Recycling and Reuse

Data on the cost of water recycling projects in California are drawn from three different sources: direct correspondence with water agencies, published documents on agency websites, and water recycling project grant proposals. While recycled water projects have been in operation for decades, complete cost information is hard to find for older projects due to changes in project ownership, lack of record keeping, and limited staff resources to go through a high volume of data. As a result, we evaluate the cost of proposed projects as well as project upgrades designed to augment water supplies. A total of 13 projects are evaluated, including seven nonpotable reuse projects and six indirect potable reuse projects. The source water for most projects in this analysis is secondary effluent from a nearby wastewater treatment plant.

Nonpotable reuse requires lower levels of treatment than other types of reuse and is distributed to customers in a separate water distribution system, which can be identified in the United States by its unique purple color. Its main applications include landscape and agricultural irrigation, habitat restoration, and certain industrial processes, such as for concrete production and cooling water. With indirect potable reuse, highly treated wastewater is put into an environmental system, such as an aquifer or reservoir, before it is treated again and put in the drinking water distribution system. Indirect potable reuse has been practiced in California since the early 1960s, and a growing number of projects are now using this approach (Crook 2010).

Table 7.2 shows cost estimates for nonpotable and indirect potable water reuse projects. Water recycling for nonpotable reuse is typically less expensive than indirect potable reuse, due to lower treatment requirements. We find that small, nonpotable reuse projects range from $0.44 to $0.93 per cubic meter, with a median cost of $0.48 per cubic meter. Expanding nonpotable reuse may require the installation or extension of a separate water distribution system, also known as a purple pipe system, at a cost of $0.77 per

Cost of Water Supply and Efficiency Options

TABLE 7.2 Water Recycling and Reuse Cost

	Sample Size	Nonpotable Reuse Facility ($ per m³)			Distribution ($ per m³)	Total Cost of Nonpotable Reuse ($ per m³)		
		Low	Median	High		Low	Median	High
Small Project (≤12 million m³)	7	$0.44	$0.48	$0.93	$0.77	$1.21	$1.25	$1.70

		Indirect Potable Reuse Facility ($ per m³)			Conveyance, Groundwater Pumping and Treatment ($ per m³)	Total Cost of Indirect Potable Reuse ($ per m³)		
		Low	Median	High		Low	Median	High
Small Project (≤12 million m³)	3	$1.21	$1.50	$1.80	$0.37	$1.59	$1.88	$2.17
Large Project (>12 million m³)	3	$0.91	$1.06	$1.28		$1.28	$1.43	$1.66

Notes: All cost estimates are rounded to the nearest cent and are shown in year 2015 dollars. Low and high costs represent the 25th and 75th percentile, respectively, of the estimated cost range. Distribution for nonpotable reuse refers to the median cost of a purple-pipe distribution system. Additional costs for distribution, pumping, and treatment for indirect potable reuse refers to the median cost of operating and maintaining finished water pumps and pipelines to transport water to an environmental buffer (e.g., a groundwater recharge basin or reservoir), plus the cost to extract and treat the groundwater.

cubic meter. Thus, the total cost for small, nonpotable reuse ranges from $1.21 to $1.70 per cubic meter, with a median cost of $1.25 per cubic meter. Project costs for large projects are not available; however, given economies of scale, they are likely to cost less than smaller projects.

We estimate that the cost of small indirect potable reuse projects, defined as those with a capacity of 12 million cubic meters per year or less, ranges from $1.21 to $1.80 per cubic meter, with a median cost of $1.50 per cubic meter. The cost of larger projects ranges from $0.91 to $1.28 per cubic meter, with a median cost of $1.06 per cubic meter. Energy is often the single largest O&M expense, accounting for 30% to 55% of the O&M costs. Prior to use, treated water is sent to an environmental buffer, such as a groundwater recharge basin or a reservoir. If the water is used to recharge groundwater, there is an additional cost of $0.37 per cubic meter to convey the water to a groundwater basin, extract it from the aquifer, and treat it to drinking water standards. Thus, the total cost for small indirect potable reuse projects ranges from $1.59 to $2.17, with a median cost of $1.88 per cubic meter. The total cost for large indirect potable reuse projects ranges from $1.28 to $1.66 per cubic meter, with a median cost of $1.43 per cubic meter.

As with storm water projects, these estimates do not include some of the potential costs and/or benefits of water reuse projects. In coastal areas, for example, recycling treated wastewater reduces pollution discharge into the ocean. Likewise, recharging groundwater aquifers with highly treated wastewater may improve groundwater quality. Integrating these benefits into the economic analysis would effectively reduce the cost

TABLE 7.3 Relative Salinity of Water

Type of Water	Relative Salinity (mg/L TDS)
Freshwater	Less than 1,000
Brackish Water	1,000 – 30,000
Seawater	30,000 – 50,000
Brine	> 50,000

of water. However, recycling water in the upper watershed could reduce water available for important downstream uses, such as fish habitat or recreation, and integrating these costs may increase the cost of water. Additional research is needed to quantify these costs and benefits.

Desalination

Desalination refers to a wide range of processes designed to remove salts from waters of different salinity levels (Table 7.3). Most desalination technologies rely on either distillation or membranes to separate salts from the product water, although most modern plants use reverse osmosis membranes. Reverse osmosis desalination typically requires pretreatment to prevent fouling of the membranes, and posttreatment to add minerals that improve taste and reduce corrosion to the water distribution system.

Interest in desalination in California began in the late 1950s. The state's first commercial desalination plant desalted brackish groundwater for residents of Coalinga in Fresno County (Crittenden et al. 2012). By 2013, there were 23 brackish groundwater desalination plants with a combined annual capacity of 170 million cubic meters (DWR 2014a). Seawater desalination has had only limited application in California, but interest remains high, with the Carlsbad desalination plant operating since December 2015 and an additional nine plants proposed along the coast (Pacific Institute 2015).

Cost of Desalination

The cost of seawater desalination is highly sensitive to regional costs for land, labor, energy, and compliance with regulatory requirements. Estimates here are based on engineering designs and plans because there are a limited number of facilities in operation along the California coast. Data on brackish water desalination facilities are more readily available because water districts have been treating brackish groundwater for several decades. However, the capital cost for facilities that have been in operation for more than 10 years is difficult to obtain and may not be relevant for estimating current costs. For these projects, we include the cost of expansion, although note that these values likely reflect the lower bound of new project costs.

We estimate that the marginal cost of a large seawater desalination plant, defined as

Cost of Water Supply and Efficiency Options

TABLE 7.4 Seawater and Brackish Water Desalination Cost

	Sample Size	Brackish Water Desalination Facility ($ per AF)			Integration ($ per AF)	Total Cost of Brackish Water Desalination Project ($ per AF)		
		Low	Median	High		Low	Median	High
Small Project (≤20 million m^3)	11	$0.73	$1.22	$1.40	$0.09	$0.83	$1.31	$1.49
Large Project (>20 million m^3)	5	$0.68	$0.82	$0.99		$0.77	$0.91	$1.08

		Seawater Desalination Facility ($ per AF)			Integration ($ per AF)	Total Cost of Seawater Desalination Project ($ per AF)		
		Low	Median	High		Low	Median	High
Small Project (≤20 million m^3)	3	$2.01	$2.13	$3.31	$0.16	$2.17	$2.29	$3.47
Large Project (>20 million m^3)	5	$1.53	$1.57	$1.90		$1.69	$1.72	$2.06

Notes: All cost estimates are rounded to the nearest cent and are shown in year 2015 dollars. Low and high costs represent the 25th and 75th percentile, respectively, of the estimated cost range. Integration cost is based on the median cost to integrate the desalinated water into the drinking water distribution system.

those with a capacity of at least 12 million cubic meters per year, ranges from $1.53 to $1.90 per cubic meter, with a median cost of $1.57 per cubic meter. The cost of smaller projects ranges from $2.01 to $3.31 per cubic meter, with a median cost of $2.13 per cubic meter. A seawater desalination plant must also be integrated into the drinking water system, which we estimate would cost an additional $0.16 per cubic meter. Thus, the total cost for a small seawater desalination project ranges from $2.17 to $3.47 per cubic meter, with a median cost of $2.29 per cubic meter. The total cost for a large seawater desalination project ranges from $1.69 to $2.06 per cubic meter, with a median cost of $1.72 per cubic meter.

Brackish water has lower salt and total dissolved solids (TDS) levels than seawater, and as a result, brackish water desalination requires less treatment to bring it to drinking water standards. We estimate that the cost of a large project with a capacity of more than 20 million cubic meters per year ranges from $0.68 to $0.99 per cubic meter, with a median cost of $0.82 per cubic meter. Smaller projects have a higher unit cost, ranging from $0.73 to $1.40 per cubic meter, with a median cost of $1.22 per cubic meter. We estimate that the cost to integrate water from a brackish water desalination facility into the drinking water distribution system is about $0.09 per cubic meter. This is less than for seawater desalination because brackish plants are typically located closer to the existing water distribution system. Thus, the total cost for a small brackish desalination project ranges from $0.83 to $1.49 per cubic meter, with a median cost of $1.31 per cubic meter. The total cost for a large brackish desalination project ranges from $0.77 to $1.08 per cubic meter, with a median cost of $0.91 per cubic meter.

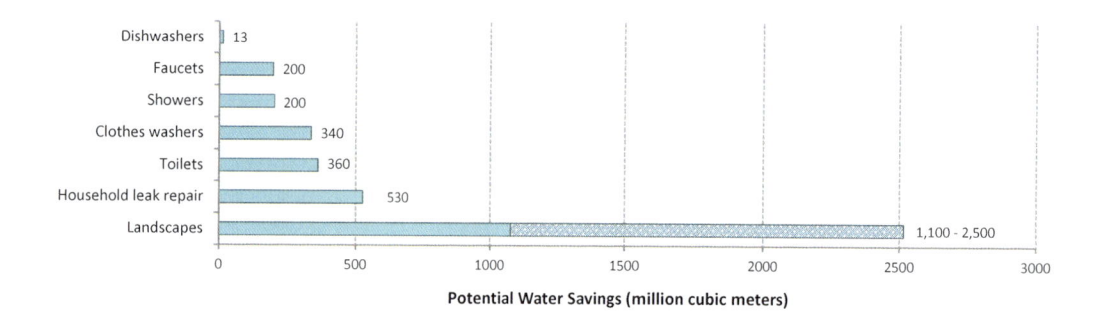

Figure 7.1 Potential Residential Water Savings, by End Use, in California.

Source: Based on data in Heberger et al. (2014)
Notes: Figure shows household water savings and does not include potential water savings from the nonresidential sector or from reducing losses in water distribution systems. Potential water savings for landscape efficiency improvements are shown as a range based on assumptions about the extent of landscape conversions.

Urban Water Conservation and Efficiency

Water conservation and efficiency are essential for meeting existing and future water needs in urban areas. California has made considerable progress in implementing water conservation and efficiency, as seen from the decline in residential water use (including both indoor and outdoor) from 620 liters per person per day (lpcd) in 2000 to under 500 lpcd in 2010 (DWR 2014c). Without these past efforts, California's current challenges would be more severe, demands on limited water supply would be higher, and ecosystem damage would be worse. Despite this progress, there is still additional potential to reduce demand for water in urban areas without affecting the services and benefits that water provides.

There are many ways to further reduce water waste and improve water efficiency in homes and businesses. A recent study by Heberger et al. (2014) found that the statewide technical potential to reduce urban water use ranged from 3.6 to 6.4 billion cubic meters per year.[4] Between 70% and 75% of the potential savings, or 2.7 to 4.4 billion cubic meters per year, are in the residential sector. As shown in Figure 7.1, water savings are possible for every end use within the home. The remainder of the savings potential (910 million to around 2.0 billion cubic meters per year) comes from efforts to improve efficiency among nonresidential users—i.e., the commercial, industrial, and institutional sectors. Finally, repairing leaks in water distribution systems reduced water losses, although insufficient data are currently available to quantify the potential water savings.

Cost of Urban Water Conservation and Efficiency Measures

The Pacific Institute analysis examined the cost of conserved water for reducing losses in the water distribution system and for various end-use efficiency measures in the

---

4. California's urban water use in 2001 to 2010 averaged over 11 billion cubic meters per year.

Cost of Water Supply and Efficiency Options

TABLE 7.5 Residential Water Conservation and Efficiency Measures

Efficiency Measure	Statewide Water Savings (1,000 m³ per year)	Measure Water Savings (liters per device per year)	Cost of Conserved Water ($ per m³) Low	Cost of Conserved Water ($ per m³) High	Notes
Toilet	360,000	18,000	-$0.51	-$0.16	13 to 4.9 lpf
		2,600	$0.95	$3.70	6.1 to 4.9 lpf
Showerhead	210,000	5,300	-$2.45	-$2.30	9.5 to 7.6 lpm
Clothes washer	330,000	27,000	-$0.61	-$0.15	
Dishwasher	14,000	1,600	$9.67	$15.66	
Landscape conversion	1,100,000 – 2,500,000	72 – 95	-$3.69	-$2.08	$22 per m²
			$0.47	$1.18	$54 per m²

Notes: All cost estimates are rounded to the nearest cent and are shown in year 2015 dollars. Measure water savings for landscape conversions are based on converting a square foot of lawn to a low water-use landscape. Because outdoor water savings are influenced by climate, we use a simplified landscape irrigation model to characterize water savings in five cities: Fresno, Oakland, Sacramento, San Diego, and Ventura.

lpf: liters per flush; lpm: liters per minute.

residential and nonresidential sectors. Data on water savings are based on the available literature, industry estimates, and expert input. The cost of the efficiency measures is based on a review of online retailers. More details about the methodology and data sources can be found in Cooley and Phurisamban (2016). Additional accurate, transparent, and consistent assessments of water efficiency measures are needed to demonstrate the performance, and ultimately the value, of these investments.

A wide variety of devices are available to reduce residential and nonresidential water use. For the residential sector, the report examined high-efficiency toilets, showerheads, clothes washers, dishwashers, and lawn conversions. Together, these end uses represent about 80% of average single-family household water use in California (DeOreo et al. 2011). For the nonresidential sector, the analysis examined a set of efficiency measures for end uses found in a wide range of businesses, such as toilets, faucet aerators, and showerheads, as well as devices for specific industries, such as modifications for medical steam sterilizers and waterless woks. There are many additional measures with high water- and energy-saving potential, such as cooling tower retrofits or specific industrial modifications that were not included in this study due to data limitations.

Residential Efficiency Measures

Table 7.5 shows the cost for conserved water for residential water conservation and efficiency measures. Efficient showerheads are among the most cost-effective measures available. Replacing older showerheads with models that use 7.6 liters per minute (lpm) would save an estimated 210 million cubic meters per year statewide at current population and use levels. These devices are relatively inexpensive and provide large financial savings over their 10-year life due to reductions in energy and wastewater costs. While replacing older showerheads that use more than 9.5 lpm would provide the greatest savings, replacing newer models is still highly cost effective.

High-efficiency toilets and clothes washers are somewhat less cost effective than showerheads but still far less costly than new supply options, and they provide much greater potential statewide water savings. High-efficiency clothes washers and toilets would save an estimated 330 million cubic meters and 360 million cubic meters per year, respectively (Heberger et al. 2014). While a new front-loading clothes washer is $340 to $460 more expensive than a standard model, this cost is more than offset by lower wastewater and energy bills, such that the cost of conserved water ranges from -$0.61 to -$0.15 per cubic meter. Similarly, the cost of conserved water for replacing older toilets that use 13 liters per flush (lpf) or more ranges from -$0.51 to -$0.16 per cubic meter saved. Replacing toilets that currently use 6 lpf is far more expensive due to lower water savings. This suggests that targeting those living in homes built before 1992, when the 6 lpf standard went into effect, would provide the greatest water savings at the lowest cost.

Table 7.5 shows the cost of reducing outdoor water use by converting lawns to low water-use landscapes. We characterize water savings in five California cities—Fresno, Oakland, Sacramento, San Diego, and Ventura—and estimate that annual water savings from landscape conversions in these cities range from 72 to 95 liters per square meter. Statewide, landscape conversions would reduce annual water use in California homes by 1.1 to 2.5 billion cubic meters (Heberger et al. 2014). We estimate that the cost of installing a low water-use landscape ranges from $32 to $54 per square meter, while installing a new lawn would cost about $11 per square meter. If the consumer is in the market for a new landscape, as may occur after a lawn dies or when buying a new home, then the incremental cost would be as low as $22 per square meter; i.e., the difference between a new lawn and a new low water-use landscape. If the customer converts an existing healthy lawn, then the cost would be $54 per square meter. At $22 per square meter, the cost of conserved water is -$3.69 to -$2.08 per cubic meter. The cost is negative due to substantial reductions in fertilizer and maintenance costs; i.e., avoided costs from reduced fertilizer use and maintenance far outweigh the cost of the landscape conversion. At $54 per square meter, the cost of conserved water is $0.47 to $1.18 per cubic meter.

Nonresidential Efficiency Measures

California's commercial, industrial, and institutional sectors (also referred to as nonresidential sectors) use approximately 3.1 billion cubic meters of water annually, accounting for about 28% of all urban water use.[5] Heberger et al. (2014) found that efficiency measures could reduce nonresidential water use by 30% to 60%, saving an estimated 910 million to 2.0 billion cubic meters per year. The estimated statewide water savings for the nonresidential sector is less than for the residential sector, which was estimated at 2.7 to 4.4 billion cubic meters per year; however, the water savings for each efficiency measure tend to be much larger for the nonresidential sector than for the residential sector. For example, a single efficient ice machine would save an estimated 49,000 liters of water per year—nearly 10 times as much water as would be saved by installing an efficient showerhead in a home. Likewise, an efficient medical steam sterilizer would save up to 2.5

---

5. Authors' calculations based on 1998–2010 average. Data from DWR (2014c).

TABLE 7.6 Nonresidential Water Conservation and Efficiency Measures

Efficiency Measure	Measure Water Savings (liters per device per year)	Cost of Conserved Water ($ per m³) Low	Cost of Conserved Water ($ per m³) High	Notes
Toilet	20,000	-$0.55	-$0.06	13 to 4.8 lpf
Toilet	2,900	$1.47	$5.29	6.1 to 4.8 lpf
Urinal	10,000	$0.79	$1.48	2.7 to 0.47 lpf
Showerhead	16,000	-$2.46	-$2.30	9.5 to 7.6 lpm
Faucet aerators	6,100	-$0.99	-$0.55	8.3 to 3.8 lpm
Pre-rinse spray valve	26,000	-$1.39	-$0.94	8.3 to 5.4 lpm
Medical steam sterilizer modification	1,700,000 – 2,500,000	-$1.03	-$0.99	
Food steamer	200,000	-$11.36	-$10.91	Boiler-based to connectionless
Ice machine	49,000	-$2.92	-$0.91	
Waterless wok	640,000	-$0.85	-$0.71	
Clothes washer	140,000	-$1.30	-$0.91	Top loader to front loader
Landscape conversion	72 – 95	-$3.69	-$2.08	Assumes $22 per m²
Landscape conversion	72 – 95	$0.47	$1.18	Assumes $54 per m²
Rotary nozzle	7,900 – 15,000	$0.16	$0.84	
Water broom	190,000	$0.13	$0.28	

Notes: All cost estimates are rounded to the nearest cent and are shown in year 2015 dollars. Water savings for landscape conversions are based on converting a square foot of lawn to a low water-use landscape. Because outdoor water savings are influenced by climate, we use a simplified landscape irrigation model to characterize water savings in five cities: Fresno, Oakland, Sacramento, San Diego, and Ventura. See Appendix B in Cooley and Phurisamban (2016) for methodology and assumptions.

lpf: liters per flush; lpm: liters per minute.

million liters per year, at least 30 times more than could be saved by retrofitting an entire home with efficient appliances and fixtures.

Table 7.6 shows the cost of conserved water for some nonresidential water conservation and efficiency measures. We find that many nonresidential measures also have a negative cost and are highly cost effective. Several efficiency measures for restaurants—such as food steamers, waterless wok stoves, and ice machines—offer significant financial savings over their lifetime. For example, an efficient connectionless food steamer, which operates as a closed system that captures and reuses steam, would save about

200,000 liters of water and 14,000 kWh of electricity per year (FSTC n.d.), resulting in a cost of conserved water of -$11.36 to -$10.91 per cubic meter. Conversely, toilet and urinal replacements are less cost effective than other measures. However, as with the residential sector, targeting high-use customers and devices would increase the cost effectiveness of these measures.

Water Loss Control

Throughout California, high-quality water is lost from the system of underground pipes that distributes water to homes, businesses, and institutions. A survey of 85 California utilities found that real water losses averaged 170 liters per service connection per day (Sturm 2013).[6] Water loss rates vary based on the age of the system, the materials used, and maintenance levels. Studies suggest that leak detection surveys could reduce annual water losses by 620,000 liters per kilometer surveyed at a cost of $190 per kilometer (Reinhard Sturm, personal communication, December 1, 2015).[7] Assuming that leak detection and repair are ongoing processes, we estimate that the levelized cost for this measure is about $0.32 per cubic meter.[8] In addition to increasing water availability and deferring or eliminating expenditures on new supply and treatment infrastructure, reducing water losses can also protect public health and reduce flood damage liabilities. While not included in this analysis, these co-benefits would further reduce the cost of conserved water from a distribution system leak-detection program.

Summary and Conclusions

Alternative water supplies and efficiency measures are being implemented around the world and there is significant opportunity to expand the implementation of these options to meet current and future water needs. Economic feasibility is an important consideration to more widespread adoption, and this chapter offers a comprehensive analysis for California of the cost of a wide range of new options—including storm water capture, recycled wastewater, seawater and brackish water desalination, and numerous urban water conservation and efficiency technologies. We provide our best estimates for the cost of these options, expressed on a dollar per unit water basis and integrate any co-benefits associated with these projects to the extent possible; however, the economic values of environmental costs and benefits are not well documented and thus not included in this analysis. While difficult to quantify, they are economically relevant and further research is needed to develop better environmental benefit and cost estimates.

Tables 7.1, 7.2, and 7.4 compare the cost of alternative water supplies. Cooley and Phurisamban (2016) find that the cost of alternative water supplies is highly varied. Large storm water capture projects are among the least expensive of the water supplies examined, with a median cost of $0.48 per cubic meter. Seawater desalination projects, by contrast, are the most expensive water supply option examined, with a median cost

---

6. Real losses are physical losses of water resulting from leaks, breaks, and overflows in the pressurized system and the utility's storage tanks.

7. Based on work with 13 California utilities (Reinhard Sturm, personal communication, December 1, 2015).

8. This estimate does not include the cost to repair the leak, as the utility would have fixed the leak regardless of when it was discovered.

of $1.72 per cubic meter for large projects and $2.29 per cubic meter for small projects. Brackish water desalination is typically much less expensive than seawater desalination due to lower energy and treatment costs. Generally, the costs of municipal recycled water projects fall between those of storm water capture and seawater desalination. Nonpotable reuse is typically less expensive than potable reuse due to the lower treatment requirements; however, the distribution costs for a nonpotable reuse system could increase the cost of that water.

Tables 7.5 and 7.6 compare the cost of water efficiency measures. Urban water conservation and efficiency offer significant water savings and are the most cost-effective ways to meet current and future water needs in a region where traditional approaches to expanding water supply are increasingly costly or unavailable. Indeed, many efficiency measures have a negative cost, which means that the financial savings over the lifetime of the device that result from lower wastewater, energy, and/or maintenance costs exceed the incremental cost of the device. Financial savings from high-efficiency showerheads and clothes washers are especially large. Landscape conversions in residential and nonresidential settings can also have a negative cost, depending on the cost of the conversion and reductions in maintenance costs. Yet, even when landscape conversions cost $54 per cubic meter, we find that the cost of conserved water is less expensive than many new water supply options. While leak detection in the water distribution system is more expensive than some of the other efficiency measures, it is also highly cost effective when compared to most traditional water supply projects.

California—and many other regions of the world—is reaching, and in many cases has exceeded, the physical, economic, ecological, and social limits of traditional supply options. Water managers must expand the way they think about both "supply" and "demand"—away from costly old approaches and toward more sustainable options for expanding supply, including improving water use efficiency, water reuse, and storm water capture. There is no "silver bullet" solution to the state's water problems, as all rational observers acknowledge. Instead, a diverse portfolio of sustainable solutions is needed. But the need to do many things does not mean we must, or can afford, to do everything. We must do the most effective things first.

References

Center for Neighborhood Technology (CNT). 2010. The Value of Green Infrastructure: A Guide to Recognizing Its Economic, Environmental, and Social Benefits. http://www.cnt.org/sites/default/files/publications/CNT_Value-of-Green-Infrastructure.pdf.

City of Pasadena. 2011. Water Integrated Resources Plan. http://ww2.cityofpasadena.net/water-andpower/WaterPlan/WIRPFinal011211.pdf.

Cooley, H., P. Gleick, and R. Wilkinson. 2014. Water Reuse Potential in California. Oakland, CA: Pacific Institute.

Cooley, H., and R. Phurisamban. 2016. The Cost of Alternative Water Supply and Efficiency Options in California. Oakland, CA: Pacific Institute.

Crittenden, J. C., R. R. Trussell, D. Hand, K. Howe, and G. Tchobanoglous. 2012. *MWH's Water Treatment: Principles and Design*, 3rd ed. Hoboken, NJ: John Wiley & Sons.

Crook, J. 2010. Regulatory Aspects of Direct Potable Reuse in California. NWRI White Paper. Fountain Valley, CA: National Water Research Institute.

DeOreo, W. B., P. W. Mayer, L. Martien, M. Hayden, A. Funk, M. Kramer-Duffield, R. Davis, et al. 2011. California Single Family Water Use Efficiency Study. California Department of Water Resources/U.S. Bureau of Reclamation CalFed Bay-Delta Program.

Department of Water Resources (DWR). 2014a. Desalination (Brackish and Sea Water). In *California Water Plan Update 2013*. http://www.water.ca.gov/waterplan/docs/cwpu2013/Final/Vol3_Ch10_Desalination.pdf.

———. 2014b. Urban Stormwater Runoff Management. In *California Water Plan Update 2013*. http://www.water.ca.gov/waterplan/docs/cwpu2013/Final/Vol3_Ch20_Urban-Stormwater-Runoff-Mgmt.pdf.

———. 2014c. Data Summary: 1998–2010, Water Balances. In *California Water Plan Update 2013*: Technical Guide.

Food Service Technology Center (FSTC). Electric Steamer Life-Cycle Cost Calculator. http://www.fishnick.com/saveenergy/tools/calculators/esteamercalc.php.

Garrison, N., J. Sahl, A. Dugger, and R. Wilkinson. 2014. Stormwater Capture Potential in Urban and Suburban California. Oakland, CA: Pacific Institute.

Geosyntec Consultants. 2015. Stormwater Capture Master Plan. Report prepared for Los Angeles Department of Water and Power. Los Angeles, CA: Los Angeles Department of Water and Power.

Gleick, P. H., and M. Palaniappan. 2010. Peak Water: Conceptual and Practical Limits to Freshwater Withdrawal and Use. *Proceedings of the National Academy of Sciences (PNAS)* 107(25): 11155–11162. http://www.pnas.org/cgi/doi/10.1073/pnas.1004812107.

Heberger, M., H. Cooley, and P. Gleick. 2014. Urban Water Conservation and Efficiency Potential in California. Oakland, CA: Pacific Institute.

Los Angeles County Flood Control District (LACFCD). 2013. IRWM Implementation Grant Proposal, Proposition 84 Round 2: Benefits and Cost Analysis. http://www.ladpw.org/wmd/irwmp/docs/Prop84Round2ImplGrantApp/Attachment%208%20Benefit%20and%20Cost%20Analysis%201%20of%2010.pdf.

Los Angeles Department of Water and Power (LADWP). 2010. Urban Water Management Plan 2010. http://www.water.ca.gov/urbanwatermanagement/2010uwmps/Los%20Angeles%20Department%20of%20Water%20and%20Power/LADWP%20UWMP_2010_LowRes.pdf.

Metropolitan Water District of Southern California (MWD). 2007. Draft Groundwater Assessment Study Report. Report Number 1308. http://edmsidm.mwdh2o.com/idmweb/cache/MWD%20EDMS/003697466-1.pdf.

———. 2016. 2015 Urban Water Management Plan. http://www.mwdh2o.com/PDF_About_Your_Water/2.4.2_Regional_Urban_Water_Management_Plan.pdf.

Newton, D., D. Balgobin, D. Badyal, R. Mills, T. Pezzetti, and H. M. Ross. 2012. Results, Challenges, and Future Approaches to California's Municipal Wastewater Recycling Survey. http://www.waterboards.ca.gov/water_issues/programs/grants_loans/water_recycling/docs/article.pdf.

Orange County Water District (OCWD). 2015. 2013–2014 Engineering's Report on Groundwater Conditions, Water Supply and Basin Utilization in the Orange County Water District. http://www.ocwd.com/media/3304/ocwd-engineers-report-2013-2014.pdf.

Pacific Institute. 2015. Existing and Proposed Seawater Desalination Plants in California. http://pacinst.org/publication/key-issues-in-seawater-desalination-proposed-facilities/.

Short, W., D. J. Packey, and T. Holt. 1995. *A Manual for the Economic Evaluation of Energy Efficiency and Renewable Energy Technologies*. Golden, CO: National Renewable Energy Laboratory.

State Water Resources Control Board (SWRCB). 2013. Policy for Water Quality Control for Recycled Water. http://www.swrcb.ca.gov/water_issues/programs/water_recycling_policy/docs/rwp_revtoc.pdf.

———. 2014. State Water Board Adopts Industrial Stormwater Permit that Enhances Pollution Reduction and Increases Reuse of Storm Water Runoff. Media Release. http://www.swrcb.ca.gov/press_room/press_releases/2014/pr040114_sw.pdf.

Sturm, R. 2013. Summary Report of BMP 1.2 Water Audit Data. California Urban Water Conservation Council. http://cuwcc.org/Portals/0/Document%20Library/Committees/Programmatic%20Committees/Utility%20Operations/Resources/BMP%201.2/CUWCC%20BMP1.2%20audit%20data%20review%20May2013.pdf.

Upper Kings Basin IRWM Authority. 2013. Attachment 8—Benefits and Costs Analysis. Project A: Fresno Irrigation District—Phase 1 Southwest Groundwater Banking Project. http://www.water.ca.gov/irwm/grants/docs/Archives/Prop84/Submitted_Applications/P84_Round2_Implementation/Upper%20Kings%20Basin%20IRWM%20Authority%20(201312340022)/Attachment%208.%20-%20Att8_IG2_BenCost_1of2.pdf.

WATER BRIEF 1

The Human Right to Water and Global Sustainability: Actions of the Vatican

Peter H. Gleick

In February 2017, Pope Francis hosted a meeting on "The Human Right to Water" at the Vatican in the Casina Pio IV, promoted by the Pontifical Academy of Sciences. This was not the first time the Pope had weighed in on the issue of water and human rights, but it was the first time the Vatican had brought together experts to discuss both the legal and ethical basis for the human right to water and the next steps needed in implementing that right. Previous volumes of *The World's Water*, as far back as the second and fourth, have addressed this question (Gleick 2000; Gleick 2004).

The 2015 Encyclical Letter *Laudato Si'* and Water

In May 2015, the Vatican released Pope Francis' much-anticipated Encyclical Letter, *Laudato Si'* (*On Care for Our Common Home*) (Vatican 2015). While considerable public attention was devoted to the portions of the Encyclical related to climate change, the letter also tackles other environmental challenges—including biodiversity, food, and especially the critical issue of freshwater. Woven throughout is attention to the social and equity dimensions of these challenges and a deep concern for the poor. Much of this focus adopts a "human rights" perspective.

Even in the 21st century, significant and unresolved disparities in access, quality, and use of water remain between the wealthier, industrialized parts of the world and poorer populations. In many parts of the world, human use and extraction of water now exceeds natural resource limits—a problem described as "peak water" (Gleick and Palaniappan 2010). Yet the global community has still failed to satisfy the basic water needs of the poorest. *Laudato Si'* addresses this in section 27, where it states: "The exploitation of the planet has already exceeded acceptable limits and we still have not solved the problem of poverty."

The Encyclical identifies several key water problems, including the lack of access to clean drinking water, "indispensable for human life and for supporting terrestrial and aquatic ecosystems" (section 28); the challenges for food production due to droughts and disparities in water availability and "water poverty" (section 28); the continued prevalence of water-related diseases afflicting the poor (section 29); contamination of groundwater (section 29); and the trend toward privatization and commodification of a resource that the Vatican describes as a "basic and universal human right" (section 30).

This framing is consistent with the formal human right to water declared by the United Nations in 2010, linking the right to water (HRW) with the right to life and well-being. Today, the UN estimates that around 2.5 billion people on the planet still lack access to safe sanitation, and 750 million do not have safe drinking water (WHO/UNICEF 2017). Worldwide, more people die from unsafe water annually than from all forms of violence, including war (Palaniappian et al. 2010).

After many years of debate, the formal declaration of the HRW shifted the focus from whether such a right existed to how to implement that right and how to understand the responsibilities of governments, institutions, the private sector, and individuals in satisfying that right.

The Encyclical also expresses concern for the inefficient and wasteful use of water in both rich and poor regions: "But water continues to be wasted, not only in the developed world but also in developing countries which possess it in abundance" and decries the risk that the "control of water by large multinational businesses may become a major source of conflict in this century" (section 31).

In the context of climate change, *Laudato Si'* notes the clear links between a warming planet and threats to water resources and other environmental conditions:

> It creates a vicious circle which aggravates the situation even more, affecting the availability of essential resources like drinking water, energy and agricultural production in warmer regions, and leading to the extinction of part of the planet's biodiversity (section 24).

Consistent with the overall theme of the Encyclical is the observation that the poorest suffer most from water problems:

> One particularly serious problem is the quality of water available to the poor. Every day, unsafe water results in many deaths and the spread of water-related diseases, including those caused by microorganisms and chemical substances. Dysentery and cholera, linked to inadequate hygiene and water supplies, are a significant cause of suffering and of infant mortality (section 29).

The Encyclical goes further and notes that "Our world has a grave social debt towards the poor who lack access to drinking water, because *they are denied the right to a life consistent with their inalienable dignity*" (section 30, italics in original).

While progress has been made in cleaning up some water pollution, especially in richer industrialized nations, many water-quality indicators are worsening, not improving (Palaniappian et al. 2010), and as populations grow, exposure to some forms of water pollution affects more people and watersheds. Even in places like California, hundreds of thousands of people—mostly in low-income communities—are at risk of exposure to water with high concentrations of nitrates because of the failure to protect and clean up groundwater systems contaminated by agricultural chemicals, animal feeding operations, and poor sewage systems (Moore et al. 2011).

To tackle these challenges, the Encyclical identifies several priorities, but especially for water:

> some questions must have higher priority. For example, we know that water is a scarce and indispensable resource and a fundamental right which conditions the exercise of other human rights. This indisputable fact overrides any other assessment of environmental impact on a region (section 185).

It also calls for reducing waste and inappropriate consumption, increasing funding to ensure universal access to basic water and sanitation, and increased education and awareness, especially in the "context of great inequity."

The world's water challenges are technical, economic, political, and social issues, but the Vatican Encyclical reminds us that ultimately, they are ethical and moral issues as well. This is a valuable and timely reminder.

The 2017 Vatican Meeting on the Human Right to Water

In February 2017, the Vatican helped to host a meeting on "The Human Right to Water." At that meeting, Pope Francis signed a formal statement prepared by the workshop participants (Gleick 2017). The Pope also offered his own statement on this subject, expanding on his words addressing water in the Encyclical Letter *Laudato Si'*. In his statement, Pope Francis notes that "all people have a right to safe drinking water" as a basic human right, and he calls on countries and non-state actors to implement that right. Interestingly, the Pope also says we must not rely on God to address this problem: "But the work is up to us, the responsibility is ours."

The following is the full text of his address:

> *I greet all of you and I thank you for taking part in this meeting concerned with the human right to water and the need for suitable public policies in this regard. It is significant that you have gathered to pool your knowledge and resources in order to respond to this urgent need and this problem of today's men and women.*
>
> *The Book of Genesis tells us that water was there in the beginning (cf. Gen 1:2); in the words of Saint Francis of Assisi, it is "useful, chaste and humble" (cf. Canticle of the Creatures). The questions that you are discussing are not marginal, but basic and pressing. Basic, because where there is water there is life, making it possible for societies to arise and advance. Pressing, because our common home needs to be protected. Yet it must also be realised that not all water is life-giving, but only water that is safe and of good quality—as St. Francis again tells us, water that "serves with humility", "chaste" water, not polluted.*
>
> *All people have a right to safe drinking water. This is a basic human right and a central issue in today's world (cf. Laudato Si', 30; Caritas in Veritate, 27). It is sad when in the legislation of a country or a group of countries, water is not considered as a human right. It is even sadder still when what is written is neglected, and this human right is denied. This is a problem that affects everyone and is a source of great suffering in our common home. It also cries out for practical solutions capable of surmounting the selfish concerns that prevent everyone from*

exercising this fundamental right. Water needs to be given the central place it deserves in the framework of public policy. Our right to water is also a duty to water. Our right to water gives rise to an inseparable duty. We are obliged to proclaim this essential human right and to defend it—as we have done—but we also need to work concretely to bring about political and juridical commitments in this regard. Every state is called to implement, also through juridical instruments, the Resolutions approved by the United Nations General Assembly since 2010 concerning the human right to a secure supply of drinking water. Similarly, non-state actors are required to assume their own responsibilities with respect to this right.

The right to water is essential for the survival of persons (cf. Laudato Si', 30) and decisive for the future of humanity. High priority needs to be given to educating future generations about the gravity of the situation. Forming consciences is a demanding task, one requiring conviction and dedication. And I wonder if, in the midst of this "piecemeal third world war" that we are experiencing, if we are not on the path towards a great world war over water.

The statistics provided by the United Nations are troubling, nor can they leave us indifferent. Each day a thousand children die from water-related illnesses and millions of persons consume polluted water. These facts are serious; we have to halt and reverse this situation. It is not too late, but it is urgent to realise the need and essential value of water for the good of mankind.

Respect for water is a condition for the exercise of the other human rights (cf. ibid., 30). If we consider this right fundamental, we will be laying the foundations for the protection of other rights. But if we neglect this basic right, how will we be able to protect and defend other rights? Our commitment to give water its proper place calls for developing a culture of care (cf. ibid., 231)—it seems to be something poetic and, indeed, Creation is a "poiesis", this culture of care that is creative—and also fostering a culture of encounter, joining in common cause all the necessary efforts made by scientists and business people, government leaders and politicians. We need to unite our voices in a single cause; then it will no longer be a case of hearing individual or isolated voices, but rather the plea of our brothers and sisters echoed in our own, and the cry of the earth for respect and responsible sharing in a treasure belonging to all. In this culture of encounter, it is essential that each state act as a guarantor of universal access to safe and clean water.

God the Creator does not abandon us in our efforts to provide access to clean drinking water to each and to all. But the work is up to us, the responsibility is ours. It is my hope that this Conference will help strengthen your convictions and that you will leave in the certainty that your work is necessary and of paramount importance so that others can live. With the "little" we have, we will be helping to make our common home a more liveable and fraternal place, better cared for, where none are rejected or excluded, but all enjoy the goods needed to live and to grow in dignity. And let us not forget the United Nations data, the figures. Let us not forget that every day a thousand children, every day, die of water-related diseases.

Thank you.

References

Gleick, P. H. 2000. The Human Right to Water. In *The World's Water 2000–2001: The Biennial Report on Freshwater Resources*, pp. 1–17. Washington, DC: Island Press.

———. 2004. The Human Right to Water: Two Steps Forward, One Step Back. In *The World's Water 2004–2005: The Biennial Report on Freshwater Resources*, pp. 204–212. Washington, DC: Island Press.

———. 2017. The Vatican Workshop and Statement on The Human Right to Water. Significant Figures, National Geographic ScienceBlogs. http://scienceblogs.com/significantfigures/index.php/2017/03/07/the-vatican-workshop-and-statement-on-the-human-right-to-water/.

Gleick, P. H., and M. Palaniappan. 2010. Peak Water Limits to Freshwater Withdrawal and Use. *Proceedings of the National Academy of Sciences (PNAS)* 107(25): 11155–11162. doi: 10.1073/pnas.1004812107.

Moore, E., E. Matalon, C. Balazs, J. Clary, L. Firestone, S. De Anda, and M. Guzman. 2011. Human Costs of Nitrate-Contaminated Drinking Water in the San Joaquin Valley. Oakland, CA: Pacific Institute. http://pacinst.org/publication/human-costs-of-nitrate-contaminated-drinking-water-in-the-san-joaquin-valley/.

Palaniappian, M., P. H. Gleick, L. Allen, M. Cohen, J. Christian-Smith, and C. Smith. 2010. Clearing the Waters: A Focus on Water Quality Solutions. Oakland, CA: Pacific Institute. http://pacinst.org/publication/clearing-the-waters-focus-on-water-quality-solutions/.

Vatican. 2015 (May). Encyclical Letter *Laudato Si'* of the Holy Father Francis, *On Care for Our Common Home*. http://w2.vatican.va/content/francesco/en/encyclicals/documents/papa-francesco_20150524_enciclica-laudato-si.html.

World Health Organization (WHO)/UNICEF. 2017. Progress on Drinking Water, Sanitation and Hygiene: 2017 Update and SDG Baselines. Geneva, Switzerland: WHO. http://www.who.int/mediacentre/news/releases/2017/launch-version-report-jmp-water-sanitation-hygiene.pdf.

WATER BRIEF 2

Access to Water through Public Drinking Fountains

Rapichan Phurisamban and Peter H. Gleick

Public drinking fountains have been documented since ancient times, with descriptions as far back as ancient Greece, where fountains were both a common sight and a public necessity. A 2nd-century Greek writer, Pausanias, wrote that a place can never rightfully be called a "city" without water fountains (Gleick 2010). Spring-fed public water fountains were typically placed in or near temples and dedicated to gods, goddesses, nymphs, and heroes.

As populations and cities grew, demand for public water systems and new water treatment and delivery technologies led to increased use of public water fountains, and by the 20th century fountains became a fixture of the urban landscape. In recent decades, however, they have been slowly disappearing from public spaces for several reasons—including the advent of commercial bottled water, decreased public investment in urban infrastructure, concern over the health risks of fountains and municipal water in general, and a *laissez-faire* attitude toward public water systems (Gleick and Phurisamban 2017).

Drinking fountains serve many purposes: they offer an alternative to bottled water or commercial soft drinks, provide easy access to public water for school children, commuters, outdoor athletes, the homeless, and tourists; and some fountains are even designed to provide water for pets. A study from the Pacific Institute, entitled "Drinking Fountains and Public Health: Improving National Water Infrastructure to Rebuild Trust and Ensure Access," discusses the state of U.S. drinking fountains and addresses concerns about their quality and links to illnesses (Phurisamban and Gleick 2017a). The report offers recommendations for improving how fountains are maintained, cleaned, and updated, and for expanding access in public places. A related assessment offers insight into new drinking water fountain technologies. Both are briefly summarized here.

Fountains and Public Health

One factor influencing the decline in public water fountains is concern about unsafe water, either because of worries about the municipal water system itself or because fountains are inadequately cleaned and maintained. The perception that water from public fountains is unhealthy is not supported by evidence. Epidemiology studies and other assessments looking at health issues associated with public water fountains have found very limited evidence of health risks. Furthermore, problems that were identified are typically traced to contamination from poor cleaning and maintenance, or from old water infrastructure in buildings rather than contamination at the point of use.

Recent reports of unsafe water from fountains show that the problem is almost never the fountain itself, but old water distribution and plumbing systems. Lead contamination, for example, is almost universally a problem associated with old piping systems and plumbing designs that no longer meet national standards. As part of any national or local water infrastructure effort, such systems should be immediately replaced. Despite the limited evidence that fountains pose any public health risk, there are specific things that can be done to both reduce the risk of contamination and help rebuild public confidence in water fountains. Phurisamban and Gleick (2017a) recommend:

- establishing comprehensive monitoring and testing of all drinking fountains;
- developing and implementing standard protocols for water fountain maintenance, repair, and replacement;
- creating broad nationwide efforts to replace old water infrastructure, especially distribution and plumbing systems, with modern piping to eliminate sources of lead, copper, and microbial contamination;
- upgrading the type and function of older drinking fountains (e.g., by installing filters);
- greatly increasing the number of fountains to improve access to municipal water in public places;
- engaging municipalities, schools, park districts, and others responsible for drinking fountains in communications efforts to help rebuild public confidence in fountains; and
- using new tools to compile and distribute information on where to find drinking fountains and to assess and report on their condition.

These efforts, combined with communications on the results of regular water testing, reports on the performance of fountains, and information on how to find and access high-quality drinking fountains, can help improve public trust in water fountains.

New Water Fountain Technology

Efforts to expand access to public water fountains can benefit from the adoption of new technologies that improve water quality, convenience, and reliability (Phurisamban and Gleick 2017b). New fountains offer features like filters, chillers, and bottle fillers. Mobile apps can make it easier to find nearby drinking fountains. Other features of modern fountains can include vandal resistance, freeze resistance, improved handicapped access, pet access, and more. Figure WB-2.1 provides a useful infographic of drinking fountain typologies (Ivanov 2015).

Outside drinking fountains should include features to help weather the elements. Vandal-resistant and durable fountains are useful in high-traffic areas. Freeze-resistant

DRINKING FOUNTAIN TYPOLOGIES

SPOUT

Spigot
Most basic fountain type. Historically, often included a drinking cup. Has now been readopted as a "bottle filler."

→ Variation: spigot with drinking hole

Upward bubbler
Invented by Luther Haws in 1906 (and perhaps by others simultaneously), still used in Portland's Benson Bubblers.

Arc bubbler
Easiest to drink from. Used in some Renaissance Roman fountains, and became widespread in the US in the late 1910's.

Arc bubbler with mouth guard
Addressing concerns about mouth contact with the water source, became widespread in the early 1920's.

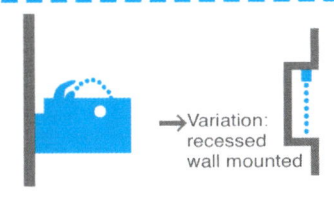

MOUNTING

Wall-mounted
Most common in indoor settings and in parks with restrooms. Saves money on pipes by double-utilizing "wet walls," but may lead to higher feelings of disgust when water is associated with bathrooms.

→ Variation: recessed wall mounted

Free-standing
Common in parks and in some urban streetscapes. Can be installed anywhere there are underground water and sewer pipes.

ON / OFF

Always on
Visually indicates that fountain is working, lets fountain double as ornamental. Still used in some very wet climates.

Pedal operated
Still found on some historic fountains. Allows fountain use while holding back hair and with one hand full.

Button operated
Evolved to conserve water and allow time for refrigeration. Most common.

Motion activated
Becoming common for no-touch bottle fillers. Can be both cool and frustrating.

ACCESS

Americans with Disabilities Act accessible
ADA compliance requires knee clearance of at least 27" and a spout no higher than 36".

Varied height access
Accommodates people of different heights and bending abilities.

Child accessible
Many fountains include a step-stool so small children can drink by themselves.

Animal accessible
Horse and oxen troughs used to fill from runoff of human use jets. Today, dog bowls are increasingly popular.

CHILL

Unchilled
Most outdoor drinking fountains are unrefrigerated. Can be unappetizing in hot climates when sited out of shade.

Ice-cooled
Early cooled fountains used ice blocks. Some drinking fountain benefactors specified with their gifts that cities had to keep drinking fountains sup- plied with ice in summer months.

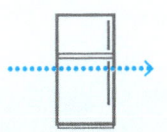

Refrigerated
Many indoor fountains pass water through refrigerated coils before dispensing. Requires electricity.

Frost-Proof
Fountains do not store water inside them, requiring a short wait once the button is pushed, but allowing fountains to stay on all winter in cold climates. Great recent innovation.

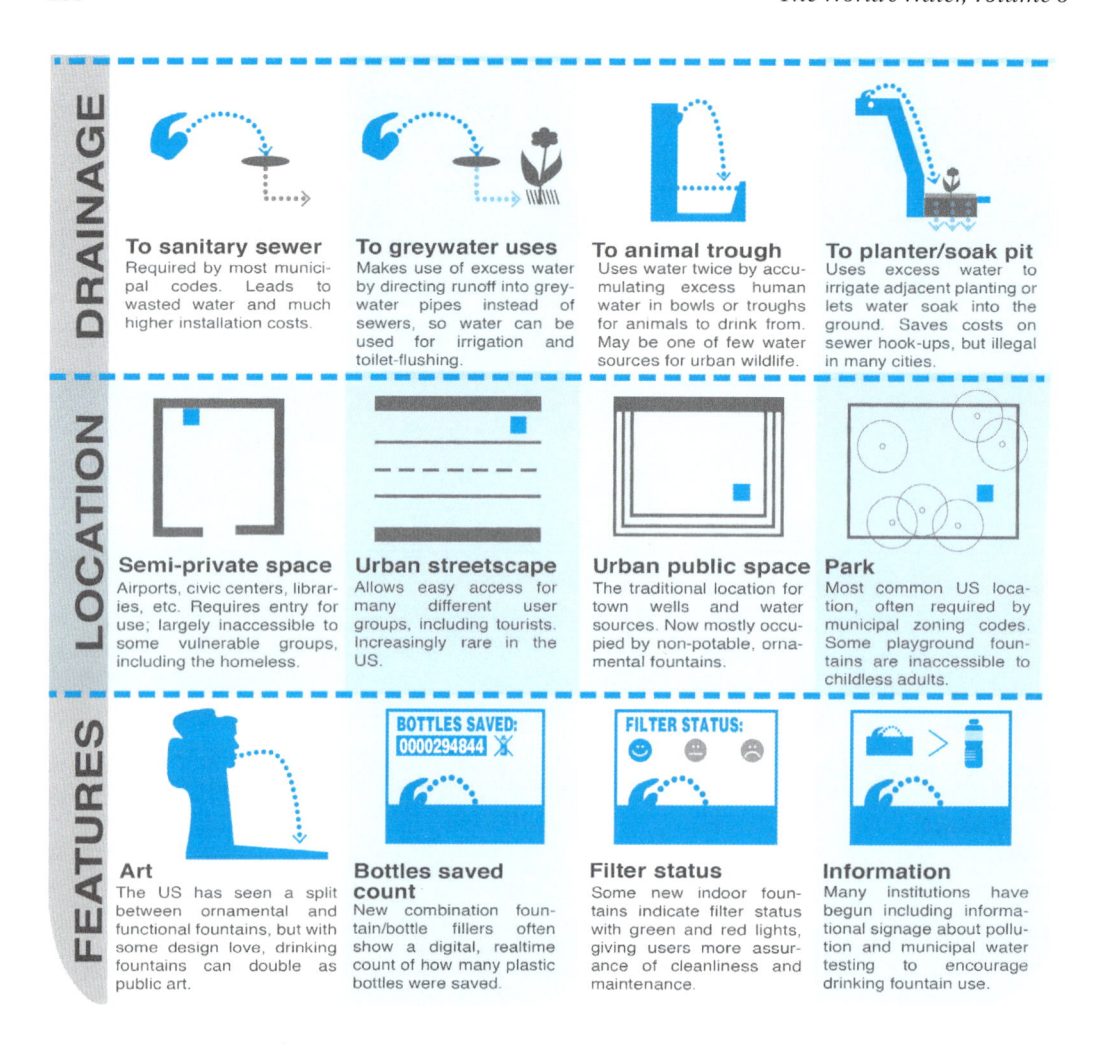

DRAINAGE	**To sanitary sewer** Required by most municipal codes. Leads to wasted water and much higher installation costs.	**To greywater uses** Makes use of excess water by directing runoff into greywater pipes instead of sewers, so water can be used for irrigation and toilet-flushing.	**To animal trough** Uses water twice by accumulating excess human water in bowls or troughs for animals to drink from. May be one of few water sources for urban wildlife.	**To planter/soak pit** Uses excess water to irrigate adjacent planting or lets water soak into the ground. Saves costs on sewer hook-ups, but illegal in many cities.
LOCATION	**Semi-private space** Airports, civic centers, libraries, etc. Requires entry for use; largely inaccessible to some vulnerable groups, including the homeless.	**Urban streetscape** Allows easy access for many different user groups, including tourists. Increasingly rare in the US.	**Urban public space** The traditional location for town wells and water sources. Now mostly occupied by non-potable, ornamental fountains.	**Park** Most common US location, often required by municipal zoning codes. Some playground fountains are inaccessible to childless adults.
FEATURES	**Art** The US has seen a split between ornamental and functional fountains, but with some design love, drinking fountains can double as public art.	**Bottles saved count** New combination fountain/bottle fillers often show a digital, realtime count of how many plastic bottles were saved.	**Filter status** Some new indoor fountains indicate filter status with green and red lights, giving users more assurance of cleanliness and maintenance.	**Information** Many institutions have begun including informational signage about pollution and municipal water testing to encourage drinking fountain use.

FIGURE WB-2.1 TYPOLOGIES OF DRINKING WATER FOUNTAINS, INCLUDING INFORMATION ON FEATURES, LOCATION, AND OTHER FOUNTAIN CHARACTERISTICS.
Source: Ivanov 2015.

features permit fountains to continue to work in winter months. A wide range of bubbler options are available to supplement the more common "arc" bubbler. The most notable improvement in drinking fountains in recent years has been incorporation of a way to refill individual bottles. Demand for bottle-filling stations began to rise significantly around 2010 during anti-bottled water campaigns and the growing availability and use of refillable bottles, the ease of use of these stations, and less perceived risk of contamination. Water chilling also makes modern fountains attractive to users.

Some cities in Europe and Australia have taken the idea of chilled water further and added carbonation capability to regular drinking fountains. The concept of sparkling water fountains originated in Italy and has spread to other regions; including France, Belgium, and Australia. In Paris, these fountains are known as "La Petillante" or "she who sparkles," and Eau de Paris, the public company responsible for providing and maintaining drinking fountains throughout the city, offers a map of all of its 725 drinking fountains, including six sparkling water fountains.

Given the public concern about tap water quality, fountains now come with a range of filters, from activated carbon and other media to sophisticated (and costly) advanced filtration systems like reverse osmosis. Adding filters also adds an additional maintenance requirement to check and replace them as needed. Other methods for removing contaminants include ceramic filters, distillers, and UV light units used to disinfect water. Since filters add to the cost of water fountains, water quality tests should be done before making any filter investments, to both ensure that filters are needed and to identify the most appropriate type to install.

Summary

There is strong support for expanding investment in the nation's water infrastructure as part of a broader infrastructure effort. One specific objective should be to expand public access to high-quality and safe municipal water by improving access to drinking fountains in schools, parks, buildings, and transit areas. Investments in drinking fountains will make access to water more widespread and offer an alternative that is far cheaper than bottled water, which has a wide range of cost and environmental liabilities. New fountain technologies, approaches for consistent cleaning and maintenance, installation of new fountains in high-traffic locations, and public conversations about water fountains can all play a role in reviving and expanding this ancient water tradition for the modern age.

References

Gleick, P. H. 2010. *Bottled and Sold: The Story Behind Our Obsession with Bottled Water*. Washington, DC: Island Press.

Gleick, P. H., and R. Phurisamban. 2017. National Water Infrastructure Efforts Must Expand Access to Public Drinking Fountains. National Geographic ScienceBlogs. http://scienceblogs.com/significantfigures/index.php/2017/03/08/national-water-infrastructure-efforts-must-expand-access-to-public-drinking-fountains/.

Ivanov, J. 2015. Drinking Fountain Typologies. http://drinkingfountains.blogspot.com/2015/03/drinking-fountain-typologies.html.

Phurisamban, R., and P. H. Gleick. 2017a. Drinking Fountains and Public Health: Improving National Water Infrastructure to Rebuild Trust and Ensure Access. Oakland, CA: Pacific Institute. http://pacinst.org/publication/drinking-fountains-public-health-improving-national-water-infrastructure-rebuild-trust-ensure-access/.

Phurisamban, R., and P. H. Gleick. 2017b. Make Public Drinking Water Fountains Great Again. Oakland, CA: Pacific Institute. http://pacinst.org/make-public-drinking-water-fountains-great/.

WATER BRIEF 3

Water and Conflict Update
Events, Trends, and Analysis

Peter H. Gleick

Introduction

The Pacific Institute continues to be the leading independent research organization tracking, analyzing, and cataloguing conflicts over water resources (see Box WB-3.1). For three decades, the Institute has maintained an online assessment of such conflicts, the *Water Conflict Chronology* (Gleick 2017). This unique database summarizes violence related to access to freshwater, attacks on water systems, the use of water as a weapon, terrorist incidents related to water, and more, going back nearly five thousand years. In mid-2017, the latest update was released, documenting recent events and noting an uptick in the number of incidents in recent years.

Figure WB-3.1 shows the average number of events per year from 1930 to 2016. As discussed in volume 7 of *The World's Water*, part of this increase could be due to improvements in reporting; new Internet tools that permit more comprehensive collection and dissemination of news, data, and information; and more awareness of the issue. But it is also possible that the growing number of water conflicts is the result of real tensions and disputes over limited freshwater resources. The data also provide evidence of a shift in the nature of these conflicts—away from water disputes between nations and toward subnational and local violence over water access. The growing risk of subnational water conflicts was noted as far back as 1998 in the first volume of *The World's Water*:

> Traditional political and ideological questions that have long dominated international discourse are now becoming more tightly woven with other variables that loomed less large in the past, including population growth, transnational pollution, resource scarcity and inequitable access to resources and their use (Gleick 1998).

Notable examples in the most recent update include a series of incidents in India associated with severe drought and protests over inadequate availability of water; persistent attacks on water systems in Syria, Iraq, and Yemen; and perhaps most disturbing, a growing number of assassinations of environmental activists who have been working to expand the voices of local communities in environmental protection around rivers and water resources.

> **BOX WB-3.1** The Pacific Institute *Water Conflict Chronology*
>
> The Pacific Institute maintains a comprehensive database, the *Water Conflict Chronology*, at http://www.worldwater.org/water-conflict/. An update is also published in each volume of *The World's Water*. Using these data, the Pacific Institute has published research papers, historical reviews, and regional case studies on water conflicts. We have organized workshops on lessons from regional water disputes in the Middle East, Central Asia, and Latin America. We have brought together experts from the fields of traditional and nontraditional arms control and helped coordinate a workshop on the role of science and religion in reducing the risks of water-related violence, which was held at the Pontifical Academy of Sciences of the Vatican.
>
> The full *Water Conflict Chronology* includes integrated Google Maps; time, location, and subject filters; and a separate searchable bibliography. The nature of entries in the chronology can be described and categorized in different ways. The Institute has split the categories or types of conflicts as follows, though other groupings and distinctions can also be useful:
>
> > *Military Tool (state actors):* where water resources, or water systems themselves, are used by a nation or state as a weapon during a military action.
> >
> > *Military Target (state actors):* where water resources or water systems are targets of military actions by nations or states.
> >
> > *Terrorism or domestic violence, including cyberterrorism (non-state actors):* where water resources, or water systems, are the targets or tools of violence or coercion by non-state actors.
> >
> > *Development Disputes (state and non-state actors):* where water resources or water systems are a major source of contention and dispute in the context of economic and social development.
>
> The *Water Conflict Chronology* has appeared in every volume of *The World's Water* since 1998. It continues to be one of the most popular and regular features of the Pacific Institute's work and the chronology is used regularly by the media and academics interested in understanding more about both the history and character of disputes over water resources (CNN/Zakaria 2013).

Access to water has often been a catalyst for tensions and violence, and water itself has long been a target and tool of war. In his history written around AD 90, Flavius Josephus describes how a few decades earlier Pontius Pilate diverted a stream to Jerusalem from the surrounding villages and then violently crushed a protest by tens of thousands of people. In 1748, an angry mob in New York burned down a ferry house on the Brooklyn shore of the East River, as revenge for what they considered unfair allocation of East River

Water and Conflict Update

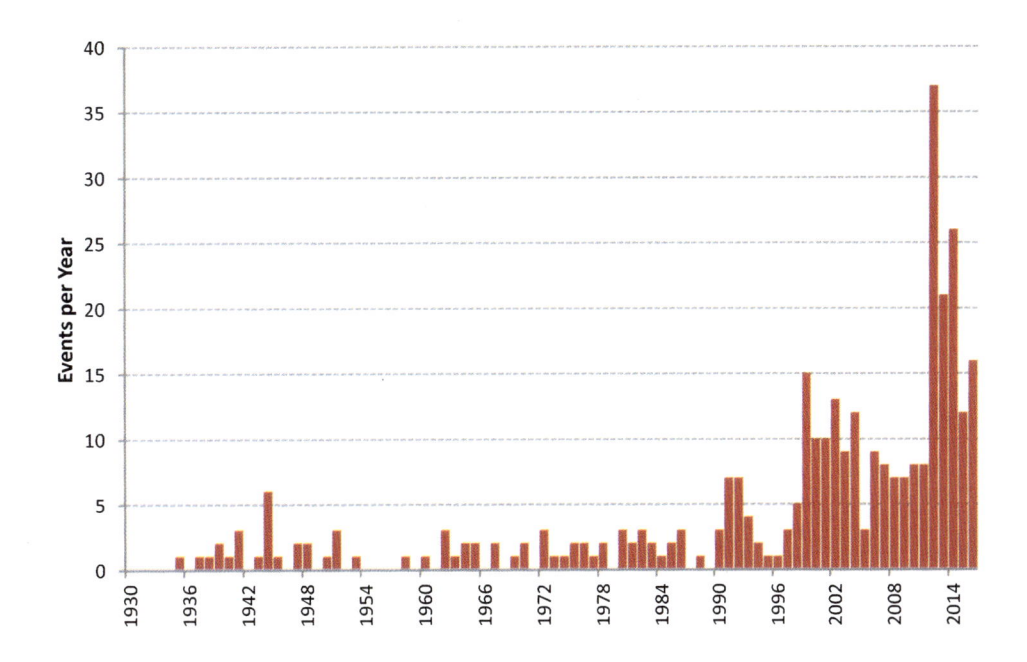

FIGURE WB-3.1 THE NUMBER OF WATER CONFLICT EVENTS REPORTED PER YEAR, 1930–2016.
Source: Gleick 2017.

water rights. (See the *Water Conflict Chronology* for additional details and full citations.)

These histories go back even further in time. Nearly four thousand years ago, Abi-Eshuh, a king in ancient Mesopotamia and grandson of Hammurabi, dammed the Tigris River to prevent the retreat of rebels who had declared independence from the Babylonian empire. In the very same region of the world, however, in 2017, the Islamic State flooded villages east of Aleppo, Syria by releasing water from a dam on the Euphrates River to halt the advance of the Syrian Arab Army, and U.S.-backed forces recaptured Syria's Tabqa Dam from ISIS.

In 2016, Berta Cáceres—a prize-winning activist opposing the Agua Zarca hydroelectric dam on the Río Gualcarque river in Honduras—was murdered after years of death threats and state persecution linked to her campaign. Two of her colleagues have also been killed (Watts 2016). In South Africa, environmental activist Sikhosiphi Radebe was murdered while opposing industrial mining development that threatened community water resources and land (Schneider 2016).

In early 2016, at least 18 people were killed and 200 injured after the Indian Army intervened to reopen the Munak canal, which supplies New Delhi with three-fifths of its freshwater supply. The canal was shut down by economic protests in Haryana State. Sabotage of the canal left more than 10 million people in India's capital without water.

Several entries describe repeated attacks on water pipelines, pumping plants, dams, and treatment systems by almost all parties in the Syria and Iraq conflicts. Since the start of the Syrian civil war—a war influenced in part by climate change, severe drought, and associated economic disruption—attacks on water pipelines, pumping plants, dams, and treatment systems have caused a 50 percent reduction in access to safe water. Similar attacks on water systems have occurred recently in Iraq and Yemen and water-related diseases like cholera are now surging (Vidal 2016).

Water and energy systems have regularly been targeted in the violence between Russia and Ukraine over the past few years. A long series of attacks have intermittently left nearly three million people without access to reliable water supplies. The attacks included repeated damage to the Donetsk Filtration Plant, the South Donbass water pipeline, energy plants that supply power to water treatment and distribution systems, and the Carbonit Water Pumping Station (Relief Web 2017).

Two new entries in the United States were also added to the chronology, including the standoff at the Malheur National Wildlife Refuge over water rights and land use, which ended with one death and several arrests, and the violence at the Standing Rock protests over the Dakota Access Pipeline, which Native Americans consider a threat to the region's water resources (including the Missouri River) and to ancient burial grounds. During the protests, hundreds of people were injured and arrested.

New historical examples have also been added, including an entry for India in AD 1260 and one for Hispaniola in 1802, both related to the use of water systems as weapons or targets during conflicts and political uprisings.

Attention to the risks of water conflicts is growing. In 2012, the U.S. National Intelligence Council (NIC) concluded: "Water challenges—shortages, poor water quality, floods—will likely increase the risk of instability and state failure, exacerbate regional tensions, and distract countries from working with the United States on important policy objectives" (ODNI 2012). The NIC noted the Middle East, Northern Africa, and South Asia already face challenges coping with water problems: "During the next 10 years, water problems will contribute to instability in states important to U.S. national security interests" and they predicted that by 2040, water shortages and contamination "probably will harm the economic performance of important trading partners." These concerns were repeated in an NIC report released in 2017 on global security threats (ODNI 2017):

> More extreme weather, water and soil stress, and food insecurity will disrupt societies … A growing number of countries will experience water stress—from population growth, urbanization, economic development, climate change, and poor water management—and tensions over shared water resources will rise.

Pressures on water resources around the world continue to grow. Researchers, water experts, diplomats, and the military need to improve their understanding of the links between water and security and work to reduce the risks of conflict.

References

CNN/Zakaria. 2013. The Coming Water Wars? Fareed Zakaria GPS, CNN World. March 22. http://globalpublicsquare.blogs.cnn.com/2013/03/22/the-coming-water-wars/comment-page-1/.

Office of the Director of National Intelligence (ODNI). 2012. Global Water Security. Intelligence Community Assessment, ICA 2012-08, February 2. https://www.dni.gov/files/documents/Newsroom/Press%20Releases/ICA_Global%20Water%20Security.pdf.

———. 2017. Global Trends: The Paradox of Progress—Trends Transforming the Global Landscape. https://www.dni.gov/index.php/global-trends/trends-transforming-the-global-landscape.

Gleick, P. H. 1998. Conflict and Cooperation over Fresh Water. In: *The World's Water 1998–1999: The Biennial Report on Freshwater Resources*. Washington, DC: Island Press.

———. 2017. The Water Conflict Chronology. http://worldwater.org/water-conflict/.

Relief Web. 2017. Humanitarian Bulletin: Ukraine. Issue 16, 1 January–28 February. http://reliefweb.int/report/ukraine/humanitarian-bulletin-ukraine-issue-16-1-january-28-february-2017.

Schneider, K. 2016. A Murder on South Africa Wild Coast Escalates Conflict Over Water, Land, Mining. *Circle of Blue Water News*, March 22. http://www.circleofblue.org/2016/south-africa/murder-south-africa-wild-coast-escalates-conflict-water-land-mining/.

Vidal, J. 2016. Water Supplies in Syria Deteriorating Fast Due to Conflict, Experts Warn. *The Guardian*. September 7. https://www.theguardian.com/environment/2016/sep/07/water-supplies-in-syria-deteriorating-fast-due-to-conflict-experts-warn?CMP=Share_iOSApp_Other.

Watts. J. 2016. Berta Cáceres, Honduran Human Rights and Environment Activist, Murdered. *The Guardian*. March 4. https://www.theguardian.com/world/2016/mar/03/honduras-berta-caceres-murder-enivronment-activist-human-rights?CMP=twt_a-environment_b-gdneco.

Water Units, Data Conversions, and Constants

Water experts, managers, scientists, and educators work with a bewildering array of different units and data. These vary with the field of work: engineers may use different water units than hydrologists; urban water agencies may use different units than reservoir operators; academics may use different units than water managers. But they also vary with regions: water agencies in England may use different units than water agencies in France or Africa; hydrologists in the eastern United States often use different units than hydrologists in the western United States. And they vary over time: today's water agency in California may sell water by the acre-foot, but its predecessor a century ago may have sold miner's inches or some other now arcane measure.

These differences are of more than academic interest. Unless a common "language" is used, or a dictionary of translations is available, errors can be made or misunderstandings can ensue. In some disciplines, unit errors can be more than embarrassing; they can be expensive, or deadly. In September 1999, the $125 million Mars Climate Orbiter spacecraft was sent crashing into the face of Mars instead of into its proper safe orbit above the surface because one of the computer programs controlling a portion of the navigational analysis used English units incompatible with the metric units used in all the other systems. The failure to translate English units into metric units was described in the findings of the preliminary investigation as the principal cause of mission failure.

This table is a comprehensive list of water units, data conversions, and constants related to water volumes, flows, pressures, and much more. Most of these units and conversions were compiled by Kent Anderson and initially published in P. H. Gleick, 1993, *Water in Crisis: A Guide to the World's Fresh Water Resources*, Oxford University Press, New York.

Water Units, Data Conversions, and Constants

Prefix (Metric)	Abbreviation	Multiple	Prefix (Metric)	Abbreviation	Multiple
deka-	da	10	deci-	d	0.1
hecto-	h	100	centi-	c	0.01
kilo-	k	1000	milli-	m	0.001
mega-	M	10^6	micro-	µ	10^{-6}
giga-	G	10^9	nano-	n	10^{-9}
tera-	T	10^{12}	pico-	P	10^{-12}
peta-	P	10^{15}	femto-	f	10^{-15}
exa-	E	10^{18}	atto-	a	10^{-18}

LENGTH (L)

1 micron (µ)	$= 1 \times 10^{-3}$ mm	**10 hectometers**	$= 1$ kilometer
	$= 1 \times 10^{-6}$ m	**1 mil**	$= 0.0254$ mm
	$= 3.3937 \times 10^{-5}$ in		$= 1 \times 10^{-3}$ in
1 millimeter (mm)	$= 0.1$ cm	**1 inch (in)**	$= 25.4$ mm
	$= 1 \times 10^{-3}$ m		$= 2.54$ cm
	$= 0.03937$ in		$= 0.08333$ ft
1 centimeter (cm)	$= 10$ mm		$= 0.0278$ yd
	$= 0.01$ m	**1 foot (ft)**	$= 30.48$ cm
	$= 1 \times 10^{-5}$ km		$= 0.3048$ m
	$= 0.3937$ in		$= 3.048 \times 10^{-4}$ km
	$= 0.03281$ ft		$= 12$ in
	$= 0.01094$ yd		$= 0.3333$ yd
1 meter (m)	$= 1000$ mm		$= 1.89 \times 10^{-4}$ mi
	$= 100$ cm	**1 yard (yd)**	$= 91.44$ cm
	$= 1 \times 10^{-3}$ km		$= 0.9144$ m
	$= 39.37$ in		$= 9.144 \times 10^{-4}$ km
	$= 3.281$ ft		$= 36$ in
	$= 1.094$ yd		$= 3$ ft
	$= 6.21 \times 10^{-4}$ mi		$= 5.68 \times 10^{-4}$ mi
1 kilometer (km)	$= 1 \times 10^5$ cm	**1 mile (mi)**	$= 1609.3$ m
	$= 1000$ m		$= 1.609$ km
	$= 3280.8$ ft		$= 5280$ ft
	$= 1093.6$ yd		$= 1760$ yd
	$= 0.621$ mi	**1 fathom (nautical)**	$= 6$ ft
10 millimeters	$= 1$ centimeter	**1 league (nautical)**	$= 5.556$ km
10 centimeters	$= 1$ decimeter		$= 3$ nautical miles
10 decimeters (dm)	$= 1$ meter	**1 league (land)**	$= 4.828$ km
			$= 5280$ yd
10 meters	$= 1$ dekameter		$= 3$ mi
10 dekameters (dam)	$= 1$ hectometer	**1 international nautical mile**	$= 1.852$ km
			$= 6076.1$ ft
			$= 1.151$ mi

Water Units, Data Conversions, and Constants *(continued)*

AREA (L²)

1 square centimeter (cm²)	$= 1 \times 10^{-4}\,m^2$	**1 square foot (ft²)**	$= 929.0\,cm^2$
	$= 0.1550\,in^2$		$= 0.0929\,m^2$
	$= 1.076 \times 10^{-3}\,ft^2$		$= 144\,in^2$
	$= 1.196 \times 10^{-4}\,yd^2$		$= 0.1111\,yd^2$
1 square meter (m²)	$= 1 \times 10^{-4}$ hectare		$= 2.296 \times 10^{-5}$ acre
	$= 1 \times 10^{-6}\,km^2$		$= 3.587 \times 10^{-8}\,mi^2$
	$= 1$ centare (French)	**1 square yard (yd²)**	$= 0.8361\,m^2$
	$= 0.01$ are		$= 8.361 \times 10^{-5}$ hectare
	$= 1550.0\,in^2$		$= 1296\,in^2$
	$= 10.76\,ft^2$		$= 9\,ft^2$
	$= 1.196\,yd^2$		$= 2.066 \times 10^{-4}$ acres
	$= 2.471 \times 10^{-4}$ acre		$= 3.228 \times 10^{-7}\,mi^2$
1 are	$= 100\,m^2$	**1 acre**	$= 4046.9\,m^2$
1 hectare (ha)	$= 1 \times 10^4\,m^2$		$= 0.40469$ ha
	$= 100$ are		$= 4.0469 \times 10^{-3}\,km^2$
	$= 0.01\,km^2$		$= 43{,}560\,ft^2$
	$= 1.076 \times 10^5\,ft^2$		$= 4840\,yd^2$
	$= 1.196 \times 10^4\,yd^2$		$= 1.5625 \times 10^{-3}\,mi^2$
	$= 2.471$ acres	**1 square mile (mi²)**	$= 2.590 \times 10^6\,m^2$
	$= 3.861 \times 10^{-3}\,mi^2$		$= 259.0$ hectares
1 square kilometer (km²)	$= 1 \times 10^6\,m^2$		$= 2.590\,km^2$
	$= 100$ hectares		$= 2.788 \times 10^7\,ft^2$
	$= 1.076 \times 10^7\,ft^2$		$= 3.098 \times 10^6\,yd^2$
	$= 1.196 \times 10^6\,yd^2$		$= 640$ acres
	$= 247.1$ acres		$= 1$ section (of land)
	$= 0.3861\,mi^2$	**1 feddan (Egyptian)**	$= 4200\,m^2$
1 square inch (in²)	$= 6.452\,cm^2$		$= 0.42$ ha
	$= 6.452 \times 10^{-4}\,m^2$		$= 1.038$ acres
	$= 6.944 \times 10^{-3}\,ft^2$		
	$= 7.716 \times 10^{-4}\,yd^2$		

(continues)

Water Units, Data Conversions, and Constants *(continued)*

VOLUME (L³)

1 cubic centimeter (cm³)	= 1 × 10⁻³ liter	**1 cubic foot (ft³)**	= 2.832 × 10⁴ cm³
	= 1 × 10⁻⁶ m³		= 28.32 liters
	= 0.06102 in³		= 0.02832 m³
	= 2.642 × 10⁻⁴ gal		= 1728 in³
	= 3.531 × 10⁻³ ft³		= 7.481 gal
1 liter (l)	= 1000 cm³		= 0.03704 yd³
	= 1 × 10⁻³ m³	**1 cubic yard (yd³)**	= 0.7646 m³
	= 61.02 in³		= 6.198 × 10⁻⁴ acre-ft
	= 0.2642 gal		= 46656 in³
	= 0.03531 ft³		
1 cubic meter (m³)	= 1 × 10⁶ cm³		= 27 ft³
	= 1000 liter	**1 acre-foot**	= 1233.48 m³
	= 1 × 10⁻⁹ km³	**(acre-ft or AF)**	= 3.259 × 10⁵ gal
	= 264.2 gal		= 43560 ft³
	= 35.31 ft³	**1 Imperial gallon**	= 4.546 liters
	= 6.29 bbl		= 277.4 in³
	= 1.3078 yd³		= 1.201 gal
	= 8.107 × 10⁻⁴ acre-ft		= 0.16055 ft³
		1 cfs-day	= 1.98 acre-feet
1 cubic decameter (dam³)	= 1000 m³		= 0.0372 in-mi²
	= 1 × 10⁶ liter	**1 inch-mi²**	= 1.738 × 10⁷ gal
	= 1 × 10⁻⁶ km³		= 2.323 × 10⁶ ft³
	= 2.642 × 10⁵ gal		= 53.3 acre-ft
	= 3.531 × 10⁴ ft³		= 26.9 cfs-days
	= 1.3078 × 10³ yd³	**1 barrel (of oil) (bbl)**	= 159 liter
	= 0.8107 acre-ft		= 0.159 m³
1 cubic hectometer (ha³)	= 1 × 10⁶ m³		= 42 gal
	= 1 × 10³ dam³		= 5.6 ft³
	= 1 × 10⁹ liter	**1 million gallons**	= 3.069 acre-ft
	= 2.642 × 10⁸ gal	**1 pint (pt)**	= 0.473 liter
	= 3.531 × 10⁷ ft³		= 28.875 in³
	= 1.3078 × 10⁶ yd³		= 0.5 qt
	= 810.7 acre-ft		= 16 fluid ounces
1 cubic kilometer (km³)	= 1 × 10¹² liter		= 32 tablespoons
	= 1 × 10⁹ m³		= 96 teaspoons
	= 1 × 10⁶ dam³	**1 quart (qt)**	= 0.946 liter
	= 1000 ha³		= 57.75 in³
	= 8.107 × 10⁵ acre-ft		= 2 pt
	= 0.24 mi³		= 0.25 gal
		1 morgen-foot (S. Africa)	= 2610.7 m³
1 cubic inch (in³)	= 16.39 cm³		
	= 0.01639 liter	**1 board-foot**	= 2359.8 cm³
	= 4.329 × 10⁻³ gal		= 144 in³
	= 5.787 × 10⁻⁴ ft²		= 0.0833 ft³
1 gallon (gal)	= 3.785 liters	**1 cord**	= 128 ft³
	= 3.785 × 10⁻³ m³		= 0.453 m³
	= 231 in³		
	= 0.1337 ft³		
	= 4.951 × 10⁻³ yd³		

Water Units, Data Conversions, and Constants *(continued)*

VOLUME/AREA (L³/L²)

1 inch of rain	= 5.610 gal/yd²	**1 box of rain**	= 3,154.0 lesh
	= 2.715 × 10⁴ gal/acre		

MASS (M)

1 gram (g or gm)	= 0.001 kg	**1 ounce (oz)**	= 28.35 g
	= 15.43 gr		= 437.5 gr
	= 0.03527 oz		= 0.0625 lb
	= 2.205 × 10⁻³ lb	**1 pound (lb)**	= 453.6 g
1 kilogram (kg)	= 1000 g		= 0.45359237 kg
	= 0.001 tonne		= 7000 gr
	= 35.27 oz		= 16 oz
	= 2.205 lb	**1 short ton (ton)**	= 907.2 kg
1 hectogram (hg)	= 100 gm		= 0.9072 tonne
	= 0.1 kg		= 2000 lb
1 metric ton (tonne or te or MT)	= 1000 kg	**1 long ton**	= 1016.0 kg
	= 2204.6 lb		= 1.016 tonne
	= 1.102 ton	**1 long ton**	= 2240 lb
	= 0.9842 long ton		= 1.12 ton
1 dalton (atomic mass unit)	= 1.6604 × 10⁻²⁴ g	**1 stone (British)**	= 6.35 kg
			= 14 lb
1 grain (gr)	= 2.286 × 10⁻³ oz		
	= 1.429 × 10⁻⁴ lb		

TIME (T)

1 second (s or sec)	= 0.01667 min	**1 day (d)**	= 24 hr
	= 2.7778 × 10⁻⁴ hr		= 86400 s
1 minute (min)	= 60 s	**1 year (yr or y)**	= 365 d
	= 0.01667 hr		= 8760 hr
1 hour (hr or h)	= 60 min		= 3.15 × 10⁷ s
	= 3600 s		

DENSITY (M/L³)

1 kilogram per cubic meter (kg/m³)	= 10⁻³ g/cm³	**1 metric ton per cubic meter (te/m³)**	= 1.0 specific gravity
	= 0.062 lb/ft³		= density of H₂O at 4°C
1 gram per cubic centimeter (g/cm³)	= 1000 kg/m³		= 8.35 lb/gal
	= 62.43 lb/ft³	**1 pound per cubic foot (lb/ft³)**	= 16.02 kg/m³

(continues)

Water Units, Data Conversions, and Constants *(continued)*

VELOCITY (L/T)

1 meter per second (m/s)	= 3.6 km/hr = 2.237 mph = 3.28 ft/s	**1 foot per second (ft/s)**	= 0.68 mph = 0.3048 m/s
1 kilometer per hour (km/h or kph)	= 0.62 mph = 0.278 m/s	**velocity of light in vacuum (c)**	= 2.9979 × 10^8 m/s = 186,000 mi/s
		1 knot	= 1.852 km/h = 1 nautical mile/hour = 1.151 mph
1 mile per hour (mph or mi/h)	= 1.609 km/h = 0.45 m/s = 1.47 ft/s		= 1.688 ft/s

VELOCITY OF SOUND IN WATER AND SEAWATER
(assuming atmospheric pressure and sea water salinity of 35,000 ppm)

Temp, °C	Pure water, (meters/sec)	Sea water, (meters/sec)
0	1,400	1,445
10	1,445	1,485
20	1,480	1,520
30	1,505	1,545

FLOW RATE (L^3/T)

1 liter per second (1/sec)	= 0.001 m^3/sec = 86.4 m^3/day = 15.9 gpm = 0.0228 mgd = 0.0353 cfs = 0.0700 AF/day	**1 cubic decameters per day (dam^3/day)**	= 11.57 1/sec = 1.157 × 10^{-2} m^3/sec = 1000 m^3/day = 1.83 × 10^6 gpm = 0.264 mgd
1 cubic meter per second (m^3/sec)	= 1000 1/sec = 8.64 × 10^4 m^3/day = 1.59 × 10^4 gpm = 22.8 mgd = 35.3 cfs = 70.0 AF/day		= 0.409 cfs = 0.811 AF/day
		1 gallon per minute (gpm)	= 0.0631 1/sec = 6.31 × 10^{-5} m^3/sec = 1.44 × 10^{-3} mgd = 2.23 × 10^{-3} cfs = 4.42 × 10^{-3} AF/day
1 cubic meter per day (m^3/day)	= 0.01157 1/sec = 1.157 × 10^{-5} m^3/sec = 0.183 gpm = 2.64 × 10^{-4} mgd = 4.09 × 10^{-4} cfs = 8.11 × 10^{-4} AF/day	**1 million gallons per day (mgd)**	= 43.8 1/sec = 0.0438 m^3/sec = 3785 m^3/day = 694 gpm = 1.55 cfs = 3.07 AF/day

Water Units, Data Conversions, and Constants *(continued)*

FLOW RATE (L³/T) (continued)

1 cubic foot per second (cfs)	= 28.3 l/sec = 0.0283 m³/sec = 2447 m³/day = 449 gpm = 0.646 mgd = 1.98 AF/day	**1 miner's inch**	= 0.02 cfs (in Idaho, Kansas, Nebraska, New Mexico, North Dakota, South Dakota, and Utah) = 0.026 cfs (in Colorado) = 0.028 cfs (in British Columbia)
1 acre-foot per day (AF/day)	= 14.3 l/sec = 0.0143 m³/sec = 1233.48 m³/day = 226 gpm = 0.326 mgd = 0.504 cfs	**1 weir** **1 quinaria (ancient Rome)**	= 0.02 garcia = 0.47–0.48 l/sec
1 miner's inch	= 0.025 cfs (in Arizona, California, Montana, and Oregon: flow of water through 1 in² aperture under 6-inch head)		

ACCELERATION (L/T²)

standard acceleration of gravity	= 9.8 m/s² = 32 ft/s²

FORCE (ML/T² = Mass × Acceleration)

1 newton (N)	= kg-m/s² = 10⁵ dynes = 0.1020 kg force = 0.2248 lb force	**1 dyne** **1 pound force**	= g·cm/s² = 10⁻⁵ N = lb mass × acceleration of gravity = 4.448 N

(continues)

Water Units, Data Conversions, and Constants *(continued)*

PRESSURE (M/L² = Force/Area)		**1 kilogram per sq. centimeter (kg/cm²)**	= 14.22 lb/in²
1 pascal (Pa)	= N/m²		
1 bar	= 1 × 10⁵ Pa		
	= 1 × 10⁶ dyne/cm²	**1 inch of water at 62°F**	= 0.0361 lb/in²
	= 1019.7 g/cm²		= 5.196 lb/ft³
	= 10.197 te/m²		= 0.0735 inch of mercury at 62°F
	= 0.9869 atmosphere		
	= 14.50 lb/in²	**1 foot of water at 62°F**	= 0.433 lb/in²
	= 1000 millibars		= 62.36 lb/ft²
1 atmosphere (atm)	= standard pressure		= 0.833 inch of mercury at 62°F
	= 760 mm of mercury at 0°C		= 2.950 × 10⁻² atmosphere
	= 1013.25 millibars	**1 pound per sq. inch (psi or lb/in²)**	= 2.309 feet of water at 62°F
	= 1033 g/cm²		= 2.036 inches of mercury at 32°F
	= 1.033 kg/cm²		
	= 14.7 lb/in²		= 0.06804 atmosphere
	= 2116 lb/ft²		= 0.07031 kg/cm²
	= 33.95 feet of water at 62°F	**1 inch of mercury at 32°F**	= 0.4192 lb/in²
	= 29.92 inches of mercury at 32°F		= 1.133 feet of water at 32°F
TEMPERATURE			
degrees Celsius or Centigrade (°C)	= (°F–32) × 5/9	**degrees Fahrenheit (°F)**	= 32 + (°C × 1.8)
Kelvins (K)	= K–273.16		= 32 + ((°K–273.16) × 1.8)
	= 273.16 + °C		
	= 273.16 + ((°F– 32) × 5/9)		

Water Units, Data Conversions, and Constants

Water Units, Data Conversions, and Constants *(continued)*

ENERGY (ML^2/T^2 = Force × Distance)

1 joule (J)	= 10^7 ergs	**1 kilowatt-hour**	= 3.6×10^6 J
	= N·m	**(kWh)**	= 3412 Btu
	= W·s		= 859.1 kcal
	= kg·m^2/s^2	**1 quad**	= 10^{15} Btu
	= 0.239 calories		= 1.055×10^{18} J
	= 9.48×10^{-4} Btu		= 293×10^9 kWh
1 calorie (cal)	= 4.184 J		= 0.001 Q
	= 3.97×10^{-3} Btu		= 33.45 GWy
	(raises 1 g H$_2$O	**1 Q**	= 1000 quads
	1°C)		≈ 10^{21} J
1 British thermal	= 1055 J	**1 foot-pound (ft-lb)**	= 1.356 J
unit (Btu)	= 252 cal (raises		= 0.324 cal
	1 lb H$_2$O 1°F)	**1 therm**	= 10^5 Btu
	= 2.93×10^{-4} kWh	**1 electron-volt (eV)**	= 1.602×10^{-19} J
1 erg	= 10^{-7} J	**1 kiloton of TNT**	= 4.2×10^{12} J
	= g·cm^2/s^2	**1 10^6 te oil equiv.**	= 7.33×10^6 bbl oil
	= dyne·cm	**(Mtoe)**	= 45×10^{15} J
1 kilocalorie (kcal)	= 1000 cal		= 0.0425 quad
	= 1 Calorie (food)		

POWER (ML^2/T^3 = rate of flow of energy)

1 watt (W)	= J/s	**1 horsepower**	= 0.178 kcal/s
	= 3600 J/hr	**(H.P. or hp)**	= 6535 kWh/yr
	= 3.412 Btu/hr		= 33,000 ft-lb/min
1 TW	= 10^{12} W		= 550 ft-lb/sec
	= 31.5×10^{18} J		= 8760 H.P.-hr/yr
	= 30 quad/yr	**H.P. input**	= 1.34 × kW input
1 kilowatt (kW)	= 1000 W		to motor
	= 1.341 horsepower		= horsepower
	= 0.239 kcal/s		input to motor
	= 3412 Btu/hr	**Water H.P.**	= H.P. required to
10^6 bbl (oil) /day	≈ 2 quads/yr		lift water at a
(Mb/d)	≈ 70 GW		definite rate to
1 quad/yr	= 33.45 GW		a given distance
	≈ 0.5 Mb/d		assuming 100%
1 horsepower	= 745.7 W		efficiency
(H.R or hp)	= 0.7457 kW		= gpm × total head
			(in feet)/3960

(continues)

Water Units, Data Conversions, and Constants *(continued)*

EXPRESSIONS OF HARDNESS[a]

1 grain per gallon	= 1 grain CaCO$_3$ per U.S. gallon	**1 French degree**	= 1 part CaCO$_3$ per 100,000 parts water
1 part per million	= 1 part CaCO$_3$ per 1,000,000 parts water	**1 German degree**	= 1 part CaO per 100,000 parts water
1 English, or Clark, degree	= 1 grain CaCO$_3$ per Imperial gallon		

CONVERSIONS OF HARDNESS

1 grain per U.S. gallon	= 17.1 ppm, as CaCO$_3$	**1 French degree**	= 10 ppm, as CaCO$_3$
1 English degree	= 14.3 ppm, as CaCO$_3$	**1 German degree**	= 17.9 ppm, as CaCO$_3$

WEIGHT OF WATER

1 cubic inch	= 0.0361 lb	**1 imperial gallon**	= 10.0 lb
1 cubic foot	= 62.4 lb	**1 cubic meter**	= 1 tonne
1 gallon	= 8.34 lb		

DENSITY OF WATER[a]

Temperature		Density
°C	°F	gm/cm^3
0	32	0.99987
1.667	35	0.99996
4.000	39.2	1.00000
4.444	40	0.99999
10.000	50	0.99975
15.556	60	0.99907
21.111	70	0.99802
26.667	80	0.99669
32.222	90	0.99510
37.778	100	0.99318
48.889	120	0.98870
60.000	140	0.98338
71.111	160	0.97729
82.222	180	0.97056
93.333	200	0.96333
100.000	212	0.95865

Note: Density of Sea Water: approximately 1.025 gm/cm^3 at 15°C.

[a]*Source:* van der Leeden, F., Troise, F. L., and Todd, D. K., 1990. *The Water Encyclopedia*, 2d edition. Lewis Publishers, Inc., Chelsea, Michigan.

Comprehensive Table of Contents

Volume 1
The World's Water 1998-1999: The Biennial Report on Freshwater Resources

Foreword by Anne H. and Paul R. Ehrlich   ix

Acknowledgments   xi

Introduction   1

ONE   The Changing Water Paradigm   5
- Twentieth-Century Water-Resources Development   6
- The Changing Nature of Demand   10
- Economics of Major Water Projects   16
- Meeting Water Demands in the Next Century   18
- Summary: New Thinking, New Actions   32
- References   33

TWO   Water and Human Health   39
- Water Supply and Sanitation: Falling Behind   39
- Basic Human Needs for Water   42
- Water-Related Diseases   47
- Update on Dracunculiasis (Guinea Worm)   50
- Update on Cholera   56
- Summary   63
- References   64

THREE   The Status of Large Dams: The End of an Era?   69
- Environmental and Social Impacts of Large Dams   75
- New Developments in the Dam Debate   80
- The Three Gorges Project, Yangtze River, China   84
- The Lesotho Highlands Project, Senqu River Basin, Lesotho   93
- References   101

FOUR   Conflict and Cooperation Over Fresh Water   105
- Conflicts Over Shared Water Resources   107
- Reducing the Risk of Water-Related Conflict   113
- The Israel-Jordan Peace Treaty of 1994   115
- The Ganges-Brahmaputra Rivers: Conflict and Agreement   118
- Water Disputes in Southern Africa   119

Summary 124
Appendix A
   Chronology of Conflict Over Waters in the Legends,
   Myths, and History of the Ancient Middle East 125
Appendix B
   Chronology of Conflict Over Water:
   1500 to the Present 128
References 132

FIVE   Climate Change and Water Resources: What Does the Future Hold? 137

What Do We Know? 138
Hydrologic Effects of Climate Change 139
Societal Impacts of Changes in Water Resources 144
Is the Hydrologic System Showing Signs of Change? 145
Recommendations and Conclusions 148
References 150

SIX   New Water Laws, New Water Institutions 155

Water Law and Policy in New South Africa: A Move Toward
   Equity 156
The Global Water Partnership 165
The World Water Council 172
The World Commission on Dams 175
References 180

SEVEN   Moving Toward a Sustainable Vision for the Earth's Fresh Water 183

Introduction 183
A Vision for 2050: Sustaining Our Waters 185

WATER BRIEFS

The Best and Worst of Science: Small Comets and the New
   Debate Over the Origin of Water on Earth 193
Water Bag Technology 200
Treaty Between the Government of the Republic of India and
   the Government of the People's Republic of Bangladesh
   on Sharing of the Ganga/Ganges Waters at Farakka 206
United Nations Conventions on the Law of the Non-
   Navigational Uses of International Watercourses 210
Water-Related Web Sites 231

DATA SECTION

Table 1   Total Renewable Freshwater Supply by Country 235
Table 2   Freshwater Withdrawal by Country and Sector 241

Comprehensive Table of Contents

Table 3	Summary of Estimated Water Use in the United States, 1900 to 1995	245
Table 4	Total and Urban Population by Country, 1975 to 1995	246
Table 5	Percentage of Population with Access to Safe Drinking Water by Country, 1970 to 1994	251
Table 6	Percentage of Population with Access to Sanitation by Country, 1970 to 1994	256
Table 7	Access to Safe Drinking Water in Developing Countries by Region, 1980 to 1994	261
Table 8	Access to Sanitation in Developing Countries by Region, 1980 to 1994	263
Table 9	Reported Cholera Cases and Deaths by Region, 1950 to 1997	265
Table 10	Reported Cholera Cases and Deaths by Country, 1996 and 1997	268
Table 11	Reported Cholera Cases and Deaths in the Americas, 1991 to 1997	271
Table 12	Reported Cases of Dracunculiasis by Country, 1972 to 1996	272
Table 13	Waterborne Disease Outbreaks in the United States by Type of Water Supply System, 1971 to 1994	274
Table 14	Hydroelectric Capacity and Production by Country, 1996	276
Table 15	Populations Displaced as a Consequence of Dam Construction, 1930 to 1996	281
Table 16	Desalination Capacity by Country (January 1, 1996)	288
Table 17	Desalination Capacity by Process (January 1, 1996)	290
Table 18	Threatened Reptiles, Amphibians, and Freshwater Fish, 1997	291
Table 19	Irrigated Area by Country and Region, 1961 to 1994	297

Index   303

Volume 2
The World's Water 2000-2001: The Biennial Report on Freshwater Resources

Foreword by Timothy E. Wirth   xiii
Acknowledgments   xv
Introduction   xvii

ONE   The Human Right to Water   1
  Is There a Human Right to Water?   2
  Existing Human Rights Laws, Covenants, and Declarations   4
  Defining and Meeting a Human Right to Water   9
  Conclusions   15
  References   15

TWO   How Much Water Is There and Whose Is It? The World's Stocks and Flows of Water and International River Basins   19
  How Much Water Is There? The Basic Hydrologic Cycle   20
  International River Basins: A New Assessment   27
  The Geopolitics of International River Basins   35
  Summary   36
  References   37

THREE   Pictures of the Future: A Review of Global Water Resources Projections   39
  Data Constraints   40
  Forty Years of Water Scenarios and Projections   42
  Analysis and Conclusions   58
  References   59

FOUR   Water for Food: How Much Will Be Needed?   63
  Feeding the World Today   63
  Feeding the World in the Future: Pieces of the Puzzle   65
  How Much Water Will Be Needed to Grow Food?   78
  Conclusions   88
  Referencess   89

FIVE   Desalination: Straw into Gold or Gold into Water   93
  History of Desalination and Current Status   94
  Desalination Technologies   98
  Other Aspects of Desalination   106
  The Tampa Bay Desalination Plant   108
  Summary   109
  References   110

SIX The Removal of Dams: A New Dimension to an Old Debate 113
Economics of Dam Removal   118
Dam Removal Case Studies: Some Completed Removals   120
Some Proposed Dam Removals or Decommissionings   126
Conclusion   134
References   134

SEVEN Water Reclamation and Reuse: Waste Not, Want Not 137
Wastewater Uses   139
Direct and Indirect Potable Water Reuse   151
Health Issues   152
Wastewater Reuse in Namibia   155
Wastewater Reclamation and Reuse in Japan   158
Wastewater Costs   159
Summary   159
References   161

WATER BRIEFS
Arsenic in the Groundwater of Bangladesh and West Bengal, India   165
Fog Collection as a Source of Water   175
Environment and Security: Water Conflict Chronology—Version 2000   182
Water-Related Web Sites   192

DATA SECTION

Table 1	Total Renewable Freshwater Supply, by Country	197
Table 2	Freshwater Withdrawal, by Country and Sector	203
Table 3	World Population, Year 0 to A.D. 2050	212
Table 4	Population, by Continent, 1750 to 2050	213
Table 5	Renewable Water Resources and Water Availability, by Continent	215
Table 6	Dynamics of Water Resources, Selected Countries, 1921 to 1985	218
Table 7	International River Basins of the World	219
Table 8	Fraction of a Country's Area in International River Basins	239
Table 9	International River Basins, by Country	247
Table 10	Irrigated Area, by Country and Region, 1961 to 1997	255
Table 11	Irrigated Area, by Continent, 1961 to 1997	264
Table 12	Human-Induced Soil Degradation, by Type and Cause, Late 1980s	266
Table 13	Continental Distribution of Human-Induced Salinization	268
Table 14	Salinization, by Country, Late 1980s	269
Table 15	Total Number of Reservoirs, by Continent and Volume	270

Table 16	Number of Reservoirs Larger than 0.1 km^3, by Continent, Time Series	271
Table 17	Volume of Reservoirs Larger than 0.1 km^3, by Continent, Time Series	273
Table 18	Dams Removed or Decommissioned in the United States, 1912 to Present	275
Table 19	Desalination Capacity, by Country, January 1999	287
Table 20	Total Desalination Capacity, by Process, June 1999	289
Table 21	Desalination Capacity, by Source of Water, June 1999	290
Table 22	Number of Threatened Species, by Country/Area, by Group, 1997	291
Table 23	Countries with the Largest Number of Fish Species	298
Table 24	Countries with the Largest Number of Fish Species per Unit Area	299
Table 25	Water Units, Data Conversions, and Constants	300

Index   311

Volume 3
The World's Water 2002-2003: The Biennial Report on Freshwater Resources

Foreword by Amory B. Lovins   xiii

Acknowledgments   xv

Introduction   xvii

ONE   **The Soft Path for Water**   1
by Gary Wolff and Peter H. Gleick

A Better Way   3
Dominance of the Hard Path in the Twentieth Century   7
Myths about the Soft Path   9
One Dimension of the Soft Path: Efficiency of Use   16
Moving Forward on the Soft Path   25
Conclusions   30
References   30

TWO   **Globalization and International Trade of Water**   33
by Peter H. Gleick, Gary Wolff, Elizabeth L. Chalecki, and Rachel Reyes

The Nature and Economics of Water, Background and Definitions   34
Water Managed as Both a Social and Economic Good   38
The Globalization of Water: International Trade   41
The Current Trade in Water   42
The Rules: International Trading Regimes   47
References   54

THREE   **The Privatization of Water and Water Systems**   57
by Peter H. Gleick, Gary Wolff, Elizabeth L. Chalecki, and Rachel Reyes

Drivers of Water Privatization   58
History of Privatization   59
The Players   61
Forms of Privatization   63
Risks of Privatization: Can and Will They Be Managed?   67
Principles and Standards for Privatization   79
Conclusions   82
References   83

FOUR   Measuring Water Well-Being: Water Indicators and Indices   87
        by Peter H. Gleick, Elizabeth L. Chalecki, and Arlene Wong

    Quality-of-Life Indicators and Why We Develop Them   88
    Limitations to Indicators and Indices   92
    Examples of Single-Factor or Weighted Water Measures   96
    Multifactor Indicators   101
    Conclusions   111
    References   112

FIVE   Pacific Island Developing Country Water Resources and Climate Change   113
        by William C.G. Burns

    PIDCs and Freshwater Sources   113
    Climate Change and PIDC Freshwater Resources   119
    Potential Impacts of Climate Change on PIDC Freshwater Resources   124
    Recommendations and Conclusions   125
    References   127

SIX   Managing Across Boundaries: The Case of the Colorado River Delta   133
        by Michael Cohen

    The Colorado River   134
    The Colorado River Delta   139
    Conclusions   144
    References   145

SEVEN   The World Commission on Dams Report: What Next?   149
        by Katherine Kao Cushing

    The WCD Organization   149
    Findings and Recommendations   151
    Strategic Priorities, Criteria, and Guidelines   153
    Reaction to the WCD Report   155
    References   172

WATER BRIEFS

    The Texts of the Ministerial Declarations from The Hague March 2000) and Bonn (December 2001)   173
    The Southeastern Anatolia (GAP) Project and Archaeology   181
        by Amar S. Mann

Water Conflict Chronology   194
by Peter S. Gleick

Water and Space   209
by Elizabeth L. Chalecki

Water-Related Web Sites   225

DATA SECTION   237

Table 1	Total Renewable Freshwater Supply, by Country (2002 Update) 237	
Table 2	Fresh Water Withdrawals, by Country and Sector (2002 Update) 243	
Table 3	Access to Safe Drinking Water by Country, 1970 to 2000 252	
Table 4	Access to Sanitation by Country, 1970 to 2000 261	
Table 5	Access to Water Supply and Sanitation by Region, 1990 and 2000 270	
Table 6	Reported Cases of Dracunculiasis by Country, 1972 to 2000 273	
Table 7	Reported Cases of Dracunculiasis Cases, Eradication Progress, 2000 276	
Table 8	National Standards for Arsenic in Drinking Water 278	
Table 9	United States National Primary Drinking Water Regulations 280	
Table 10	Irrigated Area, by Region, 1961 to 1999 289	
Table 11	Irrigated Area, Developed and Developing Countries, 1960 to 1999 290	
Table 12	Number of Dams, by Continent and Country 291	
Table 13	Number of Dams, by Country 296	
Table 14	Regional Statistics on Large Dams 300	
Table 15	Commissioning of Large Dams in the 20th Century, by Decade 301	
Table 16	Water System Rate Structures 303	
Table 17	Water Prices for Various Households 304	
Table 18	Unaccounted-for Water 305	
Table 19	United States Population and Water Withdrawals, 1900 to 1995 308	
Table 20	United States GNP and Water Withdrawals, 1900 to 1996 310	
Table 21	Hong Kong GDP and Water Withdrawls, 1952 to 2000 313	
Table 22	China GDP and Water Withdrawls, 1952 to 2000 316	

Water Units, Data Conversions, and Constants   318

Index   329

Volume 4
The World's Water 2004-2005: The Biennial Report on Freshwater Resources

Foreword by Margaret Catley-Carlson xiii

Introduction xv

ONE The Millennium Development Goals for Water: Crucial Objectives, Inadequate Commitments 1
by Peter H. Gleick

Setting Water and Sanitation Goals 2
Commitments to Achieving the MDGs for Water 2
Consequences: Water-Related Diseases 7
Measures of Illness from Water-Related Diseases 9
Scenarios of Future Deaths from Water-Related Diseases 10
Conclusions 14
References 14

TWO The Myth and Reality of Bottled Water 17
by Peter H. Gleick

Bottled Water Use History and Trends 18
The Price and Cost of Bottled Water 22
The Flavor and Taste of Water 23
Bottled Water Quality 25
Regulating Bottled Water 26
Comparison of U.S. Standards for Bottled Water and Tap Water 36
Other Concerns Associated with Bottled Water 37
Conclusions 41
References 42

THREE Water Privatization Principles and Practices 45
by Meena Palaniappan, Peter H. Gleick, Catherine Hunt, Veena Srinivasan

Update on Privatization 46
Principles and Standards for Water 47
Can the Principles Be Met? 48
Conclusions 73
References 74

FOUR Groundwater: The Challenge of Monitoring and Management 79
by Marcus Moench

Conceptual Foundations 80
Challenges in Assessment 80
Extraction and Use 81
Groundwater in Agriculture 88
The Analytical Dilemma 90
A Way Forward: Simple Data as a Catalyst for Effective Management 97
References 98

FIVE Urban Water Conservation: A Case Study of Residential Water Use in California 101
by Peter H. Gleick, Dana Haasz, Gary Wolff

The Debate over California's Water 102
Defining Water "Conservation" and "Efficiency" 103
Current Urban Water Use in California 105
A Word About Agricultural Water Use 107
Economics of Water Savings 107
Data and Information Gaps 108
Indoor Residential Water Use 109
Indoor Residential Water Conservation: Methods and Assumptions 112
Indoor Residential Summary 118
Outdoor Residential Water Use 118
Current Outdoor Residential Water Use 119
Existing Outdoor Conservation Efforts and Approaches 120
Outdoor Residential Water Conservation: Methods and Assumptions 121
Residential Outdoor Water Use Summary 125
Conclusions 126
Abbreviations and Acronyms 126
References 127

SIX Urban Water Conservation: A Case Study of Commercial and Industrial Water Use in California 131
by Peter H. Gleick, Veena Srinivasan, Christine Henges-Jeck, Gary Wolff

Background to CII Water Use 132
Current California Water Use in the CII Sectors 133
Estimated CII Water Use in California in 2000 138

Data Challenges   139
The Potential for CII Water Conservation and Efficiency
　Improvements: Methods and Assumptions   140
Methods for Estimating CII Water Use and Conservation
　Potential   143
Calculation of Conservation Potential   145
Data Constraints and Conclusions   148
Recommendations for CII Water Conservation   150
Conclusions   153
References   154

SEVEN Climate Change and California Water Resources   157
by Michael Kiparsky and Peter H. Gleick

The State of the Science   158
Climate Change and Impacts on Managed Water-Resource
　Systems   172
Moving From Climate Science to Water Policy   175
Conclusions   183
References   184

WATER BRIEFS

One　3rd World Water Forum in Kyoto: Disappointment and
　　　Possibility   189
　　by Nicholas L. Cain

　　Ministerial Declaration of the 3rd World Water Forum:
　　　Message from the Lake Biwa and Yodo River Basin   198

　　NGO Statement of the 3rd World Water Forum   202

Two　The Human Right to Water: Two Steps Forward, One Step
　　　Back   204
　　by Peter H. Gleick

　　Substantive Issues Arising in the Implementation of the International Covenant on Economic, Social, and Cultural Rights. United Nations General Comment No. 15 (2002)   213

Three　The Water and Climate Bibliography   228
　　by Peter H. Gleick and Michael Kiparksy

Four　Environment and Security: Water Conflict Chronology
　　　Version 2004–2005   234
　　by Peter H. Gleick

　　Water Conflict Chronology   236
　　by Peter H. Gleick

DATA SECTION

Data Table 1　Total Renewable Freshwater Supply by Country
　　　　　　　(2004 Update)   257
Data Table 2　Freshwater Withdrawals by Country and Sector
　　　　　　　(2004 Update)   263

Comprehensive Table of Contents

Data Table 3   Deaths and DALYs from Selected Water-Related Diseases   272
Data Table 4   Official Development Assistance Indicators   274
Data Table 5   Aid to Water Supply and Sanitation by Donor, 1996 to 2001   278
Data Table 6   Bottled Water Consumption by Country, 1997 to 2002   280
Data Table 7   Global Bottled Water Consumption by Region, 1997 to 2002   283
Data Table 8   Bottled Water Consumption, Share by Region, 1997 to 2002   285
Data Table 9   Per-capita Bottled Water Consumption by Region, 1997 to 2002   286
Data Table 10   United States Bottled Water Sales, 1991 to 2001   288
Data Table 11   Types of Packaging Used for Bottled Water in Various Countries, 1999   289
Data Table 12   Irrigated Area by Region, 1961 to 2001   291
Data Table 13   Irrigated Area, Developed and Developing Countries, 1961 to 2001   293
Data Table 14   Global Production and Yields of Major Cereal Crops, 1961 to 2002   295
Data Table 15   Global Reported Flood Deaths, 1900 to 2002   298
Data Table 16   United States Flood Damage by Fiscal Year, 1926 to 2001   301
Data Table 17   Total Outbreaks of Drinking Water-Related Disease, United States, 1973 to 2000   304
Data Table 18
      18.a Extinction Rate Estimates for Continental North American Fauna (percent loss per decade)   309
      18.b Imperiled Species for North American Fauna   309
Data Table 19   Proportion of Species at Risk, United States   311
Data Table 20   United States Population and Water Withdrawals 1900 to 2000   313
Data Table 21   United States Economic Productivity of Water, 1900 to 2000   317

WATER UNITS, DATA CONVERSIONS, AND CONSTANTS   321

COMPREHENSIVE TABLE OF CONTENTS   331

Volume 1: The World's Water 1998–1999: The Biennial Report on Freshwater Resources   331

Volume 2: The World's Water 2000–2001: The Biennial Report on Freshwater Resources   334

Volume 3: The World's Water 2002–2003: The Biennial Report on Freshwater Resources   337

COMPREHENSIVE INDEX   341

Volume 5
The World's Water 2006-2007: The Biennial Report on Freshwater Resources

Foreword by Jon Lane xiii

Introduction xv

ONE Water and Terrorism 1
by Peter H. Gleick

Introduction 1
The Worry 2
Defining Terrorism 3
History of Water-Related Conflict 5
Vulnerability of Water and Water Systems 15
Responding to the Threat of Water-Related Terrorism 22
Water Security Policy in the United States 25
Conclusion 25

TWO Going with the Flow: Preserving and Restoring Instream Water Allocations 29
by David Katz

Environmental Flow: Concepts and Applications 30
Legal Frameworks for Securing Environmental Flow 34
The Science of Determining Environmental Flow Allocations 38
The Economics and Finance of Environmental Flow Allocations 40
Making It Work: Policy Implementation 43
Conclusion 45

THREE With a Grain of Salt: An Update on Seawater Desalination 51
by Peter H. Gleick, Heather Cooley, Gary Wolff

Introduction 51
Background to Desalination 52
History of Desalination 54
Desalination Technologies 54
Current Status of Desalination 55
Advantages and Disadvantages of Desalination 66
Environmental Effects of Desalination 76
Desalination and Climate Change 80
Public Transparency 81

Summary 82
Desalination Conclusions and Recommendations 83

FOUR   Floods and Droughts 91
by Heather Cooley

Introduction 91
Droughts 92
Floods 104
The Future of Droughts and Floods 112
Conclusion 113

FIVE   Environmental Justice and Water 117
by Meena Palaniappan, Emily Lee, Andrea Samulon

Introduction 117
A Brief History of Environmental Justice in the
   United States 119
Environmental Justice in the International
   Water Context 123
Environmental Justice and International Water Issues 124
Recommendations 137
Conclusion 141

SIX   Water Risks that Face Business and Industry 145
by Peter H. Gleick, Jason Morrison

Introduction 145
Water Risks for Business 146
Some New Water Trends: Looking Ahead 150
Managing Water Risks 153
An Overview of the "Water Industry" 158
Conclusion 163

WATER BRIEFS

One   Bottled Water: An Update 169
      by Peter H. Gleick

Two   Water on Mars 175
      by Peter H. Gleick

      Introduction 175
      Background 175
      Martian History of Water 178
      Future Mars Missions 180

Three Time to Rethink Large International Water Meetings 182
      by Peter H. Gleick

      Introduction 182
      Background and History 182

	Outcomes of International Water Meetings 183
	Ministerial Statements 184
	Conclusions and Recommendations 185
	4th World Water Forum Ministerial Declaration 186
Four	Environment and Security: Water Conflict Chronology Version, 2006–2007 189
	by Peter H. Gleick
Five	The Soft Path in Verse 219
	by Gary Wolff

DATA SECTION

Data Table 1   Total Renewable Freshwater Supply by Country   221

Data Table 2   Freshwater Withdrawal by Country and Sector   228

Data Table 3   Access to Safe Drinking Water by Country, 1970 to 2002   237

Data Table 4   Access to Sanitation by Country, 1970 to 2002   247

Data Table 5   Access to Water Supply and Sanitation by Region, 1990 and 2002   256

Data Table 6   Annual Average ODA for Water, by Country, 1990 to 2004 (Total and Per Capita)   262

Data Table 7   Twenty Largest Recipients of ODA for Water, 1990 to 2004   268

Data Table 8   Twenty Largest Per Capita Recipients of ODA for Water, 1990 to 2004   270

Data Table 9   Investment in Water and Sewerage Projects with Private Participation, by Region, in Middle- and Low-Income Countries, 1990 to 2004   273

Data Table 10   Bottled Water Consumption by Country, 1997 to 2004   276

Data Table 11   Global Bottled Water Consumption, by Region, 1997 to 2004   280

Data Table 12   Per Capita Bottled Water Consumption by Region, 1997 to 2004   282

Data Table 13   Per Capita Bottled Water Consumption, by Country, 1999 to 2004   284

Data Table 14   Global Cholera Cases and Deaths Reported to the World Health Organization, 1970 to 2004   287

Data Table 15   Reported Cases of Dracunculiasis by Country, 1972 to 2005   293

Data Table 16   Irrigated Area, by Region, 1961 to 2003   298

Data Table 17   Irrigated Area, Developed and Developing Countries, 1961 to 2003   301

Data Table 18   The U.S. Water Industry Revenue (2003) and Growth (2004–2006)   303

Data Table 19   Pesticide Occurrence in Streams, Groundwater, Fish, and Sediment in the United States   305

Data Table 20   Global Desalination Capacity and Plants— January 1, 2005   308

Data Table 21 100 Largest Desalination Plants Planned, in Construction, or in Operation—January 1, 2005 310

Data Table 22 Installed Desalination Capacity by Year, Number of Plants, and Total Capacity, 1945 to 2004 314

WATER UNITS, DATA CONVERSIONS, AND CONSTANTS 319

COMPREHENSIVE TABLE OF CONTENTS 329

Volume 1: The World's Water 1998–1999: The Biennial Report on Freshwater Resources 329

Volume 2: The World's Water 2000–2001: The Biennial Report on Freshwater Resources 332

Volume 3: The World's Water 2002–2003: The Biennial Report on Freshwater Resources 335

Volume 4: The World's Water 2004–2005: The Biennial Report on Freshwater Resources 338

COMPREHENSIVE INDEX 343

Volume 6
The World's Water 2008-2009: The Biennial Report on Freshwater Resources

Foreword by Malin Falkenmark xi

Acknowledgments xiii

Introduction xv

ONE Peak Water   1
Meena Palaniappan and Peter H. Gleick
Concept of Peak Oil   2
Comparison of Water and Oil   3
Utility of the Term "Peak Water"   9
A New Water Paradigm: The Soft Path for Water   12
Conclusion   14
References   15

TWO Business Reporting on Water   17
Mari Morikawa, Jason Morrison, and Peter H. Gleick
Corporate Reporting: A Brief History   17
Qualitative Information: Water Management Policies, Strategies, and Activities   22
Water Reporting Trends by Sector   32
Conclusions and Recommendations   36
References   38

THREE Water Management in a Changing Climate   39
Heather Cooley
The Climate Is Already Changing   39
Projected Impacts of Rising Greenhouse Gas Concentrations   40
Climate Change and Water Resources   43
Vulnerability to Climate Change   44
Adaptation   45
Conclusion   53
References   54

FOUR   Millennium Development Goals: Charting Progress and the Way Forward   57
Meena Palaniappan

Millennium Development Goals   57
Measuring Progress: Methods and Definitions   60
Progress on the Water and Sanitation MDGs   62
A Closer Look at Water and Sanitation Disparities   70
Meeting the MDGs: The Way Forward   73
Conclusion   77
References   78

FIVE   China and Water   79
Peter H. Gleick

The Problems   80
Water-Related Environmental Disasters in China   82
Water Availability and Quantity   83
Groundwater Overdraft   85
Floods and Droughts   86
Climate Change and Water in China   87
Water and Chinese Politics   88
Growing Regional Conflicts Over Water   90
Moving Toward Solutions   91
Improving Public Participation   96
Conclusion   97
References   97

SIX   Urban Water-Use Efficiencies: Lessons from United States Cities   101
Heather Cooley and Peter H. Gleick

Use of Water in Urban Areas   101
Projecting and Planning for Future Water Demand   102
Per-Capita Demand   104
Water Conservation and Efficiency Efforts   106
Comparison of Water Conservation Programs   110
Rate Structures   112
Conclusion   120
References   120

WATER BRIEFS

One   Tampa Bay Desalination Plant: An Update   123
Heather Cooley

Two   Past and Future of the Salton Sea   127
Michael J. Cohen

Background   129

	Restoration 131
	Conclusion 137
Three	Three Gorges Dam Project, Yangtze River, China 139
	Peter H. Gleick
	Introduction 139
	The Project 140
	Major Environmental, Economic, Social, and Political Issues 141
	Conclusion 148
Four	Water Conflict Chronology 151
	Peter H. Gleick

DATA SECTION

Data Table 1   Total Renewable Freshwater Supply, by Country   195
Data Table 2   Freshwater Withdrawal by Country and Sector   202
Data Table 3   Access to Safe Drinking Water by Country,
              1970 to 2004   211
Data Table 4   Access to Sanitation by Country, 1970 to 2004   221
Data Table 5   MDG Progress on Access to Safe Drinking Water
              by Region   230
Data Table 6   MDG Progress on Access to Sanitation by Region   233
Data Table 7   United States Dams and Dam Safety Data, 2006   236
Data Table 8   Dams Removed or Decommissioned in the United States,
              1912 to Present   239
Data Table 9   Dams Removed or Decommissioned in the United States,
              1912 to Present, by Year and State   265
Data Table 10 United States Dams by Primary Purposes   270
Data Table 11 United States Dams by Owner   272
Data Table 12 African Dams: Number and Total Reservoir Capacity
              by Country   274
              African Dams: The 30 Highest   276
Data Table 13 Under-5 Mortality Rate by Cause and Country, 2000   279
Data Table 14 International River Basins of Africa, Asia, Europe,
              North America, and South America   289
Data Table 15 OECD Water Tariffs   312
Data Table 16 Non-OECD Water Tariffs   320
Data Table 17 Fraction of Arable Land that Is Irrigated, by Country   324
Data Table 18 Area Equipped for Irrigation, by Country   329
Data Table 19 Water Content of Things   335
Data Table 20 Top Environmental Concerns of the American Public:
              Selected Years, 1997–2008   339

WATER UNITS, DATA CONVERSIONS, AND CONSTANTS   343

COMPREHENSIVE TABLE OF CONTENTS   353

Volume 1:   The World's Water 1998–1999: The Biennial Report on
           Freshwater Resources   353

Volume 2: The World's Water 2000–2001: The Biennial Report on Freshwater Resources   356

Volume 3: The World's Water 2002–2003: The Biennial Report on Freshwater Resources   359

Volume 4: The World's Water 2004–2005: The Biennial Report on Freshwater Resources   362

Volume 5: The World's Water 2006–2007: The Biennial Report on Freshwater Resources   366

Volume 6: The World's Water 2008–2009: The Biennial Report on Freshwater Resources   370

COMPREHENSIVE INDEX   373

Volume 7
The World's Water Volume 7: The Biennial Report on Freshwater Resources

Foreword by Robert Glennon xi

Introduction xiii

ONE   Climate Change and Transboundary Waters   1

Heather Cooley, Juliet Christian-Smith, Peter H. Gleick, Lucy Allen, and Michael Cohen

Transboundary Rivers and Aquifers   2
Managing Transboundary Basins   4
Transboundary Water Management Policies and Climate Change   7
Case Studies   10
Conclusions and Recommendations   18

TWO   Corporate Water Management   23

Peter Schulte, Jason Morrison, and Peter H. Gleick

Global Trends that Affect Businesses   24
Water-Related Business Risks   25
Key Factors that Determine Extent and Type of Risk   28
Risk and Impact Assessment   30
Strategies for Improved Corporate Water Management   34
Conclusions: A Framework for Action   41

THREE   Water Quality   45

Meena Palaniappan, Peter H. Gleick, Lucy Allen, Michael J. Cohen, Juliet Christian-Smith, and Courtney Smith

Current Water-Quality Challenges   46
Consequences of Poor Water Quality   54
Moving to Solutions and Actions   65
Mechanisms to Achieve Solutions   66
Conclusion   67

FOUR   Fossil Fuels and Water Quality   73

Lucy Allen, Michael J. Cohen, David Abelson, and Bart Miller

Fossil-Fuel Production and Associated Water Use   74
Fossil Fuels and Water Quality: Direct Impacts   76
Impacts on Freshwater Ecosystems   84
Impacts on Human Communities   87
Conclusion   92

FIVE Australia's Millennium Drought: Impacts and Responses   97
Matthew Heberger

Water Resources of Australia   98
Impacts of the Millennium Drought   102
Responses to Drought   106
Conclusion   121

SIX China Dams   127
Peter H. Gleick

Dams in China   129
Dams on Chinese International Rivers   130
Exporting Chinese Dams   133
Growing Internal Concern Over Chinese Dams   135
International Principles Governing Dam Projects: The World Commission on Dams   136
Conclusions   140

SEVEN U.S. Water Policy Reform   143
Juliet Christian-Smith, Peter H. Gleick, and Heather Cooley

Background   143
International Water Reform Efforts   144
Common Themes and "Soft Path" Solutions   149
A 21st Century U.S. Water Policy   150
Conclusions   154

WATER BRIEFS   157

One   Bottled Water and Energy   157
Peter H. Gleick and Heather Cooley

Energy to Produce Bottled Water   158
Summary of Energy Uses   163
Conclusions   163

Two   The Great Lakes Water Agreements   165
Peter Schulte

History of Shared Water Resource Management   165
Conclusion   169

Three   Water in the Movies   171
Peter H. Gleick

Popular Movies/Films   171
Water Documentaries   174
Short Water Videos and Films   174

Four   Water Conflict Chronology   175
Peter H. Gleick and Matthew Heberger

DATA SECTION

Data Table 1   Total Renewable Freshwater Supply, by Country   215
Data Table 2   Freshwater Withdrawal by Country and Sector   221
Data Table 3   Access to Safe Drinking Water by Country, 1970–2008   230
Data Table 4   Access to Sanitation by Country, 1970–2008   241
Data Table 5   MDG Progress on Access to Safe Drinking Water by Region (proportion of population using an improved water source)   251
Data Table 6   MDG Progress on Access to Sanitation by Region (proportion of population using an improved sanitation facility)   254
Data Table 7   Under-5 Mortality Rate by Cause and Country, 2008   257
Data Table 8   Infant Mortality Rate by Country (per 1,000 live births)   264
Data Table 9   Death and DALYs from Selected Water-Related Diseases, 2000 and 2004   270
Data Table 10 Overseas Development Assistance for Water Supply and Sanitation, by Donating Country   273
Data Table 11 Overseas Development Assistance for Water Supply and Sanitation, by Subsector (total of all donating countries)   275
Data Table 12 Organic Water Pollutant (BOD) Emissions by Country (% from various industries), 2005   278
Data Table 13 Top Environmental Concerns of the American Public: Selected Years, 1997–2010 (% who worry "a great deal")   282
Data Table 14 Top Environmental Concerns Around the World   285
Data Table 15 Satisfaction With Local Water Quality, by Country, 2006–2007   289
Data Table 16 Extinct (or Extinct in the Wild) Freshwater Animal Species   292
Data Table 17 U.S. Federal Water-Related Agency Budgets   303
Data Table 18 Overseas Dams With Chinese Financiers, Developers, or Builders (as of August 2010)   308
Data Table 19 Per-Capita Bottled Water Consumption by Top Countries, 1999–2010 (liters per person per year)   339

WATER UNITS, DATA CONVERSIONS, AND CONSTANTS   341

COMPREHENSIVE TABLE OF CONTENTS   351

Volume 1: The World's Water 1998–1999: The Biennial Report on Freshwater Resources   351

Volume 2: The World's Water 2000–2001: The Biennial Report on Freshwater Resources   354

Volume 3: The World's Water 2002–2003: The Biennial Report on Freshwater Resources   357

Volume 4: The World's Water 2004–2005: The Biennial Report on Freshwater Resources   360

Volume 5: The World's Water 2006–2007: The Biennial Report on Freshwater Resources   364

Volume 6: The World's Water 2008–2009: The Biennial Report on Freshwater Resources   368

Volume 7: The World's Water Volume 7: The Biennial Report on Freshwater Resources   372

COMPREHENSIVE INDEX   377

Volume 8
The World's Water Volume 8: The Biennial Report on Freshwater Resources

Foreword by Ismail Serageldin xi

Introduction xiii

ONE   Global Water Governance in the Twenty-First Century   1

         Heather Cooley, Newsha Ajami, Mai-Lan Ha,
           Veena Srinivasan, Jason Morrison, Kristina Donnelly,
           and Juliet Christian-Smith

         Global Water Challenges   2
         The Emergence of Global Water Governance   6
         Conclusions   15

TWO   Shared Risks and Interests: The Case for Private Sector Engagement in Water Policy and Management   19

         Peter Schulte, Stuart Orr, and Jason Morrison

         The Business Case for Investing in Sustainable Water
            Management   21
         Utilizing Corporate Resources While Ensuring Public Interest
            Outcomes and Preventing Policy Capture   28
         Moving Forward: Unlocking Mutually Beneficial Corporate
            Action on Water   31

THREE   Sustainable Water Jobs   35

         Eli Moore, Heather Cooley, Juliet Christian-Smith, and
           Kristina Donnelly

         Water Challenges in Today's Economy   36
         Job Quality and Growth in Sustainable Water
            Occupations   54
         Conclusions   57
         Recommendations   58

FOUR   Hydraulic Fracturing and Water Resources: What Do We Know and Need to Know?   63

         Heather Cooley and Kristina Donnelly

         Overview of Hydraulic Fracturing   64
         Concerns Associated with Hydraulic Fracturing
            Operations   65
         Water Challenges   67
         Conclusions   77

FIVE   Water Footprint   83
Julian Fulton, Heather Cooley, and Peter H. Gleick

    The Water Footprint Concept   83
    Water, Carbon, and Ecological Footprints and Nexus Thinking   85
    Water Footprint Findings   87
    Conclusion   90

SIX   Key Issues for Seawater Desalination in California: Cost and Financing   93
Heather Cooley and Newsha Ajami

    How Much Does Seawater Desalination Cost?   93
    Desalination Projects and Risk   101
    Case Studies   106
    Conclusions   117

SEVEN   Zombie Water Projects   123
Peter H. Gleick, Matthew Heberger, and Kristina Donnelly

    The North American Water and Power Alliance—NAWAPA   124
    The Reber Plan   131
    Alaskan Water Shipments   133
    Las Vegas Valley Pipeline Project   136
    Diverting the Missouri River to the West   140
    Conclusions   144

WATER BRIEF

One   The Syrian Conflict and the Role of Water   147
    Peter H. Gleick

Two   The Red Sea–Dead Sea Project Update   153
    Kristina Donnelly

Three   Water and Conflict: Events, Trends, and Analysis (2011–2012)   159
    Peter H. Gleick and Matthew Heberger

Four   Water Conflict Chronology   173
    Peter H. Gleick and Matthew Heberger

DATA SECTION

Data Table 1:   Total Renewable Freshwater Supply by Country (2013 Update)   221

Data Table 2:   Freshwater Withdrawal by Country and Sector (2013 Update)   227

Data Table 3A: Access to Improved Drinking Water by Country, 1970–2008   236
Data Table 3B: Access to Improved Drinking Water by Country, 2011 Update   247
Data Table 4A: Access to Improved Sanitation by Country, 1970–2008   252
Data Table 4B: Access to Improved Sanitation by Country, 2011 Update   263
Data Table 5:   MDG Progress on Access to Safe Drinking Water by Region   268
Data Table 6:   MDG Progress on Access to Sanitation by Region   271
Data Table 7:   Monthly Natural Runoff for the World's Major River Basins, by Flow Volume   274
Data Table 8:   Monthly Natural Runoff for the World's Major River Basins, by Basin Name   292
Data Table 9:   Area Equipped for Irrigation Actually Irrigated   310
Data Table 10:   Overseas Development Assistance for Water Supply and Sanitation, by Donating Country, 2004–2011   317
Data Table 11:   Overseas Development Assistance for Water Supply and Sanitation, by Subsector, 2007–2011   320
Data Table 12:   Per Capita Water Footprint of National Consumption, by Country, 1996–2005   323
Data Table 13:   Per Capita Water Footprint of National Consumption, by Sector and Country, 1996–2005   332
Data Table 14:   Total Water Footprint of National Consumption, by Country, 1996–2005   341
Data Table 15:   Total Water Footprint of National Consumption, by Sector and Country, 1996–2005   350
Data Table 16A: Global Cholera Cases Reported to the World Health Organization, by Country, 1949–1979   359
Data Table 16B: Global Cholera Cases Reported to the World Health Organization, by Country, 1980–2011   367
Data Table 17A: Global Cholera Deaths Reported to the World Health Organization, by Country, 1949–1979   373
Data Table 17B: Global Cholera Deaths Reported to the World Health Organization, by Country, 1980–2011   381
Data Table 18A: Perceived Satisfaction with Water Quality in Sub-Saharan Africa   387
Data Table 18B: Regional Assessment of Satisfaction with Water (and Air) Quality   389
Data Table 18C: Countries Most and Least Satisfied with Water Quality   392

WATER UNITS, DATA CONVERSIONS, AND CONSTANTS   395

COMPREHENSIVE TABLE OF CONTENTS   405

    Volume 1:   The World's Water 1998–1999: The Biennial Report on Freshwater Resources   405

    Volume 2:   The World's Water 2000–2001: The Biennial Report on Freshwater Resources   408

    Volume 3:   The World's Water 2002–2003: The Biennial Report on Freshwater Resources   411

    Volume 4:   The World's Water 2004–2005: The Biennial Report on Freshwater Resources   414

    Volume 5:   The World's Water 2006–2007: The Biennial Report on Freshwater Resources   418

    Volume 6:   The World's Water 2008–2009: The Biennial Report on Freshwater Resources   422

    Volume 7:   The World's Water, Volume 7: The Biennial Report on Freshwater Resources   426

    Volume 8:   The World's Water, Volume 8: The Biennial Report on Freshwater Resources   430

COMPREHENSIVE INDEX   435

Volume 9
The World's Water Volume 9: The Report on Freshwater Resources

Foreword by Alexandra Cousteau xi

Introduction xiii

About the Authors xvii

ONE   The UN Global Compact CEO Water Mandate: History, Objectives, Strategy   1

Heather Rippman and Stefanie Woodward

Introduction: Some History on the Evolution of Corporate Social and Environmental Responsibility   1
The Formation and Role of the UN Global Compact   3
The UN CEO Water Mandate   4
Corporate Water Stewardship   6
Building Consensus: Key Water Stewardship Concepts and Terminology   8
Making Water Stewardship Accessible: Producing and Distributing Tools and Guidance   9
Managing Water-Related Impacts   10
Ensuring Integrity in Water Stewardship Initiatives   14
Making an Impact: Sustainable Management of Water and Sanitation for All   16
Conclusion   16

TWO   A Human Rights Lens for Corporate Water Stewardship: Toward Achievement of the Sustainable Development Goal for Water   21

Mai-Lan Ha

Introduction   21
The Business Case for Action on Water   22
State Recognition of the Human Rights to Water and Sanitation   25
Human Rights Responsibility of the Corporate Sector   28
Support for the Human Rights to Water and Sanitation   31
The Path Forward   35

THREE   Updating Water-Use Trends in the United States   39

Kristina Donnelly and Heather Cooley

Introduction   39
Data Collection and Organization   41
Total Water Use   42
Conclusions   49

FOUR The Water Footprint of California's Energy System, 1990–2012   55
Julian Fulton and Heather Cooley

Introduction   55
Background   55
Evaluating Water-Energy Links   57
Water for California's Total Energy System   58
Summary   61
Acknowledgements   64

FIVE The Nature and Impact of the 2012–2016 California Drought   67
Peter H. Gleick

Introduction   67
The Hydrological and Climatological Conditions behind the California Drought   67
Impacts of the California Drought   74
Overall Economic Well-Being   81

SIX Water Trading in Theory and Practice   87
Michael Cohen

Water Trading, Transfers, Markets, and Banks   88
Water Trading in Theory   88
Water Trading in Practice   90
Environmental, Economic, and Social Performance   95
Necessary, Enabling, and Limiting Conditions for Water Trades   103
Conclusion   107

SEVEN The Cost of Water Supply and Efficiency Options: A California Case   113
Heather Cooley and Rapichan Phurisamban

Introduction   113
Methods and Approach   114
Storm Water Capture   115
Water Recycling and Reuse   117
Desalination   120
Urban Water Conservation and Efficiency   122
Summary and Conclusions   126

WATER BRIEFS

One   The Human Right to Water and Global Sustainability:
Actions of the Vatican   129
Peter H. Gleick

Two   Access to Water through Public Drinking Fountains   135
Rapichan Phurisamban and Peter H. Gleick

Three   Water and Conflict Update   141
Peter H. Gleick

WATER UNITS, DATA CONVERSIONS, AND CONSTANTS   147

COMPREHENSIVE TABLE OF CONTENTS   157

Volume 1:   The World's Water 1998–1999: The Biennial Report on Freshwater Resources   157

Volume 2:   The World's Water 2000–2001: The Biennial Report on Freshwater Resources   160

Volume 3:   The World's Water 2002–2003: The Biennial Report on Freshwater Resources   163

Volume 4:   The World's Water 2004–2005: The Biennial Report on Freshwater Resources   166

Volume 5:   The World's Water 2006–2007: The Biennial Report on Freshwater Resources   170

Volume 6:   The World's Water 2008–2009: The Biennial Report on Freshwater Resources   174

Volume 7:   The World's Water Volume 7: The Biennial Report on Freshwater Resources   178

Volume 8:   The World's Water Volume 8: The Biennial Report on Freshwater Resources   182

Volume 9:   The World's Water Volume 9: The Report on Freshwater Resources   196

INDEX TO VOLUME 9   189

COMPREHENSIVE INDEX TO VOLUMES 1-8   197

Index to Volume 9

A
Access to sanitation:
   in SDG 6, 16, 17, 22
   lack of, 130
Access to water:
   historical development, 4
   in SDG 6, 16, 17, 22
   lack of, 129
   public drinking fountains, 139
Agriculture:
   historical development, 4
   water recycling and reuse, 118
   water trading, 92, 103, 108
   water use, 42–43, 87, 107
   See also California: agriculture.
Alicante Basin, 92
Alliance for Water Stewardship (AWS), 6, 9, 34
Annan, Kofi, 3
Arizona:
   Arizona Water Bank, 90
   water banks, 94, 95
   water trading, 92
Australia:
   drought, 87
   Labor Party, 99
   public drinking fountains, 138
   water banks, 94
   water trading, 88, 90–91, 97, 103, 105, 107
      buyback program, 99
      environmental performance, 98
AWS:
   See Alliance for Water Stewardship (AWS).

B
Belgium:
   HRWS, 25
   public drinking fountains, 138
Beverage Industry Environmental Roundtable (BIER), 12
Bhopal, 2
BIER:
   See Beverage Industry Environmental Roundtable (BIER).
Bioethanol:
   See Ethanol.
Blue water, definition, 57
   See also Energy-related water footprint (EWF): blue.
Bonneville Power Administration (BPA), 98
Bowen, Howard R., 1
BPA:
   See Bonneville Power Administration (BPA).
Brackish water:
   desalination, 127
      costs, 120–21
   relative salinity, 120

Brine:
   relative salinity, 120
Brown, Governor Jerry, 75, 81

C
Cáceres, Berta, 143
CALEB:
   See California Energy Balance.
California Department of Fish and Wildlife (CDFW), 81
California Department of Water Resources (CDWR), 71, 74, 92
California Energy Balance (CALEB), 57
California Energy Commission (CEC), 57
California:
   agriculture, 74–78
      employment, 78
      revenue, 75–77
         by crop, 76
   cap-and-trade program, 56
   Carlsbad, 120
   CDFW, 81
   CDWR, 71, 74, 92
   climate change, 88
   CVP, 81
   CVWD, 104
   desalination, 120
   drought, 56, 60, 87
      2012–2016, 67, 69–75
         impacts, 74–75, 77–78
            air pollution, 80
            bird populations, 81
            economic, 81–83
            ecosystem, 80–81
            fish populations, 80
            tree mortality, 81
            wildfires, 81, 82
         precipitation, 69
         temperature, 69, 71
         water use reduction, 75
      area percentage, 70
      characteristics, 67
      runoff deficit, 71, 73
   Drought Water Bank, 90, 94
   electricity generation, by source, 79
   energy system, 55, 57, 58
   energy use, 56
   energy-efficient programs, 56
   ethanol, 59, 61–63
   EWF
      blue, 57, 60–61
      consumptive use factors, 57–58, 59
      electricity production, 58, 60, 63–64
      energy policies, 63, 64
      energy types, 60
      features, 55

gray, 57, 62
green, 57, 60–61
natural gas, 59–61
oil products, 60, 61
sources, 61, 62
water types, 60–61, 64
freshwater fish, 80
Fresno, 116, 123, 124, 125
GDP increase, 56
GHGs
inventory, 56
management, 56, 63
Global Warming Solutions Act of 2006, 63
groundwater
overdraft, 74
storage, 78
hydropower generation, 55, 56, 60, 63, 78–80
IID, 92, 96, 101
Los Angeles, 99
Low Carbon Fuel Standard (LCFS), 63
MWD, 104
natural gas, 59–61
Oakland, 123, 124, 125
Owens Valley, 99
Pacific Flyway, 81
population increase, 56
precipitation, by year, 69
Rainwater Capture Act, 116
reservoir storage, 58, 72–74
river runoff, by year, 71
Sacramento, 123, 124, 125
Sacramento River, 71, 80
Sacramento Valley, 81
Sacramento–San Joaquin Delta, 71
Sacramento–San Joaquin watersheds, 71
San Diego, 123, 124, 125
San Francisco Bay, 116
San Joaquin River, 71
San Joaquin Valley, 74
SDCWA, 92, 96
shale oil, 56
solar power, 56
storm water
capture, 115–16
costs, 116–17
SWP, 81, 104
SWRCB, 81, 115, 116
temperature, by year, 70
Ventura, 123, 124, 125
water banks, 95
water conservation, 122
water contamination, 130
water efficiency, 122, 127
cost analysis, 114–15
cost savings, 126
data limitations, 115
nonresidential, 124
residential, 123, 124
water loss control, 126
water recycling and reuse, 118
water trading, 91–92, 93, 100, 107
economic effects, 96–97
environmental performance, 98
exchange agreements, 104
Imperial Valley, 101
Los Angeles, 100–101
Owens Valley, 100–101, 103
water purchases, 99
water year index, 72
water year, definition, 67
water–energy links, 57–58
Carbonit Water Pumping Station, 144
Carlsbad:

desalination plant, 120
CBWPT:
See Columbia Basin Water Transactions Program (CBWTP).
CDFW:
See California Department of Fish and Wildlife (CDFW).
CDP (Carbon Disclosure Project):
2014 Global Water Report, 7
developing corporate water tools, 9
From Water Management to Water Stewardship, 7
water questionnaire, 7
CDWR:
See California Department of Water Resources (CDWR).
CEC:
See California Energy Commission (CEC).
Central Valley Project (CVP), 81
CEO Water Mandate:
See United Nations: CEO Water Mandate.
Ceres, 9
Chernobyl, 2
Chile:
Limarí Basin, 92, 94, 100
Limarí Valley, 103
water banks, 94
water trading, 92, 100, 102, 103, 107
China:
electricity production, 64
water trading, 89, 94
Clean Water Act, 45
Climate change:
California, 88
global security crisis, 144
impacts, 53, 61, 71, 107
Laudato Si', 130
mitigation, 55, 64
water crisis, 39, 55, 64
Coachella Valley Water District (CVWD), 104
Coalinga:
desalination plant, 120
Coca-Cola Company, 4
Cold War, 2
Colorado River, 94, 104
Colorado River Basin, 87, 94, 95
Colorado Super Ditch, 90
Colorado Water Trust (CWT), 98
Colorado:
Northern Colorado Water Conservancy District, 94, 96, 97, 105
San Luis Valley, 101, 103
water banks, 94
water trading, 92, 101, 106, 107
environmental performance, 98
Columbia Basin Water Transactions Program (CBWTP), 98
Corporations:
corporate responsibility (CR), 2
corporate social responsibility (CSR), 1–2, 2
corporate stewardship, 2
corporate sustainability, 2
corporate water stewardship, 6–8, 11
advantages of good governance, 24–25
alignment with UN Guiding Principles, 30–31
assessing impacts, 32–33
building consensus, 8–10
business case for action, 22
collective action, 12–13, 16, 33
concepts and terminology, 8–10
definition, 29
elements of, 32
employee health, 24
ensuring integrity, 14–15

Index

governmental oversight, 28
HRWS integration, 31–32
human rights lens, 30, 32–33
license to operate, 24
management process, 29
managing water-related impacts, 11–14
meeting SDG goals, 22
mitigating impacts on HRWS, 34
public–private sector partnerships, 14
supply chain improvements, 33
support for HRWS, 35–36
tools and guidance, 10–11
water-related impacts, 30
social role, theories of, 1–2
UN Global Compact membership, 4
water footprint assessment, 8, 10
water-related business risk, 8–9, 12, 16, 29
See also Water risk; Water stewardship.
Costa Rica:
HRWS, 25
Cousteau, Alexandra, xi–xii
CVP:
See Central Valley Project (CVP).
CVWD:
See Coachella Valley Water District (CVWD).
CWT:
See Colorado Water Trust (CWT).

D
Dakota Access Pipeline, 144
Davis, Keith, 16
de Albuquerque, Catarina, 27
Desalination:
Carlsbad plant, 120
Coalinga plant, 120
costs, 120–21, 126, 127
technologies, 120
Donetsk Filtration Plant, 144
Driving Harmonization of Water Related Technology, 9
Drought:
Australia, 87, 97
California, 87
2012–2016, 67, 69–75
Colorado River Basin, 87
definitions, 68
Midwest, 2011–2012, 62

E
Eau de Paris, 138
Energy Star:
See United States Environmental Protection Agency (EPA): Energy Star.
Energy-related water footprint (EWF):
blue, 57, 59, 64
consumptive use factors, 57–58, 59
electricity production, 57, 58, 60
ethanol, 59, 60, 61–63
features, 55
gray, 57, 62, 64
green, 57, 59, 64
liquid fuel, 59
natural gas, 59–61
oil products, 60, 63
See also California: EWF.
England:
water trading, 93
EPA:
See United States Environmental Protection Agency (EPA).
Ethanol:
demand, increase in, 61, 62
energy policies, 63
EWF, 57, 58, 59, 60, 61–63
gray, 62
green, 60
production locations, 59
sources, 61
EWF:
See Energy-related water footprint (EWF).
Exxon Valdez, 2

F
Fleck, John, 107
Forest Stewardship Council (FSC):
sustainability certification, 11
France:
public drinking fountains, 138
Freshwater:
agricultural use, 87, 107
fish species, 80
irrigation use, 51
relative salinity, 120
sustainable management, 6
thermoelectric power, 42, 43, 45, 49
water stress, 39, 52, 87
Friedman, Milton, 1
FSC:
See Forest Stewardship Council (FSC).

G
GEMI:
water risk management questionnaire, 12
GHG:
See Greenhouse gases (GHGs).
Global Reporting Initiative (GRI), 9
Globalization:
development of, 2
opposition to, 2
problems of, 2
Gray water, definition, 57
See also Energy-related water footprint (EWF): gray.
Green water, definition, 57
See also Energy-related water footprint (EWF): green.
Greenhouse gases (GHGs):
emissions, 55, 64
management, 56, 63
GRI:
See Global Reporting Initiative (GRI).
Groundwater:
California, 74, 78
India, 99
management issues, 62
pumping and treatment costs, 117
recharge, 116, 119
water quality, 62
water trading, 99

H
Harmony Gold Mining Company, 27
Honduras:
water conflict, 143
HRWS:
See Human rights to water and sanitation (HRWS).
Human rights to water and sanitation (HRWS):
adoption by governments, 27
Belgium, 25
corporate water stewardship, integration with, 31–32
Costa Rica, 25
dimensions and definitions, 26
India, 25
Indonesia, 25
Kenya, 25
recognition by United Nations, 29
recognition by Vatican, 129

South Africa, 25, 27
    Harmony Gold Mining Company, 27
    National Water Act, 27
    Orkney Mine, 27
    state recognition of, 25–26
    water-related impacts, 30
Hydraulic fracturing, 56, 59
Hydroelectric power:
    See California: hydropower generation.

I
Idaho:
    water banks, 94
IID:
    See Imperial Irrigation District (IID).
Imperial Irrigation District (IID), 92, 96, 101
Imperial Valley, 101
India:
    groundwater
        overpumping, 99
    HRWS, 25
    water conflict, 141, 143, 144
    water trading, 93, 99
Indonesia:
    HRWS, 25
Iraq:
    water conflict, 141, 143
Irrigation:
    area, 52
    average application depth, 51
    freshwater use, annual, 51
    methods, 52
    Murray-Darling Basin, 97
    water use, decline in, 49, 52
Israel:
    water recycling and reuse, 118

K
Kazakhstan:
    water trading, 93
Kenya:
    HRWS, 25
    Water Act of 2002, 27, 28
    Water Resource Management Authority, 28
Kreps, Thomas, 1
Kyrgyzstan:
    water trading, 93

L
Läckeby Water Group, 4
LADWP:
    See Los Angeles Department of Water and Power (LADWP).
Lake Mead, 95
Laudato Si':
    climate change, 130
    human right to water, 129–30, 131
Lawrence Berkeley National Laboratories, 57
LCFS:
    See California: Low Carbon Fuel Standard (LCFS).
Levi Strauss & Co., 4
Limarí Basin, 92, 94, 100
Limarí Valley, 103
Los Angeles, 99, 100–101
Los Angeles Department of Water and Power (LADWP):
    storm water, 116

M
M&I:
    See Municipal and industrial (M&I).
Malheur National Wildlife Refuge, 144

Marine Stewardship Council (MSC):
    sustainability certification, 11
MDGs:
    See United Nations: Millennium Development Goals (MDGs).
Methyl tert-butyl ether (MTBE), 62, 63, 64
Metropolitan Water District (MWD), 104
Mexicali Valley, 92, 107
Mexico:
    National Water Law of 1992, 92
    water banks, 94
    water trading, 92
Mississippi River, 62
MSC:
    See Marine Stewardship Council (MSC).
MTBE:
    See Methyl tert-butyl ether (MTBE).
Municipal and industrial (M&I):
    water trading, 92
    water use, 41, 42–43, 45
        decline in, 45, 47, 52
Murray-Darling Basin, 91, 97, 105, 107, 108
MWD:
    See Metropolitan Water District (MWD).

N
Namibia:
    water recycling and reuse, 118
NAS:
    See National Academy of Sciences (NAS).
National Academy of Sciences (NAS), 41
National Drought Mitigation Center (NDMC), 68, 69
National Energy Policy Act of 1992, 47
National Renewable Energy Laboratory (NREL), 57
National Research Council (NRC), 89
National Water-Use Information Program (NWUIP), 41
Natural gas:
    EWF, 59–61
Nature Conservancy, The, 9
NDMC:
    See National Drought Mitigation Center (NMDC).
Nepal:
    water trading, 103
Nestlé S.A., 4, 27
Nevada:
    water banks, 94, 95
New Mexico:
    water trading, 102, 105
NGOs:
    See Non-governmental organizations (NGOs).
NIC:
    See United States National Intelligence Council (NIC).
Non-governmental organizations (NGOs), 8, 14, 28, 34
Northern Colorado Water Conservancy District, 94, 96, 97, 105
Northwest Power and Conservation Council, 98
NRC:
    See National Research Council (NRC).
NREL:
    See National Renewable Energy Laboratory (NREL).
NWUIP:
    See National Water-Use Information Program (NWUIP).

O
Ogallala aquifer, 62
Oil products:
    crude oil, sources, 59
    EWF, 57, 60–61

Index

Oman:
    water trading, 94
Omnibus Public Land Management Act of 2009, 42
Oregon:
    water trading, 94
Orkney Mine, 27
Owens Valley, 99, 100–101, 103

P

Pacific Institute:
    developing corporate water tools, 9
    Drinking Fountains and Public Health, 135
    HRWS legislation, 25
    The World's Water, 87, 129, 141
    Water Conflict Chronology, xiv, 141, 142, 143
Pakistan:
    bottling plant, 27
    water trading, 93
Peak water, 22, 113, 129
Pontifical Academy of Sciences:
    The Human Right to Water, 2017 meeting, 129, 131–32
Pope Francis:
    Laudato Si', 129–30, 131
    The Human Right to Water, 2017 meeting, 129, 131–32
Population growth, 2, 4, 39, 42, 45, 48, 49, 53, 113, 144
Pricewaterhouse Coopers, 9
Public drinking fountains:
    Ancient Greece, 135
    Australia, 138
    Belgium, 138
    France, 138
    Paris
        "La Petillante", 138
    public health concerns, 135–36
    technologies, 136, 138–39
    typologies, 138

R

Renewable Fuel Standard (RFS), 63
Reservoirs:
    California, 72–74
    evaporative losses, 58, 60, 71
RFS:
    See Renewable Fuel Standard (RFS).
Russia:
    water conflict, 144

S

SAB Miller, 4
Sacramento River, 71, 80
Sacramento Valley, 81
Sacramento–San Joaquin Delta, 71
Sacramento–San Joaquin watersheds:
    river runoff, 71
    water trading, 93
Saline water:
    definition of, 42, 120
    thermoelectric power, 43, 49
San Diego County Water Authority (SDCWA), 92, 96
San Francisco Bay, 116
San Joaquin River, 71
San Joaquin Valley, 74
San Luis Valley, 101, 103
Sanitation:
    human right to, 25–26, 35–36
    lack of access, 130
    sustainable management, 16–17
SDCWA:
    See San Diego County Water Authority (SDCWA).
SDGs:
    See United Nations: Sustainable Development Goals (SDGs).
Seawater:
    desalination, 127
        costs, 120–21
    relative salinity, 120
Secure Water Act, 42
Sierra Nevada Mountains, 55, 71, 81
Sikhosiphi, Radebe, 143
Singapore:
    water recycling and reuse, 118
South Africa:
    Department of Water Affairs and Forestry, 93
    HRWS, 25, 27
    Water Act of 1998, 93
    water conflict, 143
    water trading, 93, 102–3
South Donbass water pipeline, 144
Spain:
    Alicante Basin, 92
    water banks, 94
    water trading, 92, 100
State Water Project (SWP), 81, 104
State Water Resources Control Board (SWRCB):
    drought emergency orders, 81
    storm water, 115
    Storm Water Strategy, 116
Stockholm World Water Week, 9
Storm water:
    capture, 115–16
    costs, 116–17, 126
Suez, 4
Surface water:
    management issues, 62
SWP:
    See State Water Project (SWP).
SWRCB:
    See State Water Resources Control Board (SWRCB).
Syr Darya River, 93
Syria:
    water conflict, 141, 143

T

Texas:
    water banks, 94
Thermoelectric power:
    dry-cooling, 45
    once-through cooling, 45
    thermal pollution, 45
    water type, 44
    water use, 43–45, 46, 49, 63

U

UCSB:
    See University of California at Santa Barbara (UCSB).
Ukraine:
    water conflict, 144
UN Global Compact:
    See United Nations: Global Compact.
Unilever, 35
United Nations:
    CEO Water Mandate, 1, 4–6, 9
        building consensus, 16
        collective action, 12–13, 18
        elements of, 13
        Corporate Water Disclosure Guidelines, 13
        corporate water stewardship, 7–8, 13, 22
        Guide for Managing Integrity in Water Stewardship, 14
        Guide to Responsible Business Engagement with Water Policy, 13

Guide to Water-Related Collective Action, 13
membership growth, 7
objectives, 7
purpose of, 6
six key elements, 5
tools and guidance, 16, 18
Water Action Hub, 13
water reporting, 7
water savings, 7
Water Stewardship Toolbox, 10–11
Global Compact, 2–4, 4, 16, 35
Business for Peace, 4
Caring for Climate, 4
Leaders' Summit 2007, 4
Local Networks, 4, 5, 7, 16
Management Model, 29
member commitment, 4
principles, 3
Women's Empowerment Principles, 4
Guiding Principles for Business and Human Rights, 28–29, 30–31
HRWS, recognition of, 29
human right to water, 130
Millennium Development Goals (MDGs), 3, 4, 16, 22
Protect, Respect, and Remedy Framework, 28–29
Sustainable Development Goal 6, 16, 17, 22–23, 25
Sustainable Development Goals (SDGs), 4, 16, 22, 36
World Water Development Report, 55
United States Bureau of Reclamation (USBR), 95
United States Department of Commerce:
historical water use data, 41
United States Energy Information Administration (EIA), 57
United States Environmental Protection Agency (EPA):
Energy Star, 47
storm water, 115
WaterSense, 47
United States Geological Survey (USGS):
agriculture, 42–43
changes in water use categories, 40
consumptive use, 41
data collection, 39–42, 52
economic productivity of water, 42, 44
freshwater, 42–43
irrigation, 49, 51, 52
M&I, 41, 42–43, 45, 47, 48, 52
national water use data, 39–53
per capita, 42
residential, 48–49, 50
saline water, 42
thermoelectric, 44, 45, 46
total, 42–43
United States National Intelligence Council (NIC), 144
United States Water Resources Council, 41
United States:
water conflict, 144
University of California at Santa Barbara (UCSB), 91
USBR:
See United States Bureau of Reclamation (USBR).
USGS:
See United States Geological Survey (USGS).
Uzbekistan:
water trading, 93

V
Vatican:
Laudato Si', 129–30
Pontifical Academy of Sciences, 129
The Human Right to Water, 2017 meeting, 129, 131–32

W
WASH:
See Water, sanitation, and hygiene (WASH).
Water Conflict Chronology, xiv, 141–44
Water conflict:
Ancient Jerusalem, 142
events per year, 141, 143
Hispaniola, 144
Honduras, 143
India, 141, 143, 144
Iraq, 141, 143
ISIS, 143
Mesopotamia, 143
Russia, 144
South Africa, 143
Syria, 141, 143
Ukraine, 144
United States
Dakota Access Pipeline, 144
Malheur National Wildlife Refuge, 144
New York, East River, 142
Yemen, 141, 143
See also Water Conflict Chronology.
Water efficiency measures:
cost analysis, 114–15
cost savings, 122, 123, 126
costs, 127
data limitations, 115
nonresidential, 124–26
landscaping, 125
medical sterilizers, 125
restaurant equipment, 124–26
toilets, 125
residential, 123–24
clothes washers, 124
landscaping, 124
showerheads, 123–24
toilets, 124
water loss control, 126
Water Footprint Network, 9
Water fountains:
See Public drinking fountains.
Water quality, 6, 22, 130
Water recycling and reuse:
agriculture, 118
California, 118
costs, nonpotable, 118–19, 127
costs, potable, 119, 127
definition, 117
Israel, 118
Namibia, 118
Singapore, 118
Water resources:
business dependence on, 24
by watershed, 12
footprint assessment, 8, 10
freshwater, 6
governmental protection, 28
human right to, 25–26
peak water, 22, 113, 129
population growth, 2, 4, 39, 42, 45, 48, 49, 53, 144
prioritization of use, 27
public trusteeship, 26
reservoirs, 58
risk assessment, 8
storm water, 115–17
sustainability certification, 11
sustainable management, 1, 4, 14, 16–17, 22, 26
value to commercial activity, 24–25
See also Freshwater; Water stewardship.
Water risk, 8–10
See also Corporations: water-related business risk.

Index

Water scarcity, 6, 7, 9, 16, 53
Water stewardship:
    addressing cumulative impacts, 33
    addressing water stress, 6, 22
    corporate disclosure, 14
    definition of, 6
    ensuring integrity, 14–15
    managing water-related impacts, 11–14
    public–private sector partnerships, 14
    specific activities, 6
    supply chain, 11
        audit fatigue, 10
        codes of conduct, 11
    tools and guidance, 10–11
    See also Corporations: corporate water stewardship; United Nations: CEO Water Mandate; corporate water stewardship.
Water stress, 6, 9, 29, 39, 52, 87, 144
Water supply projects:
    cost analysis, 114
Water trading:
    agriculture, 92, 108
    Alicante Basin, 92
    allocation trades, 90–91
    Arizona, 92, 94, 95
    Arizona Water Bank, 90
    Australia, 88, 90–91, 94, 97, 103, 105, 107
        buyback program, 99
    BPA, 98
    California, 91–92, 93, 95, 96–97, 98, 100, 107
        Drought Water Bank, 90, 94
        Imperial Valley, 101
        Los Angeles, 99, 100–101
        Owens Valley, 99, 100–101, 103
        water purchases, 99
    CBWPT, 98
    Chile, 92, 94, 100, 102, 103, 107
        Limarí Basin, 92, 94, 100
        Limarí Valley, 103
    China, 89, 94
    Colorado, 94, 98, 103, 106, 107
    Colorado River, 92, 104
    Colorado Super Ditch, 90
    criticisms, 90
    CVWD, 104
    CWT, 98
    economic performance, 95–97
    enabling conditions, 104–6
    England, 93
    entitlement trades, 90–91
    environmental peformance, 98–100
    gender issues, 102, 108
    Idaho, 94
    IID, 96, 101
    India, 93, 99
    instream flows, 98
    Kazakhstan, 93
    Kyrgyzstan, 93
    limiting conditions, 106–7
    M&I, 92, 97
    market-based mechanisms, 87, 88
    Mexico, 92, 94
    Murray-Darling Basin, 91, 96, 97, 105, 107
    MWD, 104
    necessary conditions, 103–4
    Nepal, 103
    Nevada, 94, 95
    New Mexico, 102, 105
    Oman, 94
    Oregon, 94
    Pakistan, 93
    practice of, 90–95
    Sacramento–San Joaquin watersheds, 93
    San Luis Valley, 101
    SDCWA, 96
    social/equity performance, 100–103
        gender issues, 102
    South Africa, 93, 102–3
    Spain, 92, 94, 100
    SWP, 104
    terminology, 89
    Texas, 94
    theory of, 88
    Uzbekistan, 93
    Water and Choice in the Colorado Basin, 88
    water banks, 88–90, 94–95, 105
    water rights priorities, 105
    Wyoming, 92
Water use:
    agriculture, 42–43, 87
        California, 74
    blue water, 55, 57, 58, 60, 61
    consumptive use, 41
    economic productivity, increase in, 42, 44, 49
    gray water, 57
    green water, 55, 57, 58, 60, 61
    instream use, 41
    irrigation, 49, 51, 52
    M&I, 41, 42–43, 45, 47, 52
    nonresidential
        water efficiency, 125
    per capita, by state, 50
    per capita, decline in, 42, 43, 45, 47, 48–49, 52
    residential, 48–49, 50
        water efficiency, 122, 123
    thermoelectric power, 43–45, 46, 49
        water type, 44
    total, decline in, 39, 42–43, 49
    USGS data, 39–53
Water, sanitation, and hygiene (WASH), 6, 11, 35, 36
Water–energy nexus, 55, 61, 64
WaterSense:
    See United States Environmental Protection Agency (EPA): WaterSense.
Watersheds:
    evaluating water use, 12
White River, 98
World Economic Forum:
    Davos, 1999, 3
World Resources Institute (WRI):
    Aqueduct Water Risk Atlas, 12
    developing corporate water tools, 9
World Trade Organization (WTO):
    Seattle protests, 2
World Water Day, 55
Worldwide Fund for Nature (WWF):
    developing corporate water tools, 9
    Water Risk Filter, 12
WRI:
    See World Resources Institute (WRI).
WTO:
    See World Trade Organization (WTO).
WWF:
    See Worldwide Fund for Nature.
Wyoming:
    water trading, 92
Y
Yemen:
    water conflict, 141, 143

Comprehensive Index
The World's Water, Volumes 1-8

KEY (book volume in boldface numerals)
  1: The World's Water 1998-1999: The Biennial Report on Freshwater Resources
  2: The World's Water 2000-2001: The Biennial Report on Freshwater Resources
  3: The World's Water 2002-2003: The Biennial Report on Freshwater Resources
  4: The World's Water 2004-2005: The Biennial Report on Freshwater Resources
  5: The World's Water 2006-2007: The Biennial Report on Freshwater Resources
  6: The World's Water 2008-2009: The Biennial Report on Freshwater Resources
  7: The World's Water Volume 7: The Biennial Report on Freshwater Resources
  8: The World's Water Volume 8: The Biennial Report on Freshwater Resources

A
ABB, **1:**85, **7:**133
Abi-Eshuh, **1:**69, **5:**5
Abou Ali ibn Sina, **1:**51
Abu-Zeid, Mahmoud, **1:**174, **4:**193
Acceleration, measuring, **2:**306, **3:**324, **4:**331, **7:**347
Access to water. *See* Conflict/cooperation concerning freshwater; Drinking water, access; Environmental flow; Human right to water; Renewable freshwater supply; Sanitation services; Stocks and flows of freshwater; Withdrawals, water
Acidification:
  acid rain, **7:**87
  and fossil-fuel mining/processing, **7:**73–74, 86, 87
  mine drainage, **7:**52, 65
  overview, **7:**47
Adams, Dennis, **1:**196
Adaptation to climate change. *See* Climate change, adaptation
Adaptive capacity, **4:**236, **8:**5, 13
Adaptive management and environmental flows, **5:**45
Adriatic Sea, **3:**47
Afghanistan, **8:**163
Africa:
  aquifers, transboundary, **7:**3
  bottled water, **4:**288, 289, 291, **5:**281, 283
  cholera, **1:**57, 59, 61–63, 266, 269, **3:**2, **5:**289, 290
  climate change, **1:**148
  conflict/cooperation concerning freshwater, **1:**119–24, **5:**7, 9
  conflicts concerning freshwater, 2011-2012, **8:**167–68
  costs of poor water quality, **7:**63
  dams, **3:**292, **5:**151, **6:**274–78
  desalination, **2:**94, 97
  dracunculiasis, **1:**52–55, 272, **3:**274, **5:**295, 296
  drinking water, **1:**252, 262, **3:**254–55, **5:**240, 241, **6:**65
    access to, **2:**217, **6:**58, 214–15, **7:**24, 38, 41, 233–35
    progress on access to, by region, **7:**253, **8:**270
  droughts, **5:**93, 100
  economic development derailed, **1:**42
  environmental flow, **5:**32
  fog collection as a source of water, **2:**175
  Global Water Partnership, **1:**169
  groundwater, **4:**85–86
  human needs, basic, **1:**47
  human right to water, **4:**211
  hydroelectric production, **1:**71, 277–78
  irrigation, **1:**298–99, **2:**80, 85, 256–57, 265, **3:**289, **4:**296, **5:**299, **6:**324–26, 330–31

  Millennium Development Goals, **4:**7
  mortality rate
    childhood, **6:**279
    under-5, **6:**279, **7:**259–63
  Northern
    air quality, satisfaction, 2012, **8:**391
    climate change and drought, **8:**149
    conflicts concerning freshwater, 2011-2012, **8:**163–64
    drinking water, progress on access to, **7:**253, **8:**270
    sanitation, progress on access to, **7:**256, **8:**273
    water quality, satisfaction, 2012, **8:**391
    water quality, satisfaction by country, **7:**291
  population data/issues, **1:**247, **2:**214
  reclaimed water, **1:**28, **2:**139
  renewable freshwater supply, **1:**237–38, **2:**199–200, 217, **3:**239–40, **4:**263–64, **5:**223, 224
    2011 update, **7:**217–18
    2013 update, **8:**223–24
  reservoirs, **2:**270, 272, 274
  river basins in, **6:**289–96
  rivers, transboundary, **7:**3, **8:**164
  salinization, **2:**268
  sanitation services, **1:**257, 264, **3:**263–65, 271, **5:**249–50, 259, **6:**67
    access to, **7:**243–45
    progress on access to, **7:**246, **8:**273
  sub-Saharan, **6:**65–66, 76–77
    air quality, satisfaction, 2012, **8:**391
    drinking water, access, **7:**24
    drinking water, progress on access to, **7:**253, **8:**270
    sanitation, progress on access to, **7:**256, **8:**273
    schistosomiasis, **7:**58
    water quality, satisfaction 2012, **8:**391
    water quality, satisfaction by country, **7:**290 2012, **8:**387–88
  threatened/at risk species, **1:**292–93
  well-being, measuring water scarcity and, **3:**96
  withdrawals, water, **1:**242, **2:**205–7, **3:**245–47, **4:**269–70, **5:**230–31, **6:**204
    by country and sector, **7:**223–25, **8:**229–31
African Development Bank (AfDB), **1:**95–96, 173, **3:**162–63
Agreements, international:
  amendment and review process, **7:**9
  General Agreement on Tariffs and Trade (GATT), **3:**47–52
  general principles, **7:**4–5

197

Agreements, international (*continued*)
    Great Lakes–St. Lawrence River Basin, **7:**165–69
    joint institutions, role of, **7:**9–10
    North American Free Trade Agreement (NAFTA), **3:**47–48, 51–54, **8:**130
    state/provincial rights within, **7:**167–68, 169
    transboundary waters, **7:**2–7, 9, 165–69, **8:**14–15
    See also Law/legal instruments/regulatory bodies; United Nations
Agriculture:
    best management practices, **8:**53–54
    California, **4:**89, **8:**89–90
    carbon footprint, **8:**85
    cereal production, **2:**64, **4:**299–301, **7:**109–10
    conflict/cooperation concerning freshwater, **1:**111
    cropping intensity, **2:**76
    crops as a military target, **8:**165
    crop yields and food needs for current/future populations, **2:**74–76
    data problems, **3:**93
    droughts, **5:**92, 94, 98, 103
        impacts in Australia, **7:**102–5
        impacts in Syria, **8:**147–51
        management in Australia, **7:**97, 107–10
    floods, **5:**109, **8:**128
    groundwater, **4:**83, 87, 88–90
    harvesting technology, **2:**77
    irrigation
        arable land, **6:**324–28
        area equipped for, actually irrigated, by country, **6:**329–34, **8:**310–16
        basin, **2:**82
        border, **2:**82
        business/industry, water risks, **5:**162, **7:**23, 26
        and climate change, **4:**174–75, **7:**16
        conflict/cooperation concerning freshwater, **1:**110, **7:**110–11
        conflicts concerning freshwater, **8:**148, 164
        by continent, **2:**264–65
        by country and region, **1:**297–301, **2:**255–63, **4:**295–96, **5:**298–300, **6:**324–28, 329–34
        by crop type, **2:**78–80
        developing countries, **1:**24, **3:**290, **5:**301–2
        drip, **1:**23–24, **2:**82, 84, **7:**111, **8:**50
        Edwards Aquifer, **3:**74
        furrow, **2:**82, 86
        government involvement, **1:**8
        hard path for meeting water-related needs, **3:**2
        how much water is needed, **2:**81–87
        infrastructure for, **1:**6, **8:**50–53
        North American Water and Power Alliance, **1:**74, **8:**124–31
        projections, review of global water resources, **2:**45
        reclaimed water, **2:**139, 142, 145–46, **7:**54, **8:**74
        Southeastern Anatolia Project, **3:**182
        sprinkler, **2:**82, **8:**50
        surface, **2:**82, 84, 84
        total irrigated areas, **4:**297–98
        water-energy-food nexus, **8:**5–6, 85
        water quality, **2:**87
        water rights, **7:**111, 112, 113
        water-use efficiency, **3:**4, 19–20, **4:**107, **7:**109–10, 146, **8:**50–54
    land availability/quality, **2:**70–71, 73–74
        Africa, **8:**167–68
        impact of the North American Water and Power Alliance, proposed, **8:**127, 128
        Syria, **8:**148–49
    pricing, water, **1:**117, **7:**111–12
    projections, review of global water resources, **2:**45–46
    reclaimed water, **1:**28, 29
    by region, **3:**289
    runoff (*See* Runoff, agricultural)
    subsidies, **1:**24–25, 117, **7:**108, 109, 111, 152
    sustainable, **1:**187–88, **8:**10
    water assessments, **2:**46–49, 54–58
    water footprint, **8:**86, 87 (*See also* Water footprint, per-capita, of national consumption, by sector and country)
    and water quality, **7:**49–50, 62, 64, **8:**52 (*See also* Runoff, agricultural)
    well-being, measuring water scarcity and, **3:**99
    World Water Forum (2003), **4:**203
    See also Food needs for current/future populations
Aguas Argentinas, **3:**78, **4:**47
Aguas de Barcelona, **3:**63
Aguas del Aconquija, **3:**70
AIDS, **6:**58, **7:**259–63
Air quality:
    certified emission credits, **6:**51
    dust storms, **7:**105–6
    fossil-fuel combustion, **7:**77, 80, 84, 86
    methane contamination, **8:**69
    satisfaction, by region, 2012, **8:**389–91
AkzoNobel, **6:**23
Alaska, Sitka, **8:**135–36
Alaskan water shipments, proposed, **8:**133, 135–36
Albania, **1:**71, **3:**47, **7:**309
Albright, Madeleine K., **1:**106
Algae, **5:**79, **6:**83, 95
Algeria, **7:**75, 309
Alkalinization, **7:**52
Alliance for Water Efficiency, **8:**39
Alliance for Water Stewardship, **7:**26–27
Al-Qaida, **5:**15
Alstom, **7:**133
Altamonte Springs (FL), **2:**146
American Association for the Advancement of Science, **1:**149, **4:**176
American Convention on Human Rights (1969), **2:**4, 8
American Fisheries Society, **2:**113, 133
American Geophysical Union (AGU), **1:**197–98
American Rivers and Trout Unlimited (AR & TU), **2:**118, 123
American Society of Civil Engineers, **5:**24
American Water/Pridesa, **5:**62
American Water Works Association (AWWA), **2:**41, **3:**59–61, **4:**176, **5:**24
Americas:
    air quality, satisfaction, 2012, **8:**391
    cholera, **8:**359, 373
    water quality, satisfaction, 2012, **8:**391
    See also Caribbean; Central America; North America; South America
Amoebiasis, **1:**48
Amount of water. *See* Stocks and flows of freshwater
Amphibians, **1:**291–96
    effects of endocrine disruptors, **7:**49
    extinct or extinct in the wild species, **7:**56–57, 296–97
Anatolia region. *See* Southeastern Anatolia Project
Angola, **1:**119, 121, **2:**175, **3:**49, **5:**9
    dams with Chinese financiers/developers/builders, **7:**309
Anheuser-Busch, **5:**150, **6:**24
Ankara University, **3:**183
Antiochus I, **3:**184–85
Apartheid, **1:**158–59
Appleton, Albert, **4:**52–53
Aquaculture, **2:**79
Aquafina, **4:**21
Aquarius Water Trading and Transportation, Ltd., **1:**201–2, 204, **8:**135
Aquifers, **6:**10

aquifer storage and recovery (ASR), **8:**49
climate change management issues, **7:**2–3
Coca-Cola recharging, in India, **8:**21–22, 23
contamination by fossil-fuel production, **7:**51, 76
percent of global freshwater in, **7:**3
transboundary, **7:**3
Aquifers, specific:
Disi, **8:**157
Edwards, **3:**74–75, **4:**60–61
Ogallala, **3:**50, **8:**128, 141
Aral Sea, **1:**24, **3:**3, 39–41, 77, **7:**52, 131
Archaeological sites, **3:**183–89, 191. *See also*
Southeastern Anatolia Project
Area, measuring, **2:**302, **3:**320, **4:**327, **7:**345
Argentina, **3:**13, 60, 70, **4:**40, 47
arsenic in groundwater, **7:**59
Arizona, **3:**20, 138, **8:**28
Army Corps of Engineers, U.S. (ACoE), **1:**7–8, **2:**132, **3:**137, **7:**305
Reber Plan, proposed, **8:**132–33
response to drought, Missouri River and Mississippi River policy, **8:**143–44
Arrowhead bottled water, **4:**21, **7:**161
Arsenic, **2:**165–73, **3:**278–79, **4:**87, **5:**20, **6:**61, 81
in fly ash, **7:**84
from fossil-fuel production, **7:**76, 79
health effects, **7:**59
Artemis Society, **3:**218
Ascariasis, **7:**272
Ascension Island, **2:**175
Asia:
agriculture, **4:**88–90 (*See also* Asia, irrigation)
air quality, satisfaction, 2012, **8:**391
aquifers, transboundary, **7:**3
bottled water, **4:**18, 41, 291, **5:**163, 281, 283
Central (*See* Caucasus and Central Asia)
cholera, **1:**56, 58, 61, 266, 269–70, **3:**2, **5:**289, 290
climate change, **1:**147
conflict/cooperation concerning freshwater, **1:**111
dams, **3:**294–95
dracunculiasis, **1:**272–73, **3:**274–75, **5:**295, 296
drinking water, **1:**253–54, 262, **3:**257–59, **5:**243–45, **6:**66
access to, **7:**24, 237–38
progress on access to, by region, **7:**253, **8:**270
Eastern
progress on access to drinking water, **7:**24, 253, **8:**270
progress on access to sanitation, **8:**273
environmental flow, **5:**32–33
floods, **5:**108
food needs for current/future populations, **2:**75, 79
Global Water Partnership, **1:**169
groundwater, **4:**84, 88–90, 96
human needs, basic, **1:**47
hydroelectric production, **1:**71, 278–79
irrigation, **1:**299–300, **2:**80, 86, 259–61, 265, **3:**289, **4:**296, **5:**299, **6:**326–27, 331–32
population data/issues, **1:**248–49, **2:**214
pricing, water, **1:**24, **3:**69
privatization, **3:**61
renewable freshwater supply, **1:**238–39, **2:**217, **3:**240–41, **4:**264–65, **5:**225, 226
2011 update, **7:**218–19
2013 update, **8:**224–25
reservoirs, **2:**270, 272, 274
river basins, **2:**30, **6:**289, 296–301
rivers, transboundary, **7:**3
runoff, **2:**23
salinization, **2:**268
sanitation services, **1:**258–59, 264, **3:**266–68, 271, **5:**252–54, 258–61
access to, **7:**247–48
progress on access to, **7:**256, **8:**273

South-Eastern, **7:**262–63
progress on access to drinking water, **7:**24, 253, **8:**270
progress on access to sanitation, **7:**256, **8:**273
Southern
progress on access to drinking water, **7:**24, 253, **8:**270
progress on access to sanitation, **7:**256, **8:**273
water conflicts, **8:**162–63
supply systems, ancient, **1:**40
threatened/at risk species, **1:**293–94
water access, **2:**24, 217
water quality
satisfaction, 2012, **8:**391
satisfaction by country, **7:**290
well-being, measuring water scarcity and, **3:**96
Western
progress on access to drinking water, **7:**24, 253, **8:**270
progress on access to sanitation, **7:**256, **8:**273
water conflicts, 2011-2012, 162–63
withdrawals, water, **1:**243–44, **2:**208–9, **3:**248–49, **4:**272–73, **5:**233–34
by country and sector, **7:**226–28, **8:**232–33
Asian Development Bank (ADB), **1:**17, 173, **3:**118, 163, 169, **4:**7, **5:**125
reaction to World Commission on Dams report, **7:**138–39
Asmal, Kader, **1:**160, **3:**169
Assessments:
AQUASTAT database, **4:**81–82, **7:**215, 221, **8:**221, 310
Colorado River Severe Sustained Drought study (CRSSD), **4:**166
Comprehensive Assessment of the Freshwater Resources of the World, **2:**10, **3:**90
Dow Jones Indexes, **3:**167
Global Burden of Disease assessment, **7:**270
The High Efficiency Laundry Metering and Marketing Analysis project (THELMA), **4:**115
Human Development Index, **2:**165
Human Development Report, **4:**7, **6:**74, **7:**61
Human Poverty Index, **3:**87, 89, 90, 109–11, **5:**125
hydrologic cycle and accurate quantifications, **4:**92–96
International Journal on Hydropower and Dams, **1:**70
International Rice Research Institute (IRRI), **2:**75, 76
intl. river basins, **2:**27–35
measurements, **2:**300–309, **3:**318–27, **4:**325–34
Millennium Ecosystem Assessment, **7:**63
National Assessment on the Potential Consequences of Climate Variability and Change, **4:**176
Palmer Drought Severity Index, **5:**93
Standard Industrial Classification (SIC), **4:**132
Standard Precipitation Index, **5:**93
Stockholm Water Symposiums (1995/1997), **1:**165, 170
Third Assessment Report, **3:**121–23
water footprint sustainability assessment, **8:**86
Water-Global Assessment and Prognosis (WaterGAP), **2:**56
Water in Crisis: A Guide to the World's Fresh Water Resources, **2:**300, **3:**318
World Health Reports, **4:**7
World Resources Reports, **3:**88
World Water Development Report, **8:**6
See also Data issues/problems; Groundwater, monitoring/management problems; Projections, review of global water resources; Stocks and flows of freshwater; Well-being, measuring water scarcity and

Association of Southeast Asian Nations (ASEAN), **1:**169
Assyrians, **3:**184
Atlanta (GA), **3:**62, **4:**46–47
    description, **6:**103
    per-capita water demand, **6:**104–6
    population growth, **6:**103
    precipitation, **6:**104
    temperature, **6:**104
    wastewater rate structure, **6:**118–19
    water conservation, **6:**108–9, 110–12
    water rate structures, **6:**115–16
    water-use efficiency, **6:**108–9
Atlantic Salmon Federation, **2:**123
Atmosphere, harvesting water from the, **2:**175–81. *See also* Outer space, search for water in
Austin (TX), **1:**22
Australia:
    agriculture
        impact of drought, **7:**102–5
        irrigation, **2:**85, **7:**97
        production, 1960-2009, **7:**103
    bottled water, **4:**26
    conflict/cooperation concerning freshwater, **5:**10
    desalination, **5:**69, **7:**114–15, **8:**96, 97, 98, 102, 103, 118
    drought
        historical background, **7:**98, 99
        impacts, **7:**102–6
        management, **7:**106–21, 146–47
        overview, **7:**97, 98–99, 101, 121
    environmental flow, **5:**33, 35, 42
    fossil-fuel production, **7:**75
    globalization and intl. trade of water, **3:**45, 46
    legislation and policy
        Millennium Development Goals, **4:**7
        Water Efficiency Labelling and Standards Act, **7:**28, 118–19
        water policy reform, **7:**146–47
    privatization, **3:**60, 61
    reservoirs, **2:**270, 272, 274
    terrorism, **5:**16
    water resources, **2:**24, 217, **7:**98–99
    water-use efficiency, **8:**38
Austria, **3:**47
Availability, water, **2:**24–27, 215–17. *See also* Conflict/cooperation concerning freshwater; Drinking water, access; Environmental flow; Human right to water; Renewable freshwater supply; Sanitation services; Stocks and flows of freshwater; Withdrawals, water
Azov Sea, **1:**77

B
Babbitt, Bruce, **2:**124–25, 128
Babylon, ancient, **1:**109, 110
Bag technology, water, **1:**200–205, **8:**135
Baker, James, **1:**106
Bakersfield (CA), **1:**29
Balfour Beatty, **3:**166
Balkan Endemic Neuropathy (BEN), **7:**89
Bangladesh:
    agriculture, **4:**88
    Arsenic Mitigation/Water Supply Project, **2:**172
    conflict/cooperation concerning freshwater, **1:**107, 109, 118–19, 206–9
    dams with Chinese financiers/developers/builders, **7:**309
    drinking water, **3:**2
    floods, **5:**106
    groundwater, **4:**88
        arsenic in, **2:**165–73, **6:**61, **7:**55, 59, **8:**3
    Rural Advancement Committee, **2:**168

Banks, Harvey, **1:**9
Barlow, Nadine, **3:**215
Basic water requirement (BWR), **1:**44–46, **2:**10–13, **3:**101–3
Bass, **2:**123
Bath Iron Works, **2:**124
Bayer, **6:**28
Beard, Dan, **2:**129
Bechtel, **3:**63, 70
Belgium, **5:**10
Belize, **7:**309
Benin, **1:**55, **7:**309, 336
Benzene, **6:**83, **8:**72
Best available technology (BAT), **3:**18, 22–23, **4:**104, **5:**157, **6:**27
Best management practices (BMP), **8:**53–54
Best practicable technology (BPT), **3:**18, **4:**104
Beverage Marketing Corporation (BMC), **4:**18, **7:**157
BHIP Billiton, **6:**23
Biodiversity, **7:**56–57
Biofuel, **7:**25, 74, **8:**6, 167
Biological oxygen demand (BOD), **6:**28, **7:**278–81
Biologic attacks, vulnerability to, **5:**16–22
Bioretention and bioinfiltration, **8:**42
Bioswales, **8:**42
Birds:
    effects of drought, **7:**101
    effects of endocrine disruptors, **7:**49
    effects of fossil-fuel extraction, **7:**87, 91
    extinct or extinct in the wild species, **7:**297–98
    Lesotho Highlands project, **1:**98
    Siberian crane, **1:**90
    Yuma clapper rail, **3:**142
Birth, premature, **7:**259–63
Birth asphyxia, **7:**259–63
Birth defects, **7:**58, 59, 259–63
Bivalves. *See* Invertebrates, clams; Invertebrates, mussels
Black Sea, **1:**77
"Blue" water, in water footprint concept, **8:**86, 89–90, 323, 332, 341, 350
BMW, **6:**27
Bolivia, **3:**68, 69–72, **4:**54, 56–57
    conflicts concerning freshwater, 2011-2012, **8:**165–66
Bonneville Power Administration, **1:**69, **8:**129
Books, **2:**129, **5:**14
Boron, **5:**75, **7:**76
Bosnia, **1:**71
Botswana, **1:**119, 122–24, **7:**309
Bottled water:
    bottle and packaging, **4:**293–94, **7:**158–59
    brands, leading, **4:**21–22, **7:**161–62
    business/industry
        company assessments, **6:**23
        standards/rules, **4:**34–35
        water risks that face, **5:**163
    consumption
        by country, **4:**284–86, **5:**276–79, 284–86
        increase in, **7:**157
        per-capita by country, **7:**339–40
        per-capita by region, **4:**290–91, **5:**171, 282–83
        by region, **4:**287–88, **5:**169–70, 280–81
        share by region, **4:**289
        U.S., **4:**288–91, **5:**170, 281, 283, **7:**157, 158
    developing countries, **3:**44, 45
    energy considerations
        bottle manufacture, **7:**158–59, 164
        to clean, fill, seal, and label bottles, **7:**160–61, 164
        cooling process, **7:**162–63, 164
        equivalent barrels of oil used, **7:**164
        transport, **7:**161–62, 163, 164
        water processing, **7:**159–60, 164

environmental issues, **4:**41
flavor and taste, **4:**23–24
history and trends, **4:**18–22, **7:**157
hydrogeological assessments of sites for, **6:**23
intl. standards, **4:**35–36
labeling, **4:**28–31, **7:**159, 160–61, 164
overview, **4:**xvi, 17, **7:**88
price and cost, **4:**22–23
recalls, **4:**37–40, **5:**171–74
sales
    global, **3:**43, **5:**169
    and imports in, U.S., **3:**44, 343, **4:**292, **7:**157, 158
    to the poor, **4:**40–41
standards/regulations, **4:**26–27, 34–37, **5:**171, 174
    vs. tap water, **4:**36–37
summary/conclusions, **4:**41
U.S. federal regulations
    adulteration, food, **4:**32–33
    enforcement/regulatory action, **4:**34
    good manufacturing practices, **4:**32
    identity standards, **4:**27–31
    sampling/testing/FDA inspections, **4:**33–34
    water quality, **4:**31–32
U.S., consumption, **4:**288–91, **5:**170, 281, 283, **7:**157, 158
U.S., sales and imports in, **3:**44, 343, **4:**292, **7:**157, 158
water quality, **4:**17, 25–26, 31–32, 37–40
    treatment processes, **7:**159–60, 161
water sources for, **7:**159
Boundaries:
    managing across, **3:**133–34, **7:**1
    transboundary waters
        agreements, limitations and recommendations, **8:**14–15
        and climate change, **7:**1–20
        Snake Valley basin, **8:**139
    See also International river basins
Brazil:
    bottled water, **4:**40, **5:**170
    business/industry, water risks, **5:**149, 151–52, **7:**26, 89
    cholera, **1:**59
    conflict/cooperation concerning freshwater, **1:**107
    conflicts concerning freshwater, 2011-2012, **8:**165
    dams, **1:**16, **5:**134
    drought, **7:**26, **8:**22–23, 165
    energy production, **7:**26, **8:**22–23
        hydroelectric, **1:**71, **7:**129
    environmental concerns, top, **7:**287
    environmental flow, **5:**34
    human needs, basic, **1:**46
    monitoring and privatization, **3:**76–77
    privatization, **3:**76–78
    runoff, **2:**23
    sanitation services, **3:**6
    Three Gorges Dam, **1:**89
Brine, as a contaminant, **2:**107, **5:**77–80, **8:**156
British Columbia Hydro International (BC Hydro), **1:**85, 88
British Geological Survey (BGS), **2:**169
British Medical Association, **4:**63–64
Brownfield redevelopment, **8:**46
Bruce Banks Sails, **1:**202
Bruvold, William, **4:**24
Buildings:
    construction site runoff, volume, **8:**76
    green roofs, **8:**42
Burkina Faso, **1:**55, **4:**211, **8:**167
Burma (Myanmar):
    dams with Chinese financiers/developers/builders, **7:**130, 132, 133, 310–15
    Mekong River Basin, **7:**14
Burns, William, **3:**xiv

Burundi, **1:**62, **7:**11, 315–16
Business for Social Responsibility, **5:**156
Business/industry, water risks:
    assessment of, **6:**23, **7:**25–34
    China, **5:**147, 149, 160, 165
    climate change, **5:**152
    costs of poor water quality on production, **7:**64
    developing countries, **5:**150
    energy and water links, **5:**150, 151–52
    India, **5:**146, 147, 165, **8:**21–22
    management
        best available technology, **5:**157
        best management practices (BMP), **8:**53–54
        "beyond the fence line," **8:**19–20, 28
        companies, review of specific, **5:**153–55, **7:**26–27
        continuous improvement, commit to, **5:**158
        efficiency, improving water-use, **5:**161–62, **8:**20, 28
        five motivations, **7:**23
        global trends that affect, **7:**24–25
        hydrological/social/economic/political factors, **5:**153, 156, **7:**24–30
        partnerships, form strategic, **5:**158
        performance, measure and report, **5:**157–58
        public disclosure (*See* Corporate reporting)
        risks factored into decisions, **5:**157, **6:**27–28, **8:**31
        stakeholder issues, **5:**145, **6:**24, **7:**26–27, 40–41, **8:**20, 26–27
        strategies for policies/goals/targets, **5:**156–57, **7:**34–42, **8:**20–21, 31
    supply chain
        availability/reliability of, **5:**146, **7:**26
        company reporting of, **6:**24, **7:**27
        overview, **5:**145–46, **7:**28–29
        performance evaluation, **7:**35–37
        privatization, public opposition to, **5:**152–53
        public's role in water policy, **5:**150–51
        summary/conclusions, **5:**163–65
        vulnerability, **5:**149, **7:**26, **8:**23
        water quality, **5:**146–49, **8:**23
        working collaboratively with, **5:**156, **6:**24
    sustainable, **7:**23, **8:**19–22, 30, 31
    water accounting tools, **7:**32–34, 146
    water industry
        bottled water (*See* Bottled water, business/industry)
        desalination, **5:**161 (*See also* Desalination, plants)
        disinfection/purification of drinking water, **5:**159
        distribution, infrastructure for, **5:**160–61
        high-quality water, processes requiring, **5:**147–48, 160
        irrigation, **5:**162
        overview, **5:**158–59
        revenue and growth in U.S., **5:**303–4
        utilities, water, **5:**162–63, **7:**28
        wastewater treatment, **5:**153, 159–60
    organic contaminants by industry, **7:**279–81
    overview, **7:**23, 25–34, 50–51, **8:**19–31
    public perception, **7:**26–27
    water footprint, **7:**30–34, **8:**84, 87 (*See also* Water footprint, per capita, of national consumption, by sector and country)
Business sector, **3:**22–24, 169–70, **8:**27. *See also* Bottled water, business/industry; Companies; Corporate reporting; Privatization; Water conservation, California commercial/industrial water use
Bussi, Antonio, **3:**70
Byzantines, **3:**184

C
Cabot Oil & Gas Corp., **8:**70
Cadbury, **7:**37
Calgon, **3:**61
California:
  agriculture, **4:**89, **8:**89–90
  Bakersfield, **1:**29
  bottled water, **4:**24
  business/industry, water risks, **5:**158, **6:**162
  California Central Valley Project, **4:**173, **8:**131
  California Regional Assessment Group, **4:**176
  California State Water Project, **8:**131
  Carlsbad, **8:**93, 95, 98, 112–19
  climate change (*See* Climate change, California)
  conflict/cooperation concerning freshwater, **1:**109
  dams, **1:**75, **2:**120–23
  desalination, **1:**30, 32, **5:**51, 52, 63–69, 71, 73, 74
    Carlsbad plant case study, **8:**93, 95, 98, 112–19
    overview, **8:**93
    proposed and recently constructed plants, **8:**98
    Sand City plant, **8:**93
  droughts, **8:**89, 102
  East Bay Municipal Utilities District, **1:**29
  economics of water projects, **1:**16
  environmental justice, **5:**122–23
  floods, **5:**111
  fog collection as a source of water, **2:**175
  food needs for current/future populations, **2:**87
  groundwater, **4:**89
  import of water, **8:**89, 131
  industrial water use, **1:**20–21
  Irvine Ranch Water District, **4:**124–25
  Kesterson National Wildlife Refuge, **7:**47–48
  Los Angeles
    Chinatown political intrigue, **8:**138
    Los Angeles Aqueduct, **8:**131
    sustainable water jobs, **8:**39, 44, 52
  Metropolitan Water District of Southern California, **1:**22
  Monterey County, **2:**151
  North American Water and Power Alliance, **8:**124–31
  Orange County, **2:**152
  Pomona, **2:**138
  privatization, **3:**73
  projections, review of global water resources, **2:**43
  reclaimed water
    agriculture, **1:**29, **2:**142–46
    drinking water, **2:**152
    first state to attempt, **2:**137–38
    groundwater recharge, **2:**151
    health issues, **2:**154–55
    Irvine Ranch Water District, **2:**147
    Kelly Farm marsh, **2:**149
    San Jose/Santa Clara Wastewater Pollution Control Plant, **2:**149–50
    uses of, **2:**141–45
    West Basin Municipal Water District, **2:**148–49
  San Diego County, **8:**99–100
  San Francisco Bay, **3:**77, **4:**169, 183, **5:**73, **8:**132
  San Francisco Bay Project (Reber Plan), **8:**131–34
  Santa Barbara, **5:**63–64, **8:**102–3, 119
  Santa Rosa, **2:**145–46
  soft path for meeting water-related needs, **3:**20–22, 24–25
  subsidies, **1:**24–25
  toilets, energy-efficient, **1:**22, **8:**39
  twentieth-century water-resources development, **1:**9
  Visalia, **1:**29
  water conservation (*See under* Water conservation)
  water footprint, **8:**88–90, 91
  water-use efficiency, **1:**19, **8:**38

  Western Canal Water District, **2:**121
  See also Legislation, California
California-American Water Company (Cal AM), **5:**74
Cambodia, **7:**14, 59, 130, 316–17
Camdessus, Michael, **4:**195–96
Cameroon, **1:**55, **5:**32, **7:**317–18
Campylobacter jejuni, **7:**57
Canada:
  adaptation in, **6:**46
  bottled water, **4:**25, 26, 39, 288, 289, 291, **5:**281, 283
  Canadian International Development Agency, **2:**14
  cholera, **1:**266, 270
  climate change, **1:**147, 148
  conflict/cooperation concerning freshwater, **5:**6, 8
  dams, **1:**75, **3:**293
  data problems, **3:**93
  dracunculiasis, **1:**52
  drinking water, **1:**253, **3:**256, **5:**242
  environmental concerns, top, **7:**287, 288
  environmental flow, **5:**34
  Export Development Corporation, **6:**141
  fog collection as a source of water, **2:**179
  fossil-fuel production
    data, **7:**75
    natural gas, **7:**79
    tar sands, **7:**78, 79, 87, 88, 91–92
  General Agreement on Tariffs and Trade, **3:**50
  Great Lakes Basin, intl. agreements, **7:**165–69, **8:**126
  groundwater, **4:**86
  hydroelectric production, **1:**71, 74, 278, **7:**129, **8:**131
  intl. river basin, **2:**33
  irrigation, **1:**299, **2:**265, **3:**289, **4:**296, **5:**299
  James Bay Project, Quebec, **8:**131
  mortality rate, under-5, **7:**260
  North American Free Trade Agreement, **3:**47–48, 51–54, **8:**130
  North American Water and Power Alliance, **8:**124–31
  population data/issues, **1:**247, **2:**214
  renewable freshwater supply, **1:**238, **2:**200, 217, **3:**240, **4:**264
    response to North America Water and Power Alliance, **8:**129
    2011 update, **7:**218
    2013 update, **8:**224
  reservoirs, **2:**270, 272, 274
  Rocky Mountain Trench, proposed reservoir, **8:**125–26
  runoff, **2:**23
  salinization, **2:**268
  sanitation services, **1:**258, **3:**265, 272, **7:**245
  threatened/at risk species, **1:**293
  transboundary waters, **7:**3, **8:**124–31
  water availability, **2:**217
  water transfer issues, **8:**129–30
  withdrawals, water, **1:**242, **2:**207, **3:**247, **4:**271
    2011 update, **7:**225
    2013 update, **8:**231
  World Water Council, **1:**172
Canary Islands, **3:**46
Cancer, **6:**81, **7:**58, 59
Cap and trade market and environmental flows, **5:**42
Cape Verde Islands, **2:**175
Carbon dioxide, **1:**138, 139, **3:**120, 215, **4:**160, 164
  and natural gas production, **7:**80
Carbon Disclosure Project (CDP), **7:**41, **8:**22, 24
Carbon footprint, **8:**85, 86
Caribbean:
  dams, **3:**293–94
  drinking water, **1:**253, 262, **3:**255–57, **5:**241, 242
    access to, **7:**24, 235–36
    progress on access to, **7:**253, **8:**270

groundwater, **4:**86
hydroelectric production, **1:**278
irrigation, **6:**334
population data/issues, **1:**247–48, **2:**214
sanitation, **1:**258, **3:**265–66, 271, **5:**250, 251, 259
    access to, **7:**245–46
        progress on access to, **7:**256, **8:**273
threatened/at risk species, **1:**293
water quality, satisfaction by country, **7:**290–91
Caribbean National Forest, **5:**34
Carlsbad (CA), **8:**93, 95, 98, 112–17
Caspian Sea, **1:**77
Catley-Carlson, Margaret, **4:**xii–xiv
Caucasus and Central Asia:
    conflicts concerning freshwater, 2011-2012, **8:**166
    progress on access to drinking water, **8:**270
    progress on access to sanitation, **8:**273
Cellatex, **5:**15
Census of Agriculture, **8:**50
Centers for Disease Control and Prevention (CDC), **1:**52, 55, 57, **7:**305
Central African Republic, **7:**318
Central America:
    cholera, **1:**266, 270, 271
    dams, **3:**165, 293–94
    drinking water, **1:**253, **3:**255, 256, **5:**241–42, **7:**235–36
    environmental flow, **5:**34
    groundwater, **4:**86
    hydroelectric production, **1:**278
    irrigation, **1:**299, **2:**265, **4:**296, **6:**328, 334
    mortality rate, under-5, **7:**260–61
    population data, total/urban, **1:**247–48
    renewable freshwater supply, **1:**238, **2:**200, 217, **3:**240, **4:**264, **5:**224
        2011 update, **7:**218
        2013 update, **8:**224
    reservoirs, **2:**270, 272, 274
    rivers and aquifers, transboundary, **7:**3
    salinization, **2:**268
    sanitation services, **1:**258, **3:**265–66, **5:**250–51, **7:**245–46
    threatened/at risk species, **1:**293
    water availability, **2:**217
    withdrawals, water, **1:**242–43, **2:**207, **3:**247, **4:**271, **5:**231–32
        by country and sector, **7:**225, **8:**231
    See also Latin America
Centre for Ecology and Hydrology, **3:**110–11
Centro de Investigaciones Sociales Alternativas, **2:**179
CEO Water Mandate, **7:**34, **8:**9, 24
Cereal production, **2:**64, **4:**299–301
Certified emission credits (CECs), **6:**51
Chad, **1:**55, **7:**264
Chakraborti, Dipankar, **2:**167
Chalecki, Elizabeth L., **3:**xiv
Chanute (KS), **2:**152
Chemical attacks, vulnerability to, **5:**16–22
Chemical oxygen demand, **6:**28
Chiang Kai-shek, **5:**5
Childhood mortality:
    by cause, **6:**279–88, **7:**257–63
    by country, **6:**279–88, **7:**257–63, 264–69
    infant, by country, **7:**264–69
    limitations in data and reporting, **7:**257–58, 264
    under-5, **6:**279–88, **7:**257–63
    from water-related disease, **6:**58, **7:**57–58, 61–62
Children:
    infant brands of bottled water, **4:**28
    responsibility for water collection, **7:**61
    See also Birth *listings*
Chile:
    arsenic in groundwater, **7:**59

    cholera, **1:**59
    environmental concerns, top, **7:**287
    environmental flow, **5:**34, 37
    fog collection as a source of water, **2:**177–78
    General Agreement on Tariffs and Trade, **3:**49
    privatization, **3:**60, 66, 78
    Silala/Siloli River transboundary dispute, **8:**165–66
    subsidies, **4:**57–58
China:
    agriculture, **2:**86, **4:**88, 90
    algae outbreaks, **6:**83
    Beijing, **6:**90
    benzene contamination, **6:**83
    bottled water, **4:**21, 40, **5:**170
    business/industry, water risks, **5:**147, 149, 160, 165
    business/industry water use, **5:**125, **6:**81, **7:**27–28
    canal system, **8:**144
    cancer rates in, **6:**81
    climate change in, **6:**87–88
    dams, **1:**69, 70, 77, 78, 81, **5:**15–16, 133, 134, **6:**91
        construction overseas, **7:**133–35, 308–38
        hydroelectric production, **1:**71, **6:**92, **7:**129, 130
        internal concern over, **7:**135–36
        on intl. rivers, **7:**130–33
        overview, **7:**127–28, 130
        and the World Commission on Dams, **3:**170–71, **7:**128–29, 136–40
        See also under Dams, specific; Three Gorges Dam
    desalination, **6:**93
    diarrhea-related illness in, **6:**85
    diseases, water-related, **6:**85
    drinking water
        shortage of, **6:**86
        standards for, **6:**94
    droughts, **5:**97, **6:**86–87, **7:**131–32
    economic growth in, **6:**79
    economics of water projects, **1:**16, **6:**95–96
    environment
        grassroots efforts, **6:**80
        pollution in, **6:**79, 81–82 (*See also* China, water, pollution of)
        protections for, **6:**94–95
        top concerns, **7:**287, 288
        water-related disasters, **6:**82–83
    environmental flow, **5:**32
    Environmental Impact Assessment law, **6:**96
    floods, **5:**106, **6:**84–87
    food needs for current/future populations, **2:**74
    foreign investment in water markets, **6:**92
    fossil-fuel production, **7:**75, 82, **8:**63
    glaciers in, **6:**87–88
    globalization and intl. trade of water, **3:**46
    Great Wall of China, **6:**86
    groundwater, **2:**87, **3:**2, 50, **4:**79, 82, 83, 88, 90, 96–97, **5:**125, **6:**85–86
        arsenic in, **6:**81, **7:**59
        fluoride in, **6:**81
    Guangdong Province, **6:**85
    human needs, basic, **1:**46
    Jilin Province, **6:**83
    natural gas production, **8:**63
    nongovernmental organizations in, **6:**89–90, 96
    North China Plains, **6:**85–86, 90
    politics, **6:**88–90
    population, **6:**79
    privatization, **3:**59, 60
    protests in, **6:**97
    provinces, **6:**82
    public-private partnerships in, **6:**92–93
    Qinghai-Tibetan Plateau, **6:**88
    rivers, **6:**79, 81–82, 84, 88, 90, **7:**55 (*See also under* Rivers, specific)

China (*continued*)
    sanitation services, **5:**124, 147
    South-to-North Water Transfer Project, **6:**91
    State Environmental Protection Administration (SEPA), **6:**80–81, 94
    wastewater treatment plants in, **6:**92
    water
        availability of, **6:**83–85
        average domestic use of, **1:**46
        basic requirement, **2:**13
        centralized management of, **6:**89
        effects of climate change, **6:**87–88
        efficiency improvements, **6:**93–94
        expanding the supply of, **6:**91–93
        industrial use, **5:**125, **6:**81
        politics affected by, **6:**88–90
        pollution of, **6:**79, 97, **7:**64, **8:**24–25
        pricing of, **1:**25
        public participation efforts, **6:**96–97
        quality of, **6:**80–82, **8:**3
        quantity of, **6:**83–85
        regional conflicts over, **6:**90–91
        shortage of, **6:**86
        surface, **6:**81
        sustainable management of, **6:**97
        water use per unit of GDP, **6:**93
    water laws in, **6:**88–89
    wetlands in, **6:**86, 88
    withdrawals, water, **3:**316–17, **6:**85–86, **7:**27–28
    Xiluodu hydropower station, **6:**92
China International Capital Corporation (CICC), **6:**142
Chitale, Madhav, **1:**174
Chlorination, **1:**47, 60, **4:**39, **5:**159
Cholera, **1:**48, 56–63, 265–71, **3:**2, **5:**287–92
    cases reported to the WHO, by country
        1949-1979, **8:**361–66
        1980-2011, **8:**367–72
    deaths reported to the WHO, by country
        1949-1979, **8:**375–80
        1980-2011, **8:**381–86
    epidemic diarrheal diseases caused by, **7:**57
    limitations in data and reporting, **8:**359, 373–74
    See also *Vibrio; Vibrio cholerae*
Cincinnati Enquirer, **5:**171
CIPM Yangtze Joint Venture, **1:**85
Clementine spacecraft, **1:**197, **3:**212–13
Climate change:
    adaptation
        Adaptation Fund, **6:**51
        Adaptation Policy Framework, **6:**48
        assessments, **6:**47–49
        capacity for, **4:**236, **8:**5
        community participation in, **6:**49
        costs of, **6:**50–51, 53
        definition of, **6:**45
        demand-side options, **6:**46
        economic cost of, **6:**50–51
        equity issues for, **6:**51–52
        funding for, **6:**52
        general circulation models, **6:**47
        Interagency Climate Change Adaptation Task Force, **7:**153
        mainstreaming of, **6:**47
        national adaptation programs of action, **6:**48–49
        options for, **6:**45–46
        Oxfam Adaptation Financing Index, **6:**52
        participation in, **6:**49
        supply-side options, **6:**46
    air temperature increase, predicted, **7:**53
    Bibliography, The Water & Climate, **4:**xvii, 232–37
    business/industry, water risks, **5:**152
    California (*See* Climate change, California)
    changes occurring yet?, **1:**145–48
    China, **6:**87–88
    Colorado River Basin, **1:**142, 144, **4:**165–67, **7:**16–18
    costs of, **6:**50–51
    desalination, **5:**80–81
    developing countries' vulnerability to, **6:**45, 51, 54
    droughts (*See* Droughts)
    ecological effects, **4:**171–72
    environmental flow, **5:**45
    environmental justice, **5:**136–37
    floods (*See* Floods)
    food needs for current/future populations, **2:**87–88
    groundwater affected by, **4:**170, **6:**43
    hydrologic cycle, **1:**139–43, **5:**117
    hydrologic extremes, **6:**43–44
    hydrologic impacts of, **6:**48
    impacts of fossil-fuel extraction/processing, **7:**84–85
    and interbasin agreements, **8:**15
    IPCC (*See* Law/legal instruments/regulatory bodies, Intergovernmental Panel on Climate Change)
    Meking River Basin, **7:**14–15
    Nile River Basin, **7:**12–13
    overview, **1:**137–39, **7:**1, **8:**5
    precipitation, **1:**140–41, 146–47, **4:**159, 166, **6:**40, 87
    recommendations and conclusions, **1:**148–50
    reliability, water-supply, **5:**74
    renewability of resources affected by, **6:**9
    societal impacts, **1:**144–45
    summary of, **6:**53–54
    surface water effects, **6:**43, **7:**53–54
    sustainable vision for the Earth's freshwater, **1:**191
    and Syrian water conflicts, **8:**149–51
    and the Three Gorges Dam, **6:**146–47
    transboundary water management issues, **7:**1–2, 7–20
    vulnerability to, **6:**44–45, 51, 54
    water demand and, **6:**44, **7:**150
    water policy reform which addresses, **7:**153
    water quality and, **4:**167–68, **6:**44, **7:**53–54
    water resources affected by, **6:**43–44
    See also Greenhouse effect; Greenhouse gases
Climate change, California, **1:**144–45
    overview, **4:**xvii, 157–58
    policy
        economics/pricing/markets, **4:**180–81
        information gathering/reducing uncertainty, **4:**182–83
        infrastructure, existing, **4:**175–77
        institutions/institutional behaviors, new, **4:**181–82
        monitoring, hydrologic and environmental, **4:**183
        moving from science to demand management/conservation/efficiency, **4:**179–80
        new supply options, **4:**178–79
        overview, **4:**175
        planning and assessment, **4:**178
        reports recommending integration of science/water policy, **4:**176
    science
        evaporation and transpiration, **4:**159–60
        groundwater, **4:**170
        lake levels and conditions, **4:**168–69
        overview, **4:**158–59, 172–73
        precipitation, **4:**159
        sea level, **4:**169–70
        snowpack, **4:**160–61
        soil moisture, **4:**167
        the state of the ecosystems, **4:**171–72
        storms/extreme events and variability, **4:**161–63

temperature, **4:**159
  water quality, **4:**167–68
summary/conclusions, **4:**183–84
systems, managed water resource
  agriculture, **4:**174–75
  hydropower and thermal power generation, **4:**173–74
  infrastructure, water supply, **4:**173
Clinton, Bill, **2:**127, 134
Clothes, water footprint of production, **8:**84, 86
Clothes washing. *See* Laundry
CNN, Water Conflict Chronology cited by, **8:**173
Coal:
  energy content, **7:**75
  extraction and processing, **7:**73, 74, 82–84, 90
  production data, **7:**75, 82
  transport, **7:**83–84
  water consumed and energy production, **7:**25
Coastal zones:
  consequences of poor water quality, **7:**55–56
  development and desalination, **5:**80
  erosion, **7:**46
  floods, **5:**104, 108
  legislation, **5:**80
Coca-Cola, **4:**21, 38, **5:**146, 163, **6:**31–32
  beverage transport, **7:**161
  bottle development, **7:**158–59
  environmental justice, **5:**127
  and groundwater depletion in India, **8:**25
  recharging aquifers in India, **8:**21–22, 23
  water consumption and reputation, **7:**26
Cogeneration systems and desalination, **2:**107
Colombia, **1:**59, **3:**60, **5:**12, **7:**318, **8:**25–26
Colorado, **4:**95–96, **7:**52
  Missouri River diversion project, **8:**140–44
Colorado River:
  climate change, **1:**142, 144, **4:**165–67, **7:**16–18
  conflict/cooperation concerning freshwater, **1:**109, 111, **7:**6–7, **8:**91
  dams, **7:**52 (*See also* Dams, specific, Glen Canyon; Dams, specific, Hoover)
  delta characteristics, **3:**139–43, **6:**130
  fisheries, **1:**77
  hydrology, **3:**135–37, **7:**15, **8:**127
  institutional control of, **3:**134
  intl. agreements, **7:**6, 8–9, 15–16, **8:**91
  Las Vegas Valley Pipeline Project, **5:**74, **8:**136–40
  legal framework ("Law of the River"), **3:**137–39, **7:**15, 17
  restoration opportunities, **3:**143–44
  salinity, **7:**6–7
  Salton Sea inflows, **6:**129, 132
  summary/conclusions, **3:**144–45
  vegetation, **3:**134, 139–42
  water transfer, proposed, **8:**140–41
  wildlife, **1:**77, **3:**134
Columbia River Alliance, **2:**133
Comets (small) and origins of water on Earth, **1:**193–98, **3:**209–10, 219–20
Commercial sector, **3:**22–24, 169–70. *See also* Business/industry, water risks; Privatization; Water conservation, California commercial/industrial water use
Commissions. *See* International *listings;* Law/legal instruments/regulatory bodies; United Nations; World *listings*
Commodification, **3:**35
Commonwealth Development Corporation, **1:**96
Commonwealth of Independent States. *See also* Caucasus and Central Asia
  water quality, satisfaction by country, **7:**290–91
Communities:
  agricultural, loss of farms due to drought, **7:**104–5

consequences of poor water quality, **7:**61–62
empowerment through education and outreach, **8:**13–14
engagement (*See* Public participation)
impacts of fossil-fuel extraction/processing, **7:**87–88, 91–92
risks of privatization, **3:**68, 79
sustainable-water job production, **8:**59
Community structure. *See* Biodiversity; Extinct species; Introduced/invasive species; Threatened/at risk species
Companies:
  bottled water (*See* Bottled water, business/industry)
  continuous improvement commitment by, **6:**32
  current water use, **6:**22
  decision making, **6:**27–28
  environmental management system, **6:**32
  funding by, and environmental/lending standards, **8:**10
  history, **6:**17–18
  public relations and social responsibility, **8:**23, 29–30
  recommendations for, **6:**37–38, **8:**31
  small and medium enterprises (SME), **8:**26
  stakeholder issues, **5:**145, **6:**24, **7:**26–27, **8:**20, 26–27
  strategic partnerships, **6:**31–32, **7:**40, **8:**27–31
  supply chain involvement, **6:**24
  sustainable practices, **7:**23, **8:**19–22, 27–28, 30, 31
  sustainable water jobs, **8:**35–59
  water management, **6:**20
  water performance data published by, **6:**28
  water-policy statement, **6:**25–26
  See also Business/industry, water risks; Corporate reporting
Company, **2:**126
Concession models, privatization and, **3:**66–67
Conferences/meetings, international. *See* International *listings;* Law/legal instruments/regulatory bodies; United Nations; World *listings*
Conflict/cooperation concerning freshwater:
  dispute-resolution procedures, **7:**7, 168, **8:**161
  droughts, **5:**99, **7:**8–9, **8:**149–51
  economic/social development context, **7:**176, **8:**173
  environmental deficiencies and resource scarcities, **1:**105–6, **8:**161
  geopolitics and intl./transboundary waters, **2:**35–36, **7:**1, **8:**14–15, 147–51, 164
  historical (*See* Water Conflict Chronology)
  inequities in water distribution/use/development, **1:**111–13, **8:**2
  instrument/tool of conflict
    definition, **8:**173
    natural gas access as a, **8:**166
    water as a, **1:**108–10, **7:**1, 176, **8:**163, 173
  military target
    crops as, **8:**165
    definition, **8:**173
    water as a, **1:**110–11, **7:**176, **8:**149, 163, 164, 165, 167, 173
  Non-Navigational Uses of International Watercourses (*See* Convention of the Law of the Non-Navigational Uses of International Watercourses)
  number per year, 1931-2012, **8:**161
  overview, **1:**107, **3:**2–3, **5:**189–90, **7:**1
  2011-2012 events, trends, and analysis, **8:**159–69
  privatization, **3:**xviii, 70–71, 79, **4:**54, 67
  reducing the risk of conflict, **1:**113–15
  security analysis, shift in intl., **1:**105, **8:**161

Conflict/cooperation concerning freshwater (*continued*)
  subnational/intrastate *vs.* international/interstate, **8:**159, 161–62, 168–69
  summary/conclusions, **1:**124, **8:**168–69
  sustainable vision, **1:**190
  Syria, **8:**147–51
  water "wars" *vs.* water-related violence, **8:**159, 161
  See also Terrorism; *specific countries*
Congo, Democratic Republic of the, **5:**9
  dams with Chinese financiers/developers/builders, **7:**133, 318
  infant mortality rate, **7:**266
  Nile River Basin, **7:**10, 11
Congo, Republic of the:
  dams with Chinese financiers/developers/builders, **7:**133, 333
  infant mortality rate, **7:**264, 265
  location, **7:**11
Conoco-Phillips, **6:**23
Conservation. *See* Environmental flow; Land conservation; Soft path for water; Sustainable vision for the Earth's freshwater; Twenty-first century water-resources development; Water conservation; Water-use efficiency
Construction:
  building site runoff, volume, **8:**76
  major water projects, delivery methods, **8:**104–5, 107–10
Consumption/consumptive use, **1:**12, 13, **3:**103, **6:**7, 117. *See also* Projections, review of global water resources; Water footprint; Water use, consumptive/nonconsumptive; Withdrawals, water
Contracts:
  for desalination plants, 104–12, 116–19
  privatization and, **4:**65–67
Convention of the Law of the Non-Navigational Uses of International Watercourses (1997), **1:**107, 114, 124, 210–30, **2:**10, 36, 41–42, **3:**191, **5:**35
  China's vote against, **7:**129
  critical importance of, **7:**19
  overview, **7:**4–5, **8:**14
Conventions, international legal/law. *See* International *listings*; Law/legal instruments/regulatory bodies; United Nations; World *listings*
Cook Islands, **3:**118
Cooling and water use, **4:**135, **8:**86
Copper, **7:**59–60, **8:**166
Corporate issues. *See also* Bottled water, business/industry; Business/industry, water risks; Companies; Corporate reporting; Privatization; Water conservation, California commercial/industrial water use
Corporate reporting:
  bottled water, **6:**23
  continuous improvement, **6:**32
  current water use, **6:**22
  history, **6:**17–18
  non-financial reports, **6:**18–20
  recommendations for, **6:**37–38
  stakeholder consultations and engagement, **6:**24
  strategic partnerships, **6:**31–32
  summary of, **6:**36–37
  supply chain involvement, **6:**24
  sustainability reports, **6:**18, 20
  water management, **6:**20, **7:**40–41
  water performance (*See* Water performance reporting)
  water-policy statement, **6:**25–26
  water risk-assessment programs, **6:**23–24
Costa Rica, **5:**34
Costs. *See* Economy/economic issues
Côte d'Ivoire, **1:**55, **4:**65–67, **7:**319
Councils. *See* Law/legal instruments/regulatory bodies
Court decisions and conflict/cooperation concerning freshwater, **1:**109, 120. *See also* Law/legal instruments/regulatory bodies; Legislation
Covenant, the Sword, and the Arm of the Lord (CSA), **5:**21–22
Covenants. *See* Law/legal instruments/regulatory bodies; United Nations
Crane, Siberian, **1:**90
Crayfish, **7:**56
Crime:
  corruption, **8:**12–13, 76, 148
  tax fraud, **8:**131
  vandalism, **8:**76
Critical Trends, **3:**88
Croatia, **1:**71
Crocodiles, **7:**56
Crowdsourcing, **8:**12
Cryptosporidium, **1:**48, **2:**157, **4:**52, **5:**2, 159, **7:**47
Cucapá people, **3:**139
Cultural importance of water, **3:**40
Curaçao, **2:**94–95
Current good manufacturing practice (CGMP), **4:**27
CW Leonis, **3:**219–20
Cyanide, **5:**20
Cyberterrorism, **5:**16, **6:**152, **7:**176
Cyprus, **1:**202–4, **2:**108, **8:**135
Cyrus the Great, **1:**109

D
Dams:
  Africa, **3:**292, **5:**151, **6:**274–78
  business/industry, water risks, **5:**151
  Central Asia, **8:**166
  China (*See* China, dams)
  by continent and country, **3:**291–99
  debate, new developments in the, **1:**80–83
  economic issues, **1:**16, **2:**117–19, 122–24, 127, 129–30
    development and construction by China, **7:**133–35
    private-sector funding, **1:**82
  environmental flow, **5:**32–34
  environmental impacts
    impact statements, **7:**134–35
    overview, **7:**130
    threatened/at risk species due to, **3:**3, **7:**154, **8:**128–29
  environmental/social impacts, **1:**15, 75–80, 83
  and floods, **5:**106, **6:**144, **8:**128
  Gabcikovo-Nagymaros project, **1:**109, 120
  grandiose water-transfer plans, **1:**74–75
  Korean peninsula, **1:**109–10
  large
    historical background, **1:**69
    North American Water and Power Alliance, proposed, **8:**127
    San Francisco Bay, proposed, **8:**132
    total worldwide, **7:**127
    U.S. begins construction of, **1:**69–70
  opposition to, **1:**80–82, **7:**127, 132, 134–36
  by owner, **6:**272–73
  power generation (*See* Hydroelectric production)
  primary purposes of, **6:**270–71
  removal (*See* Dams, removing/decommissioning)
  runoff, humanity appropriating half of the world's, **5:**29
  safety data, **6:**236–38
  social impacts
    displaced people, **1:**77–80, 85, 90, 97–98, 281–87, **5:**134, 151

Sudan, **7:**133, 134
    U.S., **7:**154
    environmental justice, **5:**133–36
    schistosomiasis outbreaks, **1:**49, **7:**58
specific sites (*See* Dams, specific)
Syria, strategic attacks on, **8:**149
by the Tennessee Valley Authority, **7:**153
terrorism risks, **5:**15–16
twentieth-century water-resources development, **1:**6
World Water Forum (2003), **4:**193
See also World Commission on Dams
Dams, removing/decommissioning:
    case studies, completed removals
        Edwards, **2:**xix, 123–25
        Maisons-Rouges and Saint-Etienne-du-Vigan, **2:**125
        Newport No. 11, **2:**125–26
        Quaker Neck, **2:**126
        Sacramento River valley, **2:**120–23
    economics, **2:**118–19, **8:**46
    employment, **8:**46, 48
    hydroelectric production, **1:**83, **2:**114–15
    1912 to present, **2:**275–86, **6:**239–69
    overview, **2:**xix, 113–14
    proposed
        Elwha and Glines Canyon, **2:**127–28
        Glen Canyon, **2:**128–31
        Pacific Northwest, **2:**131–34, **8:**129
        Peterson, **2:**128
        Savage Rapids, **2:**128
        Scotts Peak, **2:**126–27
    purpose for being built no longer valid, **2:**117
    renewal of federal hydropower licenses, **2:**114–15
    safety issues, **2:**130
    by state, **6:**265–69
    states taking action, **2:**117–18
    summary/conclusions, **2:**134
    twentieth century by decade, **3:**301–2
    by year, **6:**265–69
Dams, specific:
    American Falls, **7:**153
    Aswan, **7:**12
    Ataturk, **1:**110, **3:**182, 184, 185
    Auburn, **1:**16
    Bakun, **1:**16
    Balbina, **5:**134
    Banqiao, **5:**15–16
    Batman, **3:**187
    Belinga, **7:**134–35
    Belo Monte, **8:**165
    Birecik, **3:**185, 186–87
    Bonneville, **1:**69, **8:**129
    Chixoy, **3:**13
    Cizre, **3:**189–90
    Condit, **2:**119
    Edwards, **1:**83, **2:**xix, 119, 123–25
    Elwha, **1:**83
    Farakka Barrage, **1:**118–19
    Fort Peck, **1:**69
    Fort Randall, **1:**70
    Garrison, **1:**70, **5:**123
    Gezhou, **6:**142
    Gibe, **7:**134
    Glen Canyon, **1:**75–76, **2:**128–31
    Glines Canyon, **1:**83
    Gorges, **5:**133
    Grand Coulee, **1:**69
    Grand Ethiopian Renaissance (Hidase; Millennium), **8:**164
    Hetch Hetchy, **1:**15, 80–81
    Hoa Binh, **5:**134
    Hoover, **1:**69, **3:**137, **8:**127
    Ice Harbor, **2:**131–34
    Ilisu, **3:**187–89, 191
    Imperial, **7:**7
    Itaipu, **1:**16
    Kabini, **8:**163
    Kalabagh, **1:**16
    Karakaya, **3:**182, 184
    Kariba, **5:**134
    Katse, **1:**93, 95
    Keban, **3:**184
    Kenzua, **7:**153
    Koyna, **1:**77
    Krishna Raja Sagar, **8:**163
    Laguna, **3:**136
    Little Goose, **2:**131–34
    Lower Granite, **2:**131–34
    Lower Monumental, **2:**131–34
    Machalgho, **8:**163
    Manitowoc Rapids, **2:**119
    Merowe, **7:**133–34
    Morelos, **3:**138, 142, **7:**7
    Myitsone, **7:**132–33
    Nam Theun I, **1:**16
    Nujiang, **7:**135
    Nurek, **1:**70, **8:**127
    Oahe, **1:**70
    Pak Mun, **5:**134
    Peterson, **2:**128
    Pubugou, **7:**136
    Quaker Neck, **2:**126
    Rogun, **8:**166
    Sadd el-Kafara, **1:**69
    St. Francis, **5:**16
    Salling, **2:**119
    Sandstone, **2:**119
    Sapta Koshi High, **1:**16
    Sardar Sarovar, **5:**133
    Savage Rapids, **2:**119, 128
    Sennâr, **7:**58
    Shasta, **1:**69, **5:**123, **7:**153
    Shimantan, **5:**15–16
    Snake River, **2:**131–33
    Tabqa (al-Thawrah), **8:**149
    Ta Bu, **5:**134
    Tantangara, **7:**114
    Three Gorges Dam (*See* Three Gorges Dam)
    Tiger Leaping Gorge, **6:**89, **7:**135
    Tishrin, **8:**149
    Welch, **2:**118
    Woolen Mills, **2:**117–19
    Wullar, **8:**163
    Xiaowan, **7:**131
    Yacyreta, **3:**13
    Yangliuhu, **7:**135
    Yangtze River, **7:**46, 129
    See also China, dams, construction overseas; Lesotho Highlands project; Southeastern Anatolia Project
Dasani bottled water, **4:**21
Data issues/problems:
    centralized data portal, **8:**12
    climate change, **4:**183
    collection, recommendations, **8:**12
    conversions/units/constants, **2:**300–309, **3:**318–27, **4:**325–34, **5:**319–28, **7:**75, 341–50, **8:**395–404
    global water footprint, **8:**87, 88
    global water resources, projections, **2:**40–42
    groundwater, **4:**97–98
    mobile connectivity, **8:**12
    need for data and water policy reform, **7:**152
    open access to information, **4:**70–73, **8:**11
    polls, **7:**283, 286, **8:**387, 389, 392

Data issues/problems (*continued*)
    sustainable water jobs, **8:**58
    urban commercial/industrial water use in California, **4:**139–40, 148–50, 152–53
    urban residential water use in California, **4:**108–9
    well-being, measuring water scarcity and, **3:**93
da Vinci, Leonardo, **1:**109
Dead Sea:
    mineral extraction from evaporation ponds, **8:**155, 156
    overview, **8:**153, 154–55
    water transfer from the Red Sea to, **8:**144, 153–57
Deaths. *See* Mortality
Decision making:
    joint, for transboundary waters, **7:**13, 167, **8:**14–15
    open/democratic, **4:**70–73, **8:**11
    precautionary principle, **7:**167
    water risks factored into, **6:**27–28
Declarations. *See* Law/legal instruments/regulatory bodies; United Nations
Deer Park, **4:**21
Deforestation, **8:**26
Deltas, river, **3:**xix
    Colorado River, **3:**139–43, **6:**130
    Sacramento River, **8:**133
Demand management, **4:**179–80, **7:**153–54. *See also* Water-use efficiency; Withdrawals, water
    with desalination projects, **8:**102, 106, 115–16, 119
Demographic health surveys (DHS), **6:**61
Demographics of sustainable water job employees, **8:**56
Dengue fever, **3:**2
Denmark, **1:**52
Density, measuring, **2:**304, **3:**327, **4:**329, 334, **7:**345, 350
Desalination:
    advantages and disadvantages, **5:**66–76, **8:**108–9
    Australia, **5:**69, **7:**114–15, **8:**96, 97, 98, 102, 103, 118
    business/industry, water risks, **5:**161
    California (*See* California, desalination)
    capacity by country/process/source of water, **1:**131, 288–90, **2:**287–90, **5:**58–60
    capacity statistics, **5:**56–57, 59–60, 308–17
    Carlsbad (CA), **8:**93, 95, 98, 112–19
    case studies, **8:**106–17
    China, **6:**93
    climate change, **5:**80–81, **7:**16
    concentrate disposal, **2:**107
    economic issues (*See* Economy/economic issues, desalination)
    energy use/reuse, **2:**107, **5:**69–71, 75–76
    environmental effects of, **5:**76–80
    global status of, **5:**55–58
    health, water quality and, **5:**74–75
    history and current status, **2:**94–98, **5:**54
    intakes, water, impingement/entrainment, **5:**76–77
    Israel, **5:**51, 69, 71, 72, **8:**157
    Nauru, **3:**118
    oversight process, regulatory and, **5:**81–82
    overview, **1:**29–30, **2:**93–94, **5:**51–53
    plants
        capacity of actual/planned, **5:**308–17
        Carlsbad (CA), **8:**93, 95, 98, 112–19
        costs, **8:**93–101, 117–18
        project delivery methods, **8:**104–5, 107–10
        projects and risks, **8:**101–6, 118–19
        project structure, **8:**119
        proposed and recently constructed, **8:**98, 103, 155–56
        Sand City (CA), **8:**93
        Tampa Bay (FL), **2:**108–9, **5:**61–63, **6:**123–25, **8:**98, 103, 106–12, 118
    processes, **5:**54–55
        freezing, **2:**104
        ion-exchange methods, **2:**104
        membrane
            costs, **8:**94
            electrodialysis, **2:**101–2
            overview, **2:**101
            reverse osmosis (*See* Reverse osmosis)
        membrane distillation, **2:**104–5
        overview, **2:**103–4
        solar and wind-driven systems, **2:**105–6
        thermal
            multiple-effect distillation, **2:**99–100
            multistage flash distillation, **2:**96, 100
            overview, **2:**98–99
            vapor compression distillation, **2:**100–101
    production statistics, monthly, **8:**107
    Red Sea, proposed, **8:**154, 155–56
    reliability value of, **5:**73–74
    salt concentrations of different waters, **2:**94, **5:**53
    source of water/process, capacity by, **5:**56–57, 59–60
    summary/conclusions/recommendations, **2:**109–10, **5:**82–86, **8:**117–18
    Tampa Bay (FL), **2:**108–9, **5:**61–63, **6:**123–25, **8:**98, 103, 106–12, 118
    U.S., **5:**58–63
Desertification, Australia, **7:**105–6
Deutsche Morgan Grenfell, **1:**96
Developing countries:
    agriculture, **1:**24
        irrigation, **3:**290, **4:**297–98, **5:**301–2
    bottled water, **4:**40–41
    business/industry, water risks, **5:**150
    cholera, **1:**56
    climate change vulnerability of, **6:**44–45, 51, 54
    dams, **1:**82
    diseases, water-related, **5:**117
    dracunculiasis, **1:**51
    drinking water, **1:**261–62
    economic development derailed, **1:**42
    education and expertise in water quality, **7:**66
    efficiency, improving water-use, **1:**19
    food needs for current/future populations, **2:**69
    industrial water use, **1:**21
    knowledge and technology transfer, **8:**10–11
    Pacific Islands developing countries (PIDC), **5:**136
    pollution sources, **8:**3
    population increases and lack of basic water services, **3:**2
    privatization, **3:**79, **4:**46
    sanitation services, **1:**263–64
    toilets, energy-efficient, **1:**22
    unaccounted for water, **4:**59
    See also Environmental justice; *specific countries*
Development:
    assistance (*See* Overseas Development Assistance)
    economic/social, as context for water conflicts, **7:**176, **8:**173 (*See also* Water Conflict Chronology)
    the right to, **2:**8–9
    technology (*See* Technology development; Technology transfer)
Development Assistance Committee (DAC). *See* Overseas Development Assistance
Diageo's Water of Life, **7:**38
Diarrhea, **1:**48, **4:**8, 11, **6:**58, 75, 85
    disability adjusted life year (DALY), **7:**272
    morbidity, **7:**58
    mortality, **7:**57–58, 272
        childhood, under-5, **7:**259–63
Dioxins, **7:**48, 60
Diptheria, **7:**272
Disability-adjusted life year (DALY), **4:**9, 276–77, **7:**57, 270–72

Diseases, water-related, **1:**186–87
  amoebiasis, **1:**48
  ascariasis, **7:**272
  Balkan Endemic Neuropathy, **7:**89
  Campylobacter jejuni, **7:**57
  in China, **6:**85
  cholera (*See* Cholera)
  dams, removing/decommissioning, **2:**130
  death, **6:**58, 73
    and disability-adjusted life year from, **4:**276–77, **7:**57, 270–72
    limitations in data and reporting, **7:**270–71
    projected, **4:**9–10, 12–13
  dengue fever, **3:**2
  diarrhea (*See* Diarrhea)
  diptheria, **7:**272
  dysentery, **1:**42, **4:**64
  emerging diseases/pathogens, **2:**155, **7:**49
  encephalitis, **7:**47
  environmental justice, **5:**128–29
  failure, **3:**2, **5:**117
  fecal coliform bacteria, **7:**52–53 (*See also* Fecal contamination)
  Guinea worm (*See* Dracunculiasis)
  hepatitis, **7:**58
  hookworm, **7:**61, 272
  malaria, **1:**49–50, **6:**58, **7:**259–63
  meningitis, **7:**47
  outbreaks in U.S., **4:**308–12
  overview, **1:**47–50, 274–75, **7:**47, 57–58
  poliomyelitis, **7:**272
  roundworm, **7:**61–62
  schistosomiasis, **1:**48, 49, **7:**58, 272
  Shigella, **7:**57
  trachoma, **1:**48, **7:**272
  trichuriasis, **1:**48, **7:**272
  trypanosomiasis, **7:**272
  typhoid, **1:**48, **7:**57, 58
  waterborne, **1:**47–49, 274–75, **4:**8, **7:**57–58
  waterborne *vs.* water-based, **7:**57–58
  whipworm, **7:**61
  See also Millennium Development Goals
Dishwashers, **4:**109, 116, **6:**106
Displaced people, **8:**148, 164, 167. *See also* Dams, social impacts, displaced people
Dolphins, **1:**77, 90, **3:**49, 50, **7:**56
Dow Jones Indexes, **3:**167
Downstream users, **5:**37. *See also* Human right to water
Dracunculiasis (Guinea worm), **1:**39, 48–56, 272–73, **3:**273–77, **5:**293–97
  host zooplankton, **7:**58
  overview, **7:**47
Drinking water, access:
  collection distance, **7:**231, **8:**237
  collection the responsibility of children/women, **7:**61, 89, **8:**168
  in conflict areas (*See* Conflict/cooperation concerning freshwater, military target, water as a)
  corporate efforts to improve, **7:**38, 41
  costs of, **6:**73
  by country, **1:**251–55, **3:**252–60, **5:**237–46, **6:**211–20
    urban and rural, 1970-2008, **7:**230–40, **8:**236–46
    urban and rural, 2011, **8:**236–38, 247–51
  deaths due to lack, **8:**164
  defining terms, **4:**28, **7:**230–31, **8:**236–37
  developing countries, **1:**261–62, **7:**62
  disinfection and purification, **5:**159
  fluoride, **4:**87
  funding of, **6:**73, **8:**9
  "improved," use of term, **7:**230–31, 251–52, **8:**236–37, 239
  infrastructure for tap water, **8:**35–36
  intl. organizations, recommendations by, **2:**10–11
  limitations in data and reporting, **6:**61–62, **7:**231, 251–52, **8:**237–38, 268–69
  Overseas Development Assistance, **7:**273–77
    by donating country
      2004-2011, **8:**317–19
    by subsector
      2007-2011, **8:**320–22
  reclaimed water, **2:**151–52
  by region, MDG progress on, **6:**65–67, 230–32, **7:**24, 251–53, **8:**268–70
  rural areas, **6:**70–71 (*See also* Drinking water, access, by country)
  shortages, **6:**44, 53, 86, **8:**1, 164, 167
  statistics regarding, **6:**44, **7:**24, 61, **8:**1
  twentieth-century water-resources development, **3:**2
  urban areas, **6:**70–71, **7:**97 (*See also* Drinking water, access, by country)
  violence during long walks to water source, **8:**168
  well-being, measuring water scarcity and, **3:**96–98
  World Health Organization, **4:**208
  World Water Forum (2003), **4:**202
  See also Health, water issues; Human right to water; Millennium Development Goals; Soft path for water; Water quality; Well-being, measuring water scarcity and; *specific contaminants*
Droughts, **1:**142, 143, **4:**163, 203–4, **6:**44
  agricultural effects, **5:**92, 94, 98, 103, **7:**102–5, **8:**148
  Atlanta (GA), **6:**108
  Australia, **7:**97–121, 146–47
  beginning of, determination, **5:**93
  Brazil, **7:**26, **8:**22–23, 165
  California, **8:**89, 102
  causes, **5:**95–96, **7:**8, 98, 100
  China, **5:**97, **6:**86–87, **7:**131–32
  Colorado River basin, **8:**138
  defining terms, **5:**92, **7:**99–100
  disturbances promoting ecosystem diversity, **5:**91
  due to climate change, **5:**112–13, **6:**44, **7:**8–9, 54, 101–2, **8:**149–51
  ecological effects, **5:**92, **7:**54, 100–101
  economy/economic issues, **5:**91–92, 98, 103
  effects of, **5:**95–99
  fires, **5:**98, 102, **7:**105
  forecasting, **7:**109, 146
  future of, **5:**112–13
  management
    agricultural, **7:**107–14
    crisis management, **5:**99, 111, **7:**8–9, 97, 100, 106–7
    impact and vulnerability assessment, **5:**100–101
    mitigation and response, **5:**101–3
    monitoring and early warning, **5:**99–100, **7:**20
    national policy development, **7:**106–7, 112
    public participation, **7:**116–17
    risk management, **5:**99, **7:**109
    water market/water trading, **7:**110, 111–14, 146
  National Drought Mitigation Center, **5:**94
  overview, **5:**91–92
  short-lived or persistent, **5:**93
  summary/conclusions, **5:**113–14
  Syria, **8:**147–51
  transboundary agreements, **7:**8–9, 12–13, 15
  urban areas, **5:**98, **7:**114–21
  U.S., **5:**93, **6:**44, **8:**130
DuPont, **1:**52
Dust Bowl (U.S.), **5:**93
Dutch Water Line strategy, **5:**5
Dynamics, water, **2:**218
Dysentery, **1:**42, **4:**64

E
Early Warning Monitoring to Detect Hazardous Events in Water Supplies, **5:**2, 20
Earth, origins of water on, **1:**93–98, **3:**209–12, **6:**5
Earthquakes. *See* Seismic activity
Earth Water, **5:**163
East Bay Municipal Utilities District (EBMUD), **5:**73
East Timor. *See* Timor Leste
Economy/economic issues:
    access to water, **5:**125, **7:**67
    bag technology, water, **1:**200, 204–5
    bottled water, **4:**17, 22–23
    budgets, U.S. federal agency water-related, **7:**303–7
    cholera, **1:**60
    climate change, **6:**50–51
    Colorado River, **3:**143
    conflict/cooperation concerning freshwater, **5:**15
    cost effectiveness, **4:**105, **8:**38
    cost of water (*See* Pricing, water)
    costs of water in San Diego County, **8:**100
    credit rating for project, **8:**117
    dams, **1:**82, **2:**117–19, 122–24, 127, 129–30, 132
    desalination, **1:**30, **2:**95, 105–9, **5:**62–63, 66, 68–73
        cost comparisons, **8:**99–101, 118
        cost estimates, **8:**95–99, 110–11, 112–13, 118
        cost terminology, **8:**94–95
        overview, **8:**93–94, 117
    developing countries, **1:**42, **3:**127 (*See also* Overseas Development Assistance)
    disadvantages of corporate sustainability, **8:**30
    disincentives for corporate sustainability, **8:**30
    droughts, **5:**91–92, 98, 103
    economic development, **7:**176, **8:**27
    economic good, treating water as an, **3:**xviii, 33–34, 37–38, 58, **4:**45
    economies of scale, and hard path for meeting water-related needs, **3:**8
    efficiency, economic, **4:**104, 105, **8:**100
    environmental flow, **5:**32, 40–43
    financial assistance, **6:**74
    fishing, **2:**117
    floods, **5:**91–92, 108, 109
    Global Water Partnership, **1:**171
    human needs, basic, **1:**46–47
    human right to water, **2:**13–14, **8:**29
    incentives and rebates for water conservation, **6:**110–12, **7:**119, 121
    incentives for corporate sustainability, **8:**23, 31
    industrial water use, **1:**21, **7:**67
    infrastructure projects, funding, **8:**105
    intl. water meetings, **5:**183
    investment issues, **1:**6–7, **5:**273–75, **8:**12–13, 23
    jobs, sustainable water, **8:**35–59
    Lesotho Highlands project, **1:**95–97, 99
    Millennium Development Goals, **4:**6–7, **7:**64
    overruns, water-supply project, **3:**13
    poor water quality, **7:**63–65
    privatization, **3:**77, **4:**53–60
    productivity of water, U.S., **4:**321–24
    property values and green space, **8:**42
    recession, U.S., **8:**130
    reclaimed water, **2:**156, 159, **7:**121
    recycled water, **8:**11
    revenue/growth of the water industry, **5:**303–4
    sanitation services, **5:**273–75
    soft path for meeting water-related needs, **3:**5, 6–7, 12–15, 23–25, **7:**150
    Southeastern Anatolia Project, **3:**182, 187, 190–91
    subsidies
        agriculture, **1:**24–25, 117, **7:**108, 109, 152
        desalination, **5:**69
        engineering projects, large-scale, **1:**8
    government and intl. organizations, **1:**17, **7:**67, 108, 109, 111, 152, **8:**35
    privatization, **3:**70–72, **4:**50, 53–60
    twenty-first century water-resources development, **1:**24–25
    supply-side solutions, **1:**6
    tariffs, water, **6:**312–23
    terrorism, **5:**20
    Three Gorges Dam, **1:**86–89
    transfer of water, **8:**136, 140, 142
    transparency and accountability, **8:**12–14, 112
    twentieth-century water-resources development, end of, **1:**16–17
    urban commercial/industrial water use in California, **4:**140–43, 151–52
    urban residential water use in California, **4:**107–8
    wages for sustainable water jobs, **8:**56
    water policy reform, **7:**152–53
    water scarcity effects on, **6:**45
    water use *vs.* Gross Domestic Product, **8:**142
    withdrawals, water, **3:**310–17
    World Commission on Dams, **3:**158, 162–64, 167–69
    World Water Forum (2003), **4:**195–96
    See also Business/industry, water risks; Environmental justice; Globalization and international trade of water; Privatization; World Bank
Ecosystems:
    bioproductivity, **8:**86
    classification of impacts, three-tier, **7:**52
    climate change, **4:**171–72
    community structure (*See* Biodiversity; Extinct species; Introduced/invasive species; Threatened/at risk species)
    costs of poor water quality, **7:**63
    ecological footprint, **8:**85, 86
    environmental flow, **5:**30–31
    impacts of dams and water withdrawals, **3:**3, **8:**37
    impacts of desalination, **8:**94
    impacts of fossil-fuel extraction/processing, **7:**84–87
    impacts of water quality degradation factors, **4:**168, **7:**46–49, 52, 54–57
    privatization, **4:**51–53, **7:**146
    reclaimed water, **2:**149–50
    reserve, and nature's right to water, **7:**145
    restoration/protection, **2:**149–50, **7:**66, 146, **8:**44–48
    World Water Forum (2003), **4:**203
    See also Environmental flow; Environmental issues; Fish; Soft path for water; Sustainable vision for the Earth's freshwater
Ecuador, **1:**59, 71, **2:**178–79, **3:**49, **5:**124
    dams with Chinese financiers/developers/builders, **7:**319
    fossil-fuel production, **7:**88
Education:
    effects of inadequate sanitation on, **6:**58, **7:**61
    empowerment through, **8:**13–14
    Save Water and Energy Education Program, **4:**114
    and sustainable water jobs, **8:**55, 57, 58, 59
    in water quality, **7:**66
    in water supply and sanitation, **7:**277
    in water supply and sanitation, Overseas Development Assistance
        by donating country 2004-2011, **8:**317–19
        by subsector 2007-2011, **8:**320–22
Edwards Manufacturing Company, **2:**124

Eels, **2:**123
Efficiency (*See* Water-use efficiency)
Egypt, **1:**118, **2:**26, 33, **4:**18, 40, **5:**163
    conflicts concerning freshwater, 2011-2012, **8:**164
    decline in infant mortality rate, **7:**264
    Nile River Basin, **7:**11, **8:**164
    Suez Canal, **8:**153
    tourism, **8:**153, 164
Eighteen District Towns project, **2:**167
El Niño/Southern Oscillation (ENSO), **1:**139, 143, 147, **3:**119, **4:**162, **5:**95, 106
    and the Millennium Drought in Australia, **7:**98
Employees, demographics of sustainable water jobs, **8:**56
Employment, sustainable water jobs, **8:**35–59
Encana Oil and Gas Inc., **8:**71, 73
Encephalitis, **7:**47
Endocrine disruptors, **7:**49, 59, 60
End use of water as a social concern, **3:**7, 8–9
Energy issues:
    bottled water, **7:**157–64
    business/industry, water risks, **5:**150, 151–52
    desalination, **2:**107, **5:**69–71, 75–76, **8:**95
    droughts, **5:**98, **8:**22–23
    energy efficiency, **3:**xiii, **7:**84, 162
    measuring energy, **2:**308, **3:**326, **4:**333, **7:**75, 349
    power generation
        by fossil fuel type, **7:**75
            renewable technologies (*See* Hydroelectric production; Nuclear power; Solar power; Wind power)
            retirement of aging power plants, **7:**84–85
            thermal pollution from, **7:**85
            water consumption, **7:**25, 26, 31, 73, 74–75
            water footprint, **8:**5–6, 84
    tap water, **7:**163
    transfer of water, **8:**142
    transportation method, **7:**162
    unit conversions, **7:**75
    U.S. Department of Energy, **1:**23, **4:**115
    water-energy-food nexus, **8:**5–6
    water treatment, **7:**60
Engineering projects, large-scale. *See* Dams; Transfers, water; Twentieth-century water-resources development; Zombie water projects
England:
    cholera, **1:**56
    desalination, **2:**94
    droughts, **5:**92
    eutrophication, **7:**63
    human right to water, **4:**212
    Lesotho Highlands project, **1:**96
    Office of Water Services, **4:**64–65
    privatization, **3:**58, 60, 61, 78, **4:**62–65
    sanitation services, **5:**129–31
    Southeastern Anatolia Project, **3:**191
    World Commission on Dams, **3:**162, 170
    See also United Kingdom
Enron, **3:**63
Environment, **2:**182
Environmental flow:
    characteristics of hydrologic regimes, **5:**31
    economics/finance, **5:**40–43
    General Accounting Office, **5:**119
    legal framework, **5:**34–37
    policy implementation, **5:**43–45
    projects in practice, **5:**32–34
    science of determining, **5:**38–40
    summary/conclusions, **5:**45–46
    water quality link, **5:**40
    World Commission on Dams' recommendations, **5:**30

Environmental Impact Assessment (EIA), **6:**96, **7:**31–32
Environmental issues:
    bottled water, **4:**41
    change, global, **1:**1
    cleanup (*See* Habitat restoration; Remediation)
    contaminants in water (*See* Wastewater; Water quality; *specific contaminants*)
    as context for business/industry water risk, **7:**29, **8:**24
    dams/reservoirs, **1:**15, 75–80, 83, 91, **8:**128
    desalination, **5:**76–80
    ecological impacts (*See* Ecosystems)
    environmental flow (*See* Environmental flow)
    environmental justice (*See* Environmental justice)
    green infrastructure (low-impact development), **8:**39, 41
    green jobs, **8:**35–59
    Lesotho Highlands project, **1:**98
    nature's right to water, **7:**145
    reclaimed water, **2:**149–50
    shrimp/tuna and turtle/dolphin disputes, **3:**49, 50
    stormwater, pollution by (*See* Stormwater runoff)
    sustainable vision for the Earth's freshwater, **1:**188–90
    Three Gorges Dam, **1:**89–9, **6:**142
    top concerns around the world, **7:**285–88
    top concerns of U.S. public, **7:**282–84
    twentieth-century water-resources development, end of, **1:**12, 15–16
    U.S., **6:**339–41, **7:**282–84
Environmental justice:
    climate change, **5:**136–37
    Coca-Cola, **5:**127
    dams, **5:**133–36
    discrimination, environmental, **5:**118–19
    economic issues, **2:**13–14, **8:**29
    environmentalism of the poor, **5:**123–24
    Environmental Justice Coalition for Water, **5:**122–23
    good governance, **5:**138–41
    history of movement in U.S., **5:**119–20, 122–23
    intl. context, **5:**118
    overview, **5:**117–18
    principles of, **5:**120–22
    privatization, **5:**131–33
    right to water (*See* Human right to water)
    sanitation services, **5:**127–31
    summary/conclusions, **5:**141–42
    water access, **5:**124–25, 127, **6:**44, **8:**29
    water quality, **5:**127–29
    women and water, **5:**126, 134, **7:**61, 89, **8:**168
Environmental management system, **6:**32
Eritrea, **5:**95, **7:**11
Ethanol, **7:**74
Ethical issues:
    land ethic of Leopold, **8:**129
    public relations and social responsibility, **8:**23, 29–30
Ethiopia, **1:**55, **4:**211, **5:**95, 97
    conflicts concerning freshwater, 2011-2012, **8:**164
    dams
        with Chinese financiers/developers/builders, **7:**133, 134, 319–20
        Grand Ethiopian Renaissance (Hidase; Millennium), **8:**164
    drinking water, access to, **6:**71–73, **7:**41
    Nile River Basin, **7:**11, **8:**164
    sanitation, **6:**71–73
Ethos Water, **5:**163
Europa Orbiter, **3:**218

Europe:
    air quality, satisfaction, 2012, **8:**391
    aquifers, transboundary, **7:**3
    bottled water, **4:**22, 25, 288, 289, 291, **5:**163, 281, 283
        per-capita consumption, **7:**340
    cholera, **1:**56, 267, 270, **5:**289, 290
    dams, **3:**165, 292–93
    drinking water, **5:**159, 245–46, **7:**239–40
    Eco-Management and Audit Scheme, **6:**18
    environmental flow, **5:**33
    European Convention on Human Rights (1950), **2:**4, 8
    European Union Development Fund, **1:**96
    European Union Water Framework Directive (2000), **7:**25, 147–49
    food needs for current/future populations, **2:**69
    globalization and intl. trade of water, **3:**45, 47
    Global Water Partnership, **1:**169
    groundwater, **4:**84–85
    human right to water, **4:**209, 214
    hydroelectric production, **1:**71, 279–80
    irrigation, **1:**301, **2:**261–62, 265, **3:**289, **6:**332–33
    Lesotho Highlands project, **1:**96
    mortality rate, under-5, **7:**261–62
    population data/issues, **1:**249–50, **2:**214
    privatization, **3:**58, 60, 61
    renewable freshwater supply, **1:**239–40, **2:**201–2, 217, **3:**241–42, **4:**265–66, **5:**225–26
        2011 update, **7:**219–20
        2013 update, **8:**225–26
    reservoirs, **2:**270, 274
    river basins, **6:**289, 301–06
        transboundary, **2:**29–31, **7:**3
    salinization, **2:**268
    sanitation services, **3:**272, **5:**255, **7:**249–50
    threatened/at risk species, **1:**294–95, **7:**56
    water availability, **2:**217
    waterborne diseases, **1:**48
    water quality, satisfaction
        by country, **7:**290–91
        2012, **8:**391
    withdrawals, water, **1:**244, **2:**209–11, **3:**249–50, **4:**273–75, **5:**235–36
        by country and sector, **7:**228–29, **8:**233–35
Evaporation of water into atmosphere, **1:**141, **2:**20, 22, 83, **4:**159–60, **6:**43, 107
    reduction in drought management, **7:**110
    reduction with aquifer recharge, **8:**49
Evaporation ponds, potash extraction from, **8:**155
Everglades, **8:**48
Evian bottled water, **7:**162
Excreta. *See* Fecal contamination
Ex-Im Bank, **1:**88–89, **3:**163
Export credit agencies (ECAs), **3:**191
Extinct species:
    freshwater animal species, **7:**292–302
    rates for fauna from continental North America, **4:**313–14
    See also Threatened/at risk species

F
Faucets, **4:**117, **6:**106, **7:**28, **8:**38
Fecal contamination, **1:**47–48, **7:**52–53, 57–58
Fertilizer. *See also* Runoff, agricultural
Field flooding, **2:**82, 86
Fiji, **3:**45, 46, 118, **7:**320
Fiji Spring Water, **7:**162
Films:
    and portrayals of terrorism, **5:**14
    water in, **7:**171–74
Filtration, water, **1:**47, **5:**55, **7:**159–60, 161
Finn, Kathy, **4:**69–70

Fire, and drought, **5:**98, 102, **7:**105
First Peoples. *See* Indigenous populations
Fish:
    aquaculture, **2:**79
    bass, **2:**123
    carp, **2:**118
    climate change, **4:**168
    Colorado River, **3:**142
    dams, removing/decommissioning, **2:**118, 123, 126, 128, 131–33
    dams/reservoirs affecting, **1:**77, 83, 90, 98, **2:**117, **6:**142–43
        North American Water and Power Alliance, **8:**128–29
        percent of North American species threatened, **7:**154
    desert pupfish, **3:**142
    droughts, **5:**98, 102
    eels, **2:**123
    endocrine disruptors, **7:**49
    extinct or extinct in the wild species, **7:**294–96
    floods, **5:**109
    food needs for current/future populations, **2:**79
    fossil-fuel extraction affecting, **7:**87, 91
    herring, **2:**123
    impacts of desalination, **8:**94
    largest number of species, countries with, **2:**298–99
    pesticides, **5:**305–7
    Sacramento River, **2:**120–21, **8:**128
    salmon (*See* Salmon)
    San Francisco Bay, **8:**132
    sturgeon, **1:**77, 90, **2:**123
    threatened/at risk species, **1:**291–96, **3:**3, 39–40, 142
        due to dams, **7:**154, **8:**37, 128–29
        due to mining drainage, **7:**52
        percent endangered, **8:**3
        percent of U.S. species, **7:**56, **8:**37
    tuna, **3:**49, 50
Floods, **1:**142–43, **4:**162–63, 203–4, 302–7, **5:**104–9
    Australia, **7:**97
    causes, **5:**106, **7:**9
    China, **6:**84–87, 144
    control, **5:**110–12
    and dams, **5:**106, **6:**144, **8:**128
    Dead Sea, **8:**155
    definition, **5:**104–5
    disturbances promoting ecosystem health, **5:**91, **7:**134
    economy/economic issues, **5:**91–92
    effects of, **5:**106–10
    flash, **5:**104
    frequency, calculation, **5:**105
    future of, **5:**112–13
    Johnstown Flood of 1889, **5:**16
    management, **5:**110–12
    Mekong River Basin, **7:**14–15
    overview, **5:**91–92
    summary/conclusions, **5:**113–14
    transboundary agreements, **7:**9
Florida:
    Altamonte Springs, **2:**146
    reclaimed water, **2:**146–47
    St. Petersburg, **2:**146–47
    Tampa Bay (*See under* Desalination, plants)
Flow-limited resources, **6:**6
Flow rates, **2:**305–6, **3:**104, 324, **4:**168, 172, 331. *See also* Hydrologic cycle; Stocks and flows of freshwater
    measuring, **7:**346–47
Fluoride, **4:**87, **6:**81
Fog collection as a source of water, **2:**175–81
Food. *See also* Agriculture, water footprint

access, impacts of fossil-fuel extraction/
    processing, **7:**91
adulteration, **4:**32–33
BOD emissions by country, **7:**279–81
diets, regional, **2:**64–66
fish, **2:**79
genetically-modified, **7:**110
grains (*See* Grains)
meat consumption, **2:**68–69, 72, 79–80
water footprint to produce, **8:**83–84
Food needs for current/future populations:
    agricultural land (*See* Agriculture, land availability/
        quality)
    climate change, **2:**87–88
    cropping intensity, **2:**76
    crop yields, **2:**74–76
    eaten by humans, fraction of crop production,
        2:76–77
    inequalities in food distribution/consumption,
        2:64, 67–70
    kind of food will people eat, what, **2:**68–70
    need and want to eat, how much food will people,
        2:67–68
    overview, **2:**65
    people to feed, how many, **2:**66–67
    production may be unable to keep pace with
        future needs, **2:**64
    progress in feeding Earth's population, **2:**63–64
    summary/conclusions, **2:**88
    water-energy-food nexus, **8:**5–6
    water needed to grow food (*See* Agriculture,
        irrigation)
Footprint. *See* Carbon footprint; Ecosystems,
    ecological footprint; Water footprint
Force, measuring, **2:**306, **3:**324, **4:**331, **7:**347
Foreign Affairs, **3:**xiii
Forestry, urban, **8:**42
Fossil fuels. *See* Petroleum and fossil fuels
Fossil groundwater, **6:**9–10
France:
    conflict/cooperation concerning freshwater, **5:**15
    dams
        Three Gorges Dam, **1:**89
        World Commission on Dams, **3:**159
    dracunculiasis, **1:**52
    environmental concerns, top, **7:**287
    globalization and intl. trade of water, **3:**45
    Global Water Partnership, **1:**171
    human right to water, **4:**209
    Lesotho Highlands project, **1:**96
    privatization, **3:**60, 61
Frank, Louis A., **1:**194–98, **3:**209–10
French Polynesia, **3:**118
Freshwater:
    percent of global in aquifers, **7:**3
    percent of global in the Great Lakes, **7:**165
    See also Drinking water; Lakes; Renewable
        freshwater supply; Rivers; Streams;
        Surface water; Withdrawals, water
Freshwater Action Network, **8:**13
Fuels. *See* Biofuel; Petroleum and fossil fuels
Furans, **7:**48, 60
Fusion Energy Foundation, **8:**130, 131
Future, the. *See* Projections, review of global water
    resources; Soft path for water; Sustainable
    vision for the Earth's freshwater; Twenty-first
    century water-resources development

G
Gabon, **7:**133, 134–35, 321
Galileo spacecraft, **3:**217, 218
Gallup Organization, surveys. *See* Air quality,
    satisfaction; Water quality, satisfaction

Gambia, **4:**211, **7:**321
Gap, Inc., **5:**156, **8:**22
Gardens, **1:**23, **4:**122–23, **7:**115, 116, **8:**42
Garrison Diversion Project, **8:**140
Gases. *See* Greenhouse gases; Natural gas
Gaza Strip, **8:**154
Gaziantep Museum, **3:**186, 187
GEC Alsthom, **1:**85
GE Infrastructure, **5:**159–61
General circulation models (GCMs), **3:**121–23, **4:**158,
    159, 162, 167, **6:**41–42, 47
General Electric, **1:**85, **7:**133
Geophysical Research Letters, **1:**194
Georgia (country), **7:**321
Georgia (state). *See* Atlanta
Germany:
    conflict/cooperation concerning freshwater, **5:**5, 7
    dams
        Three Gorges Dam, **1:**88, 89
        World Commission on Dams, **3:**159, 170
    environmental concerns, top, **7:**287
    fossil-fuel production, **7:**75
    intl. river basin, **2:**29
    Lesotho Highlands project, **1:**96
    privatization, **3:**61
    terrorism, **5:**20
Ghana, **1:**46, 52, 54, **7:**321–22
Giardia, **1:**48, **2:**157, **4:**52, **7:**47, 57
Glen Canyon Institute, **2:**130
Glennon, Robert, **7:**xiii–xiv
Global action networks (GAN), **8:**9
Global Environmental Facility (GEF), **6:**51–52
Global Environmental Management Initiative (GEMI),
    7:33, 34
Global Environmental Outlook, **3:**88
Global Environment Monitoring System/Water
    Programme (GEMS/Water), **7:**50, 53
Globalization and international trade of water:
    Alaskan water shipments, proposed, **8:**133, 135–36
    business/industry, water risks, **5:**151
    defining terms
        commodification, **3:**35
        economic good, **3:**37–38
        globalization, **3:**34–35
        private/public goods, **3:**34
        privatization, **3:**35
        social good, **3:**36–37
        "virtual water," **8:**4, 83, 91
    General Agreement on Tariffs and Trade, **3:**48–51
    governance issues, **8:**6–16
    North American Free Trade Agreement, **3:**47–48,
        51–54, **8:**130
    overview, **3:**33–34, 41–42, **8:**4
    raw or value-added resource, **3:**42–47
    rules, intl. trading regimes, **3:**47–48
    social and economic good, water managed as
        both, **3:**38–40
    water footprint issues, **8:**84, 86, 88, 89, 90, 91
    World Water Forum (2003), **4:**192, 193
Global Reporting Initiative (GRI), **5:**158, **6:**18
    G3 Guidelines, **7:**41
    Sustainability Reporting Guidelines, **6:**28–29
    Water Protocol, **6:**28, 36
Global Runoff Data Centre, **8:**274, 292
Global Water Partnership (GWP), **1:**165–72, 175, 176,
    5:183, **6:**73
The Goddess of the Gorges, **1:**84
Goh Chok Tong, **1:**110
Goodland, Robert, **1:**77
Good manufacturing practice (GMP), **4:**32, 33
Good manufacturing practices, current (CGMP), **4:**27
Gorbachev, Mikhail, **1:**106
Gorton, Slade, **2:**134

Government/politics:
　business/industry water risk and, **7:**30
　desalination, **8:**101
　droughts, **5:**92, 94–95, 103, **7:**106–7, **8:**147–51
　environmental justice, **5:**138–41
　governance issues, **8:**6–16
　human right to water, **2:**3, **5:**137–38, **8:**29
　infrastructure damage due to political violence, **8:**164–65
　irrigation, **1:**8
　military targets (*See under* Conflict/cooperation concerning freshwater; Terrorism)
　military tools (*See under* Conflict/cooperation concerning freshwater)
　North American Water and Power Alliance issues, **8:**124–25, 131
　"policy capture," **8:**28–31
　policy reform, **7:**144, **8:**59
　privatization, **3:**68, **4:**60–73
　Red Sea-Dead Sea projects, **8:**154, 156
　subnational/intrastate *vs.* international, **8:**159, 161–62, 168–69
　subsidies, **1:**17, **7:**67, 108, 109, 111, 152, **8:**35
　twentieth-century water-resources development, **1:**7–8, 17
　World Commission on Dams report, **3:**170–71, **7:**139–40
　See also Climate change, California, policy; Conflict/cooperation concerning freshwater; Human right to water; Law/legal instruments/regulatory bodies; Legislation; Stocks and flows of freshwater; *specific countries*
Grains:
　cereal, **2:**64, **4:**299–301, **7:**109–10
　production, **2:**64, 299–301
　rice, **2:**74–79
　wheat, **2:**75, **4:**89
Grand Banks, **1:**77
Grand Canyon, **1:**15, **2:**138, 146
Granite State Artesian, **4:**39
Grants Pass Irrigation District (GPID), **2:**128
Great Lakes. *See under* Lakes, specific
Greece:
　ancient water systems, **2:**137
　bag technology, water, **1:**202, 204, 205, **8:**135
　conflict/cooperation concerning freshwater, **5:**5
　hydroelectric production, **1:**71
　supply systems, ancient, **1:**40
Greenhouse effect, **1:**137, 138, **3:**126, **4:**171, **6:**39. *See also* Climate change *listings;* Greenhouse gases
Greenhouse gases, **6:**40–43, 53, **7:**80, 84, **8:**85
Greenroof & Greenwall Projects Database, **8:**42
"Green" water, in water footprint concept, **8:**85–86, 90, 323, 332, 341, 350
Greywater (reclaimed water), **7:**120, **8:**38, 48–49
Gross national product (GNP) and water withdrawals, **3:**310–17
Groundwater:
　arsenic in, **2:**165–73, **3:**278–79, **4:**87, **6:**61, **7:**59
　as "blue water" in water footprint concept, **8:**86, 89–90, 323, 332, 341, 350
　climate change, **4:**170, **6:**43
　consequences of poor water quality, **4:**83, 87, **7:**55
　contamination by fossil-fuel production, **7:**51, 76, 79, **8:**66, 68, 69–71
　data problems, **3:**93
　desalination of brackish, **8:**114
　food needs for current/future populations, **2:**87
　fossil, **6:**9–10
　General Agreement on Tariffs and Trade, **3:**49–50
　hard path for meeting water-related needs, **3:**2
　monitoring/management problems
　　agriculture, **4:**88–90
　　analytical dilemma, **4:**90–97
　　challenges in assessments, **4:**80–81
　　conceptual foundations of assessments, **4:**80
　　data and effective management, **4:**97–98
　　extraction and use, **4:**81–88, **8:**2
　　overview, **4:**79
　overextraction, **5:**125, 128, **7:**54, 55
　　India, **8:**25
　　Las Vegas (NV) area, **8:**137
　　overview, **8:**2
　　Syria, **8:**148–49
　Pacific Island developing countries, **3:**116–18
　pesticides, **5:**307
　privatization, **3:**77, **4:**60–61
　public ownership rights and privatization, **3:**74
　recharge, **8:**49
　reclaimed water, **2:**150–51, **8:**49
　reliability, water-supply, **5:**74, **7:**55
　and sinkhole formation, **8:**155
　stocks and flows of freshwater, **2:**20
　water quality, **4:**83, 87, **8:**114
　well-being, measuring water scarcity and, **3:**104
　See also Aquifers
Groupe DANONE, **7:**41
Guatemala, **3:**13, **8:**166
Guidelines for Drinking-Water Quality (WHO), **4:**26–27, 31
Guinea, **3:**76, **7:**322
Guinea worm. *See* Dracunculiasis
Gulf of California, **3:**141, 142
Gulf of Mexico, **1:**77, **7:**73
Guyana, **7:**323
Gwembe Tonga people, **5:**134

H
Habitat degradation/loss:
　from drought, **5:**98, 103, 109
　from fossil-fuel extraction/processing, **7:**85, 87
　wetlands, **7:**56, 63, **8:**4
Habitat restoration:
　Dead Sea, **8:**155–56
　ecosystems, **2:**149–50, **7:**66, 146, **8:**44–48
　rivers, **2:**xix, 127, **3:**143–44, **8:**45–46, 48
　Salton Sea, **6:**131–37, **8:**144
　streams, **8:**45–46
Habitat simulation, as environmental flow methodology, **5:**39
Haiti, **1:**46, **8:**359, 373
Halogen Occultation Experiment (HALOE), **1:**196
Hamidi, Ahmed Z., **1:**110
Harcourt, Mike, **1:**88
Hardness, measuring, **2:**309, **3:**327, **4:**334, **7:**350
Hard path for meeting water-related needs, **3:**xviii, 2, **6:**13–14. *See also* Soft path for water; Twentieth-century water-resources development
Harran, **3:**185
Harvard School of Public Health, **4:**9
　Global Burden of Disease assessment, **7:**270
Hazardous waste landfills, **5:**119, 124
Health:
　and high concentrations of metals, **7:**59–60
　and high concentrations of nutrients, **7:**58–59
　hunger and malnutrition, **2:**70, **6:**58, **7:**57–58
　maternal, **6:**58
　and persistent organic pollutants, **7:**48, 60
　sanitation issues (*See* Sanitation services)
　toxins (*See* Water quality, contaminants)
　water issues
　　arsenic (*See* Groundwater, arsenic in)
　　costs of poor water quality, **7:**63–64

desalination, **5**:74–75
diseases (*See* Diseases, water-related)
droughts, **5**:102
floods, **5**:109
fluoride, **4**:87
human needs for water, basic, **1**:42–47
privatization, **4**:47
reclaimed water, **2**:152–56
summary/conclusions, **1**:63–64
The Heat Is On (Gelbspan), **5**:136
Helmut Kaiser, **5**:161
Hepatitis, **7**:58
Herodotus, **1**:109
Herring, **2**:123
Historic flow as environmental flow methodology, **5**:39
Hittites, **3**:184
HIV, **6**:58, **7**:259–63
Hoecker, James, **2**:124
Holistic approaches to environmental flow methodology, **5**:39
Holland. *See* Netherlands
Holmberg, Johan, **1**:166, 175
Honduras, **1**:71, **4**:54–55
Hong Kong, **3**:46, 313–15
Hookworm, **7**:61, 272
Hoppa, Gregory, **3**:217
Human Development Report, **4**:7, **6**:74, **7**:61
Human rights and international law, **2**:4–9
Human right to adequate sanitation, **8**:271
Human right to water, **2**:1–2
barriers to, **4**:212–13
defining terms, **2**:9–13
economic issues, **2**:13–14, **8**:29
environmental flow, **5**:37
and environmental justice, **5**:137–38
failure to meet, consequences of the, **2**:14–15
is there a right?, **2**:2–3
laws/covenants/declarations, **2**:4–9, **7**:36, 251, **8**:268
legal obligations, translating rights into, **2**:3, 13–14
overview, **4**:207–8
Prior Appropriation Doctrine, **5**:37
progress toward acknowledging, **4**:208–11
summary/conclusions, **2**:15
why bother?, **4**:214
See also Environmental justice; Law/legal instruments/regulatory bodies, International Covenant on Economic, Social, and Cultural Rights
Hungary, **1**:109, 120
Hunger. *See* Health, hunger and malnutrition
Hurricane Katrina, **5**:24, 110
Hydraulic geometry as environmental flow methodology, **5**:39
Hydroelectric production:
California, **4**:173–74
capacity, countries with largest installed, **1**:72, 276–80, **7**:129
Central Asia, **8**:166
China, **6**:92
Colorado River, **4**:165
dams, removing/decommissioning, **1**:83
Dead Sea, proposed, **8**:154, 156
electricity generation data, **7**:73–74, 130
Glen Canyon Dam, **2**:129–30
grandiose water-transfer schemes, **1**:74–75
James Bay Project, Quebec, **8**:131
North American Water and Power Alliance, **8**:128
percentage of electricity generated with hydropower, **1**:73–74
by region, **1**:70–71
Snake River, **2**:132–33, **8**:129

Southeastern Anatolia Project, **3**:182
Syria, **8**:149
Three Gorges Dam, **1**:84, **6**:140
transboundary water agreements, **7**:6, 132
water consumption (water footprint), **7**:25, **8**:84
well-being, measuring water scarcity and, **3**:103
Hydrogen sulfide, **7**:76, 80
Hydrologic cycle:
climate change, **1**:139–43, **4**:183, **5**:117
desalination, **2**:95, **5**:52
droughts, **5**:94
quantifications, accurate, **4**:92–96
stocks and flows of freshwater, **2**:20–27
See also Environmental flow
Hydrologic extremes, **6**:43–44
Hydro-Quebec International, **1**:85

I
Iceland, **1**:71
Idaho Rivers United, **2**:133
Identity standards and bottled water, **4**:27–31
Illinois, Cache River basin, **8**:46, 48
India:
agriculture, **4**:88, 89, **8**:10
irrigation, **2**:85, 86
basic water requirement, **2**:13
bottled water, **4**:22, 25, 40
business/industry, water risks that face, **5**:146, 147, 165
Chipko movement, **5**:124
cholera, **1**:61
conflict/cooperation concerning freshwater, **1**:107, 109, 118–19, 206–9, **5**:13, 15
Cauvery River Basin, **7**:3
PepsiCo and Coca-Cola, **7**:26
conflicts concerning freshwater 2011-2012, **8**:163
dams, **1**:70, 78, 81, **5**:133
displaced people due to, **1**:78
Kabini, **8**:163
Krishna Raja Sagar, **8**:163
World Commission on Dams, **3**:159, 170, **7**:139
Wullar, and conflict with Pakistan, **8**:163
Dhaka Community Hospital, **2**:170
dracunculiasis, **1**:53, 55
economics of water projects, **1**:16, 17
environmental concerns, top, **7**:287, 288
environmental justice, **5**:124, 127
floods, **5**:106
fossil-fuel production, **7**:75, 88, 89
groundwater, **3**:2, 50, **4**:82, 83, 88–90, 92–95, **5**:125, 128
arsenic in, **2**:165–73, **4**:87, **7**:59
and Coca-Cola, **8**:21–22, 23, 25
overextraction, **7**:55, **8**:25
human right to water, **4**:211
hydroelectric production, **7**:5, 129
industrial water use, **1**:21
intl. river basin, **2**:27, **8**:163
New Delhi, **8**:163
renewable water availability in, **6**:83
sanitation services, **5**:128
Tamil Nadu state, **8**:163
water quality, **8**:3
water use, domestic, **1**:46
Indian Ocean Dipole, **7**:98
Indicators/indices, water-related, **3**:87. *See also* Well-being, measuring water scarcity and
Indigenous populations, **5**:123, 124, **7**:91–92. *See also* Environmental justice
Africa, **8**:167
Brazil, **8**:165
Canada, **8**:128

Indonesia:
    bottled water, **5:**170
    cholera, **1:**58
    climate change, **1:**147
    conflicts concerning freshwater, **8:**168
    dams with Chinese financiers/developers/builders, **7:**323
    fossil-fuel production, **7:**75
    General Agreement on Tariffs and Trade, **3:**49
    human needs, basic, **1:**46
    pricing, water, **1:**25, **3:**69
Industrial sculptures, **5:**219, 220
Industrial water treatment, **5:**160
Industrial water use, **1:**20–21, **5:**124–25. *See also* Business/industry, water risks; Projections, review of global water resources; Water conservation, California commercial/industrial water use; Water footprint, per capita, of national consumption, by sector and country
Infrared Space Observatory, **3:**220
Insects:
    extinct or extinct in the wild species, **7:**293–94
    stone fly, **7:**56
    as vectors for water-related diseases, **1:**49–50, **4:**8–9
Institute of Marine Aerodynamics, **1:**202
Integrated water planning, **1:**17, **3:**21, **8:**59. *See also* Global Water Partnership
Integrated water resource management (IWRM), **8:**27
Intel, **8:**28
Intensity, water, **3:**17–19
Inter-American Development Bank, **3:**163
Interferometry, **3:**221
International alliances/conferences/meetings, time to rethink large, **5:**182–85. *See also* Law/legal instruments/regulatory bodies
International Association of Hydrological Sciences (IAHS), **5:**183
International Bottled Water Association (IBWA), **4:**26, 34, **5:**174
International Council of Bottled Water Association (ICBWA), **4:**26
International Drinking Water Supply and Sanitation Decade (1981-90), **3:**37
International Food Policy Research Institute (IFPRI), **2:**64
International Freshwater Conference in Bonn (2001), **3:**xviii
International Hydrological Program (IHP), **5:**183
International Law Association (ILA), **5:**35, **7:**4
International Law Commission, **1:**107
International Maize and Wheat Improvement Center, **2:**75
International river basins:
    Africa, **6:**289–96, **7:**11–13, **8:**164
    Asia, **2:**30, **6:**289, 296–301
    assessments, **2:**27–35, **7:**2–3
    Central Asia, **8:**166
    climate change and management issues, **7:**2–10, **8:**15
    by country, **2:**247–54, **6:**289–311
    Europe, **6:**289, 301–06
    fraction of a country's area in, **2:**239–46
    geopolitics, **2:**35–36, **8:**15
    North America, **6:**289, 306–08, **7:**165–69 (*See also* Colorado River)
    runoff (*See* Rivers, runoff)
    South America, **6:**289, 308–11
    use of water footprint data for management, **8:**91
    of the world, **2:**219–38
International Rivers Network, **7:**308
International trade. *See* Globalization and international trade of water

International Union for Conservation of Nature (IUCN), **7:**292–302. *See also* World Conservation Union
International Water Association (IWA), **5:**182
International Water Ltd., **3:**70
International Water Management Institute (IWMI), **3:**197, **4:**88, 108, **8:**2
International Water Resources Association (IWRA), **1:**172, **5:**183, **7:**19
Internet, **1:**231–34, **2:**192–96, **3:**225–35. *See also* Websites, water-related
    increased conflict reporting due to, **8:**159
    WikiLeaks and conflict transparency, **8:**164
Introduced/invasive species, **7:**48
Invertebrates:
    clams, **3:**142–43
    crayfish, **7:**56
    effects of acid rain, **7:**87
    extinct or extinct in the wild species, **7:**292, 298–302
    mercury in, **7:**59
    mussels, **7:**56
    shrimp, **3:**49, 50, 141, 142
Iran, **1:**58, **5:**8
    conflict concerning freshwater, **8:**163
    dams with Chinese financiers/developers/builders, **7:**323
    fossil-fuel production, **7:**75, 79
Iraq, **1:**59, 110–11, 118, **5:**13, 15–16
Irrigation. *See* Agriculture, irrigation; Gardens; Lawns
ISO 14001, **6:**32
Israel:
    conflict/cooperation concerning freshwater, **1:**107, 109, 110–11, 115–16, **5:**6, 7, 10, 14–15
    conflicts concerning freshwater, 2011-2012, **8:**165
    desalination, **5:**51, 69, 71, 72, **8:**157
    drip irrigation, **1:**23
    environmental flow, **5:**33
    globalization and intl. trade of water, **3:**45
    reclaimed water, **1:**25, 29, **2:**138, 142
    Red Sea-Dead Sea projects, **8:**144, 153–57
    terrorism, **5:**21
    tourism, **8:**155
    well-being, measuring water scarcity and, **3:**98
Italy, **3:**47, 61, **5:**11

J
Japan:
    conflict/cooperation concerning freshwater, **5:**5
    dracunculiasis, **1:**52, 53
    environmental flow, **5:**42
    industrial water use, **1:**20
    reclaimed water, **2:**139, 140, 158–59
    soft path for meeting water-related needs, **3:**23
    World Commission on Dams, **3:**159
Jarboe, James E., **5:**4
Jefferson, Thomas, **2:**94
Jerusalem Post, **5:**71
Jobs. *See* Employees; Employment
Joint Monitoring Programme (WHO), **6:**60, 73, **7:**230, 241, **8:**236, 252
Jolly, Richard, **2:**3, **4:**196
Jordan, **1:**107, 109, 115–16, **2:**33, **5:**12, 33
    Disi Water Conveyance Project, **8:**157
    Red Sea-Dead Sea projects, **8:**144, 153–57
JPMorgan, **7:**27, 134
Jupiter, search for water on moons, **3:**217–18

K
Kansas:
    Chanute, **2:**152
    Missouri River diversion project, **8:**140–44
Kantor, Mickey, **3:**51–52

Kazakhstan, **7:**75, 323
Kennebec Hydro Developers, **2:**124
Kennedy, John F., **2:**95
Kenya:
    conflicts concerning freshwater, 2011-2012, **8:**167
    dams
        with Chinese financiers/developers/builders, **7:**323–24
        effects of Lake Turkana dam, **7:**134
    dracunculiasis, **1:**53, 55
    droughts, **5:**92, 99
    environmental concerns, top, **7:**287
    fog collection as a source of water, **2:**175
    food needs for current/future populations, **2:**76–77
    Lake Naivasha, **8:**26, 27
    Nairobi, **8:**167
    Nile River Basin, **7:**11 (*See also* Rivers, specific, Nile)
    public-private strategies for shared risk, **8:**26, 27
    sanitation services, **1:**42
Kerogen, **7:**79
Khan, Akhtar H., **1:**39, **4:**71
King, Angus, **2:**123
Kiribati, **3:**118
Kirin, **6:**27
Kitchens and CII water use, **4:**135, 137
Knowledge transfer, **8:**10–11
Kokh, Peter, **3:**218
Korea, **7:**274
Korean peninsula, **1:**53, 109–10
Korean War, **1:**110
Kosovo, **5:**9
Kruger National Park, **1:**120–23, **5:**32
Kurdish Workers' Party (PKK), **5:**22
Kuwait, **1:**111, **2:**94, 97, **4:**18, **5:**69, 160, 163
    fossil-fuel production, **7:**75
Kyoto Protocol, **6:**51
Kyrgyzstan, **7:**324, **8:**166

L
Labeling and bottled water, **4:**28–31, **7:**159, 160–61, 164
Lagash-Umma border dispute, **5:**5
Lakes, **4:**168–69, **7:**55
Lakes, specific:
    Cahuilla, **6:**129
    Chad, **1:**111, 148
    Chapala, **3:**77
    Dead Sea, **8:**153–57
    Great Lakes, **1:**111, **3:**50, **7:**165–69, **8:**126, 128
    Kostonjärvi, **5:**33
    Mead, **3:**137, 140, **7:**8–9, 17, **8:**137, 138
    Mono, **5:**37, 41
    Naivasha, **8:**26, 27
    Oulujärvi, **5:**33
    Powell, **1:**76, **2:**129, 130, **7:**8–9, 17
    Taihu, **6:**95
    Turkana, **7:**134
Land, agricultural. *See* Agriculture, land availability/quality
Land conservation, **8:**42
Landfills, hazardous waste, **5:**119, 124
Landscape design, **1:**23, **4:**122–23, 135, 137–38, **6:**107
Land-use management and floods, **5:**111–12
La Niña, **3:**119, **5:**95, **7:**98
Lao People's Democratic Republic, **7:**14, 324–27
Laos, **1:**16, 71, **7:**130
La Paz-El Alto, **3:**68, 69–72
Laser leveling, agriculture and, **3:**20
Las Vegas (NV):
    description, **6:**103
    Las Vegas Pipeline Project, **8:**136–40
    per-capita water demand, **6:**104–6
    population growth, **6:**103
    precipitation, **6:**104
    temperature, **6:**104
    wastewater rate structure, **6:**118–19
    water conservation, **6:**107–8, 110–12
    water rate structures, **6:**115–17
    and water supply reliability, **5:**74
    water-use efficiency, **6:**107–8
Latin America:
    bottled water, **4:**40
    cholera, **1:**56, 57, 59–61, **3:**2, **8:**359, 373
    climate change, **1:**147
    conflicts concerning freshwater, 2011-2012, **8:**165–66
    dams, **1:**77, 81
    drinking water, **1:**262, **6:**65
        access to, **7:**24, 253, **8:**270
    human needs, basic, **1:**47
    hydroelectric production, **1:**71
    irrigation, **2:**86
    population, **2:**214
    sanitation, **1:**264, **3:**271, **5:**259
        progress on access to, **7:**256, **8:**273
    water quality, satisfaction by country, **7:**290–91
    See also Central America; South America
Laundry:
    emerging technologies, **5:**219, 220
    The High Efficiency Laundry Metering and Marketing Analysis project (THELMA), **4:**115
    laundry water and CII water use, **4:**138
    washing machines, **1:**23, **4:**114–16, **5:**219, 220
Lavelin International, **1:**85
Law/legal instruments/regulatory bodies:
    Agenda 21, **5:**34
    Agreement on Technical Barriers to Trade (TBT), **4:**35
    Agreement on the Application of Sanitary and Phytosanitary Measures (SPS), **4:**35
    Appalachian Regional Commission, **7:**307
    Beijing Platform of Action, **2:**8
    Berlin Conference Report (2004), **5:**35
    Berlin Rules (2004), **7:**5
    Bonn Declaration (2001), **3:**173, 178–80
    Boundary Waters Treaty (1909), **7:**167, 168
    Budapest Treaty, **7:**5
    Bureau of Government Research (BGR), **4:**69–70
    Cairo Programme of Action, **2:**8
    California Bay-Delta Authority, **4:**181
    California Coastal Commission (CCC), **5:**2
    California Department of Water Resources, **1:**9, 29, **3:**11, **4:**170, 232
    California Energy Commission, **4:**157, 176, 232, **5:**76, 232
    Climate and Water Panel, **1:**149
    Climate Change and Water Intra-governmental Panel, **7:**153
    Code of Federal Regulations (CFR), **4:**31–32
    Codex Alimentarius Commission (CAC), **4:**26, 35–36
    Colorado River, **3:**137–39
    Consortium for Energy Efficiency (CEE), **4:**115
    Consultative Group on International Agricultural Research (CGIAR), **3:**90
    Convention of the Rights of the Child (1989), **2:**4, 9, **4:**209
    Convention on Biological Diversity (CBD), **3:**166, **5:**34
    Copenhagen Declaration, **2:**8
    Corporate Industrial Water Management Group, **5:**157
    Declaration on the Right to Development (1986), **2:**4

Law/legal instruments/regulatory bodies (*continued*)
  Dublin Conference (1992), **1:**24, 165–66, 169, **3:**37, 58, 101, **5:**34
  Earth Summit (1992), **3:**38, 88, 101, **5:**137
  Emergency Management and Emergency Preparedness Office, **5:**24
  environmental flow, **5:**34–37
  Environmental Modification Convention (1977), **1:**114, **5:**4
  European Convention on Human Rights (1950), **2:**4, 8
  European Union Water Framework Directive (2000), **7:**25, 147–49
  Federal Bureau of Investigation (FBI), **5:**24
  Federal Emergency Management Agency (FEMA), **5:**24, 96–97
  Federal Energy Regulatory Commission (FERC), **1:**83, **2:**123–24, 126, **5:**36
  Federal Maritime Commission, **7:**307
  First National People of Color Environmental Leadership Summit (1991), **5:**120
  Food and Agricultural Organization (*See under* United Nations)
  Ganges Water Agreement (1977), **1:**119
  General Agreement on Tariffs and Trade, **3:**47–52
  Geneva Conventions, **1:**114, **5:**4
  Global Water Partnership (GWP), **1:**165–72, 175, 176, **5:**183
  Great Lakes–St. Lawrence River Basin Sustainable Water Resources Agreement (2005), **7:**165, 167–68
  Great Lakes–St. Lawrence River Basin Water Resources Compact (2008), **7:**165, 167–69
  Great Lakes–St. Lawrence River Basin Water Resources Council, **7:**168
  groundwater, **4:**95–96
  Hague Declaration (2000), **3:**173–77, **4:**2, **5:**139, 140
  Harmon Doctrine, **7:**4
  Helsinki Rules (1966), **1:**114, **7:**4
  human rights and intl. law, **2:**4–9
  India-Bangladesh, **1:**107, 119, 206–9
  Indus Water Treaty, **8:**163
  Interagency Climate Change Adaptation Task Force, **7:**153
  intergovernmental, limitations and recommendations, **8:**7–8
  Intergovernmental Panel on Climate Change (IPCC), **1:**137, 138, 140, 145, 149, **3:**120–23, **5:**81, 136, **6:**39–40, 45
    Fourth Assessment Report, **7:**7
    2008 report, **8:**5
  International Boundary and Water Commission, **7:**17, 307
  International Commission on Irrigation and Drainage (ICID), **5:**183
  International Commission on Large Dams (ICOLD), **1:**70
  International Court of Justice, **7:**5, 7
  International Covenant on Economic, Social, and Cultural Rights, **2:**4, **4:**208
    actors other than states, obligations of, **4:**231
    Article 2 (1), **2:**6–7
    Article 11, **2:**7
    Article 12, **2:**7
    Declaration on the Right to Development, **2:**9
    implementation at national level, **4:**228–30
    introduction, **4:**216–18
    normative content of the right of water, **4:**218–21
    special topics of broad application, **4:**220–21
    states parties' obligations, **4:**222–26
    violations, **4:**226–28
  International Joint Commission, **3:**50, **7:**167, 168, 307
  intl. law, role of, **1:**114–15
  intl. waters, **5:**182–85, **7:**9–10, 165–69
  Israel-Jordan Peace Treaty (1994), **1:**107, 115–16
  Joint Declaration to Enhance Cooperation in the Colorado River Delta, **3:**144
  Kyoto Protocol, **5:**137
  Kyoto Third World Water Forum (2003), **5:**183
  Mar del Plata Conference (1977), **1:**40, 42, **2:**8, 10, 47, **4:**209, **5:**183, 185
  Massachusetts Water Resources Authority, **3:**20
  Mekong River Basin Agreement (1995), **7:**13
  Mekong River Commission (MRC), **1:**82, **3:**165, **5:**35, **7:**13, 15
  Minute 306, **3:**144–45
  Multilateral Working Group on Water Resources, **8:**154
  National Rainwater and Graywater Initiative, **7:**120
  National Water Commission (Australia), **7:**107, 113, 118
  National Water Commission (U.S.), **7:**152
  Natural Resources Council of Maine, **2:**123
  Nile Basin Initiative, **7:**12
  Nile River Basin Commission, **7:**12
  Nile Waters Treaty (1959), **7:**11–13
  Non-Navigational Uses of International Watercourses (*See* Convention of the Law of the Non-Navigational Uses of International Watercourses)
  North American Free Trade Agreement (NAFTA), **3:**47–48, 51–54, **8:**130
  North American Water and Power Alliance (NAWAPA), **1:**74, **8:**124–31
  OECD (*See* Organisation for Economic Cooperation and Development)
  Okavango River Basin Commission (OKACOM), **1:**122, 124
  Organisation for African Unity (OAU), **1:**120
  overview, **1:**155, **7:**66–67
  Ramsar Convention, **3:**166, **5:**34, **7:**110
  Russian Federation Water Code, **7:**149
  Secretariat, Global Water Partnership's, **1:**170–71
  Snake River Dam Removal Economics Working Group, **2:**132
  South African Department of Water Affairs and Forestry, **1:**96
  South Asian Association for Regional Cooperation (SAARC), **1:**118
  Southern Africa Development Community (SADC), **1:**156–58, 169, **3:**165
  Southern Nevada Water Authority, **8:**136–40, 141
  Southwest Florida Water Management District (SWFWMD), **2:**108, **5:**62
  Stockholm Convention on Persistent Organic Pollutants, **7:**60
  Surface Transportation Board, **7:**307
  Surface Water Treatment Rule (SWTR), **4:**52
  Swedish International Development Agency (SIDA), **1:**165, 170, 171, **2:**14, **3:**162
  Third World Centre for Water Management, **5:**139
  Upper Occoquan Sewage Authority, **2:**152
  U.S. Agency for International Development (USAID), **1:**44, **2:**10, 14, **6:**51, **7:**306, **8:**7
  U.S. Bureau of Land Management (BLM), **8:**138
  U.S. Bureau of Reclamation (BoR), **1:**7, 69, 88, **2:**128, **3:**137, **4:**10, **6:**135
    Colorado River Basin policy, **7:**17, **8:**140
    Mekong River Basin policy, **7:**13
    North American Water and Power Alliance, **8:**128
    Reber Plan, **8:**133
    water policy reform, **7:**152
  U.S. Congress, **7:**307
  U.S. Department of Agriculture, **7:**305

U.S. Department of Commerce, **7:**305
U.S. Department of Defense, **7:**305
U.S. Department of Energy, **1:**23, **4:**115
U.S. Department of Health and Human Services, **5:**24
U.S. Department of Homeland Security, **5:**23, **7:**305
U.S. Department of Housing and Urban Development, **4:**117, **7:**305
U.S. Department of Interior, **2:**123, 127, **7:**305–6, **8:**69
U.S. Department of Justice, **7:**306
U.S. Department of Labor, **7:**306
U.S. Department of State, **7:**306
U.S. Department of the President, **7:**306
U.S. Department of Transportation, **7:**306
U.S. Environmental Protection Agency (EPA), **2:**123, 152, **4:**37, 52, **5:**23, 24
    infrastructure reports, **8:**36, 39
    reclaimed water reuse, **8:**49
    water-related budget, **7:**307
    water withdrawals for natural gas extraction, **8:**67
    well monitoring, **8:**73
U.S. Fish and Wildlife Service, **2:**126, 128, **8:**46
U.S. Food and Drug Administration (FDA), **4:**26, 37, 40, **5:**171
    water-related budget, **7:**305
U.S. Mexico Treaty on the Utilization of the Colorado and Tijuana Rivers, **3:**138
U.S. National Park Service, **2:**117, 127
U.S. National Primary Drinking Water Regulation (NPDWR), **3:**280–88
Vienna Declaration, **2:**8
Water Aid and Water for People, **2:**14
Water Environment Federation (WEF), **3:**78, **4:**62, **5:**182–83, **8:**10
Water Law Review Conference (1996), **1:**161
Water Sentinel Initiative, **5:**23
Water Supply and Sanitation Collaborative Council (WSSCC), **2:**3, 13–14, **5:**138, **7:**230, 241, **8:**236, 252
See also See also American *listings;* Bottled water, U.S. federal regulations; International *listings;* Environmental justice; National *listings;* World *listings*
Lawns, **1:**23, **4:**122–23, **7:**115, 116
Law of Conservation of Energy, **6:**7
Lead, **7:**59
League of Nations, **8:**359, 373
Leak rates, **4:**109, 117–18
Leases and environmental flows, **5:**42
Leasing contracts, **3:**66, **5:**42
Least Developed Countries Fund (LDCF), **6:**51–52
Lebanon, **1:**115
Lechwe, Kafue, **5:**32
Lecornu, Jacques, **1:**174
Legislation:
    Australia. Water Efficiency Labelling and Standards Act, **7:**28, 118–19
    California
        Central Valley Project Improvement Act, **5:**36, **7:**152
        Coastal Act, **5:**80
        Water Conservation in Landscaping Act of 1990, **4:**120–21
    Israel. Water Law of 1959, **5:**35
    Japan. River Law of 1997, **5:**35
    South Africa
        Act 54 of 1956, **1:**93, 160
        Apartheid Equal Rights Amendment (ERA), **1:**158–59
        National Water Act of 1998, **7:**145
        National Water Law of 1998, **5:**35, 37
    Switzerland. Water Protection Act of 1991, **5:**35
    U.S.
        American Recovery and Reinvestment Act of 2009, **8:**35
        Bioterrorism Act of 2002, **5:**23
        Clean Air Act, **4:**68
        Clean Water Act of 1972, **1:**15, **4:**68, **5:**36, **7:**24, 144, **8:**77
        Electric Consumers Protection Act, **5:**36
        Elwha River Ecosystem and Fisheries Restoration Act of 1992, **2:**127
        Endangered Species Act of 1973, **1:**15, **5:**37
        Federal Energy Policy Act of 1992, **1:**21, **7:**28
        Federal Food, Drug, and Cosmetic Act, **4:**27
        Federal Power Act, **5:**36
        Federal Reclamation Act of 1902, **1:**8
        Federal Wild and Scenic Rivers Act of 1968, **1:**15, **8:**129
        Flood Control Act of 1936, **1:**16
        National Environmental Policy Act of 1969/1970, **2:**130, **5:**36, **8:**129
        National Water Commission Act, **7:**143–44
        National Wild and Scenic Rivers Act of 1997, **2:**120, **5:**36
        Nutrition Labeling and Education Act of 1990, **4:**28
        Public Health, Security, and Bioterrorism Preparedness and Response Act of 2002, **5:**23
        Safe Drinking Water Act of 1974, **1:**15, **3:**280, **4:**27
        Saline Water Conversion Act of 1952, **2:**94, 95
        Secure Water Act of 2009, **7:**152
        Water Desalination Act, **2:**95
        Water Resources Development Act, **7:**166, 167
Le Moigne, Guy, **1:**172, 174
Lesotho, **3:**159–60
Lesotho Highlands project:
    chronology of events, **1:**100
    components of, **1:**93, 95
    displaced people, **1:**97–98
    economic issues, **1:**16
    financing the, **1:**95–97
    impacts of the, **1:**97–99
    Kingdom of Lesotho, geographical characteristics of, **1:**93, 94
    Lesotho Highlands Development Authority, **1:**96, 98
    management team, **1:**93
    opposition to, **1:**81, 98, 99
    update, project, **1:**99–101
Levees and flood management, **5:**111
Levi Strauss, **5:**156
Li Bai, **1:**84
Liberia, **1:**63, **7:**264
Libya, **8:**164
Licenses for hydropower dams, **2:**114–15, 123–24
Life cycle assessment (LCA), **7:**32–34, 158
Linnaeus, **1:**51
Living with Water (Netherlands), **6:**49
Lovins, Amory B., **3:**xiii–xiv
Low-energy precision application (LEPA), **1:**23
Lunar Prospector spacecraft, **1:**197, **3:**213

M
Macedonia, **1:**71, **5:**10
Machiavelli, **1:**109
Madagascar, **7:**327
Madres del Este de Los Angeles Santa Isabel, **8:**39
Malaria, **1:**49–50, **6:**58, **7:**259–63
Malawi, **4:**22

Malaysia:
   conflict/cooperation concerning freshwater, **1:**110
   dams with Chinese financiers/developers/
      builders, **7:**327–28
   data, strict access to water, **2:**41
   disputes with Singapore, **1:**22
   economics of water projects, **1:**16
   floods, **5:**106
   globalization and intl. trade of water, **3:**46
   hydroelectric production, **1:**71
   prices, water, **3:**69
   privatization, **3:**61
Maldives, **5:**106, **7:**264
Mali, **1:**52–53, 55, **7:**328, **8:**167
Mallorca, **3:**45, 46
Malnutrition. *See* Health, hunger and malnutrition
Mammals:
   dolphins, **1:**77, 90, **3:**49, 50, **7:**56
   extinct or extinct in the wild species, **7:**297–98
Manila Water Company, **4:**46
Mao Tse-tung, **1:**85
Mariner 4, **5:**175, 177
Marion Pepsi-Cola Bottling Co., **4:**38
Mars, water on:
   exploration, **3:**214–17
   future Mars missions, **5:**180
   history, **5:**178–80
   instrumental analyses, **5:**178
   missions to Mars, **5:**175–78
   overview, **5:**175
   visual evidence of, **5:**177–78
Mars Climate Orbiter, **2:**300
Mars Express, **5:**178
Mars Global Surveyor (MGS), **3:**214, **5:**177
Marshall Islands, **3:**118, **5:**136
Mars Odyssey, **3:**xx
Mars Orbital Camera (MOC), **3:**215, **5:**178
Mars Reconnaissance Orbiter, **5:**180
Maryland, Montgomery County, **8:**44
Mass, measuring, **2:**304, **7:**345
Mauritania, **1:**55, **4:**82, **8:**167
Maximum available savings (MAS), **3:**18, **4:**105
Maximum cost-effective savings (MCES), **3:**18, 24, **4:**105
Maximum practical savings (MPS), **3:**18, 24, **4:**105
Maytag Corporation, **1:**23
McDonald's, **6:**24
McKernan, John, **2:**123
McPhee, John, **2:**113
Measurements, water, **2:**25, 300–309, **3:**318–27, **4:**325–34. *See also* Assessments; Well-being, measuring water scarcity and
Media. *See* Books; Films
Mediterranean Region:
   climate change and drought, **8:**150
   Eastern, mortality rate, under-5, **7:**261
Mediterranean Sea, **1:**75, 77, **8:**153–57
Medusa Corporation, **1:**204
Meetings/conferences, international. *See*
      International *listings;* Law/legal instruments/
      regulatory bodies; United Nations; World
      listings
Meningitis, **7:**47
Merck, **6:**27
Mercury:
   as a contaminant from energy production, **7:**50, 52
   and fossil fuel production, **7:**82, 83, 84, 88, 91
   health impacts, **7:**59
   in measurement of pressure, **7:**348
   terrorism, water contamination with, **7:**194
Metals, as contaminants, **4:**168
   ecological effects, **7:**47–48
   fossil-fuel production, **7:**76, 79, 86, **8:**72
   health effects, **7:**59–60
   industrial wastewater, **7:**50
   mining, **7:**52
   road runoff, **7:**53
Meteorites, water-bearing, **3:**210–12, 216
Methane, **1:**138, 139, **7:**80, 81–82, 89
   coal bed methane, **8:**63
   contamination of drinking water, **8:**69, 70–71, 73
Methemoglobinemia (blue-baby syndrome), **7:**58
Mexico:
   bottled water, **4:**40, **5:**170
   Colorado River, **3:**134, 137, 138, 141, 144–45
      intl. agreements, **7:**6, 8–9, 15–16
   environmental concerns, top, **7:**287, 288
   environmental flow, **5:**37
   fossil-fuel production, **7:**75
   groundwater, **4:**82, 83, 96, **5:**125
      arsenic in, **7:**59
   hydroelectric production, **1:**71
   irrigation, **3:**289, **7:**16
   monitoring efforts, **3:**76–77
   North American Free Trade Agreement, **3:**47–48, 51–54, **8:**130
   North American Water and Power Alliance, **8:**124–31
   privatization, **3:**60
   sanitation services, **3:**272
   surface water, effects of climate change, **6:**43
   water-use efficiency, improving, **1:**19
Michigan, Watervliet dam removal, **8:**48
Middle East:
   air quality, satisfaction, 2012, **8:**391
   bottled water, **4:**288, 289, 291, **5:**281, 283
   climate change and drought, **8:**149–51
   conflict/cooperation concerning freshwater, **1:**107–18, **2:**182, **5:**15
   conflicts concerning freshwater
      effects of Syrian conflict, **8:**147–51, 154
      2011-2012, **8:**163–64
   desalination, **2:**94, 97, **5:**54, 55, 57, 58, 68–69
   dracunculiasis, **1:**272–73, **3:**274–75, **5:**295, 296
   environmental flow, **5:**33
   groundwater, **4:**82, 85, **5:**125
   irrigation, **2:**87
   reclaimed water, **1:**28, **2:**139
   water quality, satisfaction
      by country, **7:**290–91
      2012, **8:**391
   See also Mediterranean Region, Eastern; Southeastern Anatolia Project; *specific countries*
Military targets. *See under* Conflict/cooperation concerning freshwater; Terrorism
Military tools. *See under* Conflict/cooperation concerning freshwater; Terrorism
Millennium Development Goals (MDGs), **6:**57–78
   commitments to achieving the goals, **4:**2, 6–7
   creation of, **6:**57
   diseases, water-related
      classes of, four, **4:**8–9
      future deaths from, **4:**10, 12–13
      measures of illness/death, **4:**9–11
      mortality from, **4:**9–10, **6:**58, 73
      overview, **4:**7–8
   drinking water, access to
      baseline conditions, **6:**62
      description of, **6:**58
      Ethiopia, **6:**71–73
      goals, **6:**211, **7:**230, **8:**236
      limitations in data and reporting, **6:**61–62, **7:**231, 251–52, **8:**237–38, 268–69
      need for, **6:**63
      population growth effects on, **6:**63

progress by region, **6:**65–67, 230–32, **7:**251–53, **8:**268–70
progress on, **6:**62–70
targets for, **4:**2–5, **6:**62
economic return on meeting, **7:**64
funding of, **6:**73
future of, **6:**73–75
overview, **4:**xv, 1, **6:**57, **8:**3–4
progress measurements, **6:**60–62
projections for meeting, **4:**13–14
sanitation
baseline conditions, **6:**62
description of, **6:**58
Ethiopia, **6:**71–73
limitations in data and reporting, **6:**61–62, **7:**242, 254–55, **8:**253, 271–72
need for, **6:**63
population growth effects on, **6:**63
prioritizing of, **6:**75
progress by region, **3:**270–72, **5:**256–61, **6:**67–70, 233–35, **7:**254–56, **8:**271–73
targets for, **4:**2–5, **6:**62
in urban areas, **6:**70–71
summary/conclusions, **4:**14, **6:**77
targets for, **6:**57–58, 233
technology improvements, **6:**60
within-country disparities, **6:**77
Minas Conga, **8:**166
Mineral extraction, from evaporation ponds, **8:**155, 156
Mineral water, **4:**29
Mining:
copper, **8:**166
fossil fuels, **6:**33, **7:**73–74, 83, 85–86, 89
gold, **8:**166
processes, **7:**73–74
protests and violence, Latin America, **8:**166
wastewater, **8:**74
water footprint, **7:**31, **8:**84
See also Petroleum and fossil fuels
Ministerial statements/declarations at global water conferences, **5:**184–85. *See also* Law/legal instruments/regulatory bodies
Minoan civilization, **1:**40, **2:**137
Missouri River diversion project, **8:**140–44
Mohamad, Mahathir, **1:**110
Mokaba, Peter, **1:**123
Moldavia, **5:**8
Mongolia, **7:**328
Monitoring:
drought, joint intl., **5:**99–100, **7:**20
environmental, of desalination plants, **8:**94
and privatization, **3:**75–77, 81–82, **4:**59–60, 62–65 (*See also* Groundwater, monitoring/management problems)
recommendations, **8:**12
water quality, technology, **8:**49
The Monkey Wrench Gang (Abbey), **2:**129, **5:**14
Monterey County (CA), **2:**151
Moon, search for water on the, **3:**212–14
Morocco, **7:**328
Mortality:
childhood (*See* Childhood mortality)
from cholera (*See* Cholera, deaths reported to the WHO, by country)
due to lack of water (*See* Drinking water, access, deaths due to lack)
from water-related disease (*See* Diseases, water-related, death)
Mothers of East Los Angeles, **1:**22
Mount Pelion, **4:**39
Movies. *See* Films
Mozambique, **1:**63, 119–21, **5:**7, **7:**328–29

Mueller, Robert, **5:**4
Muir, John, **1:**80–81
Municipal water, **1:**29, **4:**29, **5:**73, **6:**101
infrastructure projects, project delivery, **8:**104
pipeline from desalination plant to, **8:**115
Myanmar. *See* Burma (Myanmar)

N
Nalco, **3:**61
Namibia:
conflict/cooperation concerning freshwater, **1:**119, 122–24
fog collection as a source of water, **2:**175
Lesotho Highlands project, **1:**98–99
reclaimed water, **1:**28, **2:**152, 156–58
Narmada Project, **1:**17
National Academy of Sciences, **1:**28, **2:**155, 166
National adaptation programs of action (NAPAs), **6:**48–49
National Aeronautics and Space Administration (NASA), **5:**96, 178. *See also* Outer space, search for water in
National Arsenic Mitigation Information Centre, **2:**172
National Council of Women of Canada, **3:**78, **4:**62
National Drought Policy Commission, **5:**95
National Environmental Protection Agency of China, **1:**92
National Fish and Wildlife Foundation, **2:**124
National Geographic, **3:**89
National Institute of Preventative and Social Medicine, **2:**167
National Marine Fisheries Service, **2:**128, 132
National Oceanic and Atmospheric Administration (NOAA):
Earth System Research Laboratory, **8:**151
National Pollutant Discharge Elimination System (NPDES), **8:**77
National Radio Astronomy Observatory, **3:**221
Native Americans. *See* Indigenous populations
Natural gas:
access as a military goal, **8:**166
coal bed methane, **8:**63
consumption in the U.S., **7:**80
energy content, **7:**75
extraction/processing, **7:**74, 75, 79–80, 81, **8:**63–78
hydraulic fracturing, **8:**63–78
overview, **8:**64–65, 77–78
production by country, **7:**75
shale gas, **8:**63, 67, 68
tight gas, **8:**63
unconventional reservoirs, **7:**80–82, **8:**63
water consumption (water footprint), **7:**25
Natural Springs, **4:**38
The Nature Conservancy, **8:**26
Nature's right to water, **7:**145
Nauru, **3:**45, 46, 118
NAWAPA Foundation, **8:**124
Needs, basic water, **1:**185–86, **2:**10–13, **4:**49–51. *See also* Drinking water, access; Health, water issues; Human right to water; Sanitation services; Well-being, measuring water scarcity and
Negev desert, **4:**89
Nepal:
arsenic in groundwater, **4:**87
bottled water, **4:**22–23
conflict/cooperation concerning freshwater, **5:**11
dams, **1:**16, 71, 81
with Chinese financiers/developers/builders, **7:**329–30
World Commission on Dams, **3:**160
hydropower potential, **7:**6
irrigation, **2:**86

Nestle, **4:**21, 40, 41, **5:**163
  bottled water, **7:**158, 161–62
Netherlands, **5:**5
  agriculture and external water resources, **8:**88
  arsenic in groundwater, **2:**167
  climate change, **4:**158, 232
  dracunculiasis, **1:**52
  fossil-fuel production, **7:**75
  Global Water Partnership, **1:**169
  Living with Water strategy, **6:**49
  Millennium Development Goals, **4:**7
  open access to information, **4:**72–73
  public-private partnerships, **3:**66
Neufeld, David, **1:**197
Nevada. *See also* Las Vegas
  North American Water and Power Alliance, **8:**124–31
  Snake Valley, **8:**139
The New Economy of Water (Gleick), **4:**47
New Hampshire, **2:**118
New Mexico, North American Water and Power Alliance, **8:**126
New Orleans (LA), **4:**67–70
New Orleans City Business, **4:**69
New Orleans Times Picayune, **4:**69
Newton Valley Water, **4:**38
New York (NY), **4:**51–53
New Yorker, **2:**113
New York Times, **2:**113, **5:**20
New Zealand:
  bottled water, **4:**26
  climate change, **5:**136
  environmental flow, **5:**33
  globalization and intl. trade of water, **3:**45, 46
  privatization, **3:**61
  reservoirs, **2:**270, 272, 274
Nexus thinking, **8:**5–6, 85–86
Niger, **7:**264, 331
Nigeria, **1:**52, 55, **5:**41
  dams with Chinese financiers/developers/builders, **7:**331
  environmental concerns, top, **7:**287
  fossil-fuel production, **7:**74, 77, 91
Nike, **5:**156, **6:**24, 27
Nitrogen compounds, effects on human health, **7:**58–59. *See also* Nutrients; Runoff, agricultural
Nitrous oxide, **1:**138, 139
Nongovernmental organizations (NGOs), **1:**81, **3:**157, **4:**198, 205–6, **6:**31, 52, 74, 89–90, 96. *See also specific organizations*
  corporate partnerships with, **7:**40
  expanding role and recommendations for, **8:**8–9
Nonrenewable resources, **6:**6–7, 15
Nordic Water Supply Company, **1:**202–5, **8:**135
North Africa. *See* Africa, Northern
North America:
  irrigation, **6:**328, 334
  river basins, **6:**289, 306–08
  transboundary waters, **7:**3, 165–69, **8:**124–31, 139
North American Water and Power Alliance (NAWAPA), **1:**74, **8:**124–31
North Dakota, **8:**140
Northstar Asset Management, **7:**36
Norway, **1:**52, 89, **3:**160
  fossil-fuel production, **7:**75
  hydroelectric production, **7:**129
Nuclear power:
  Dead Sea project, proposed, **8:**154
  water consumption and energy generation, **7:**25
Nutrients:
  cycling/loading, **1:**77, **4:**172, **5:**128
  effects of high concentrations on human health, **7:**58–59
  enrichment/eutrophication, **7:**46, 49–50, 58–59, 63

O
Oak Ridge Laboratory, **1:**23, **4:**115
Oberti Olives, **3:**22–23
Occupational Information Network (O*NET), **8:**55
Oceania:
  bottled water, **4:**288, 289, 291, **5:**281, 283
  cholera, **1:**267, 270, **5:**289, 290
  dams, **3:**295
  drinking water, **1:**254–55, 262, **3:**259–60, **5:**245, **6:**65
    access to, **7:**24, 239
    progress on access to, **7:**253, **8:**270
  groundwater, **4:**86
  hydroelectric production, **1:**71, 280
  irrigation, **1:**301, **2:**263, 265, **3:**289, **4:**296, **5:**299, **6:**328, 334
  population data/issues, **1:**250, **2:**214
  privatization, **3:**61
  renewable freshwater supply, **1:**240, **2:**202, 217, **3:**242, **4:**266, **5:**227
    2011 update, **7:**220
    2013 update, **8:**226
  salinization, **2:**268
  sanitation, **1:**259–60, 264, **3:**268–69, 272, **5:**254–55, 261
    progress on access to, **7:**249, **8:**273
  threatened/at risk species, **1:**295–96
  water access, **2:**24, 217
  withdrawals, water, **1:**244, **2:**211, **3:**251, **4:**275, **5:**236
    by country and sector, **7:**229, **8:**235
  See also Pacific Island developing countries
Ogoni people, **5:**124
Ohio, Euclid Creek dam removal, **8:**46, 48
Oil:
  extraction and refining (*See* Petroleum and fossil fuels)
  oil production by country, **7:**75
  peak (*See* Peak oil)
  spills, **7:**73, 74, 77, 87, 90
  substitutes for, **6:**8–9, 12
  transport of, **6:**14–15, **7:**74, 91
  vs. water, **6:**3–9, 14
Oil shale, **7:**75, 78–79
Olivero, John, **1:**196
Oman, the Sultanate of, **2:**179–80, **7:**331
Ontario Hydro, **1:**88
Orange County (CA), **2:**152
Order of the Rising Sun, **5:**20
Oregon, **2:**128, **8:**129
Organic contaminants:
  bioaccumulation and bioconcentration, **7:**48, 60
  BOD emissions by country/industry, **7:**278–81
  emerging, **7:**48
  health effects, **7:**60
  from hydraulic fracturing for natural gas, **8:**72
  industrial wastewater, **7:**50, 51
  See also Persistent organic pollutants; Pesticides
Organisation for Economic Cooperation and Development (OECD):
  description of, **2:**56, **3:**90, 91, 164, **4:**6
  water tariffs, **6:**312–19
  See also Overseas Development Assistance
Orion Nebula, **3:**220
Outer space, search for water in:
  clouds, interstellar, **3:**219–20
  Earth's water, origin of, **1:**93–98, **3:**209–12, **6:**5
  exploration plans, **3:**216–17
  Jupiter's moons, **3:**217–18
  Mars, **3:**214–17, **5:**175–80
  moon, the, **3:**212–14

solar system, beyond our, **3:**218–19
summary/conclusions, **3:**221–22
universe, on the other side of the, **3:**220–21
Overseas Development Assistance (ODA), **4:**6, 278–83, **5:**262–72
   water supply and sanitation
      by donating country, **7:**273–74
         2004-2011, **8:**317–19
      by subsector, **7:**275–77
         2007-2011, **8:**320–22
Oxfam Adaptation Financing Index, **6:**52
Ozguc, Nimet, **3:**185
Ozone, for bottled water, **7:**159, 161

P
Pacific Environment, **8:**10
Pacific Institute:
   bottled water, **4:**24
   climate change, **4:**232, 234, 236
   Colorado River and the Basin Study, **8:**141
   desalination, **8:**93
   hydraulic fracturing, **8:**65, 77
   privatization, **4:**45–46, 193–94, **5:**132–33
   sustainable water jobs research, **8:**35–36
   Water Conflict Chronology (*See* Water Conflict Chronology)
   water use in California, **4:**105, **8:**38, 89
Pacific Island developing countries (PIDCs):
   climate change
      overview, **3:**xix
      precipitation, **3:**115–16, 124–25
      projections for 21st century, **3:**121–23
      science overview, **3:**119–21
      sea-level rise, **3:**124
      severe impacts of, **5:**136
      storms and temperatures, **3:**125
   freshwater resources
      description and status of, **3:**115–18
      overview, **3:**113–14
      threats to, **3:**118–19
   profile of, **3:**115
   summary/conclusions, **3:**125–27
   terrain of, **3:**116
   See also Oceania
Pacific Region, Western:
   mortality rate, under-5, **7:**263
Packard Humanities Institute, **3:**187
Pakistan:
   agriculture, **4:**88, 89
   bottled water, **4:**40
   conflicts concerning freshwater, 2011-2012, **8:**163
   conflicts/cooperation concerning freshwater, **5:**10, 13, 15
   dams
      with Chinese financiers/developers/builders, **7:**331–33
      World Commission on Dams, **3:**160
      Wullar, construction by India, **8:**163
   dracunculiasis, **1:**53
   groundwater, **4:**82, 88, 89
   Orangi Pilot Project, **4:**71–72
   sanitation services, **4:**71–72
Palau, **3:**118
Palestine:
   conflicts concerning freshwater, 2011-2012, **8:**165
   Gaza Strip, **8:**154
   West Bank, **8:**165
Palestinian National Authority:
   Red Sea-Dead Sea projects, **8:**144, 153–57
Palestinians, **1:**109, 118, **5:**6, 10, 13–15
Panama, **3:**160
Papua New Guinea, **7:**333
Paraguay, **3:**13

Parasites. *See under* Diseases, water-related
Partnerships:
   with community-based organizations, **8:**59
   public-private (*See* Public-private partnerships)
   strategic corporate, **6:**31–32
Pathogenic organisms. *See* Diseases, water-related
Peak oil, **6:**1–3, 14
Peak water:
   description of, **6:**8, **8:**159
   ecological, **6:**10–12, 15
   fossil groundwater, **6:**9–10
   limitations of term, **6:**15
   summary of, **6:**15
   utility of, **6:**9–14
Pennsylvania, **2:**118
   Dimock Township, methane contamination from hydraulic fracturing, **8:**69, 70–71
   disposal of hydraulic fracturing wastewater, **8:**74
   illegal wastewater disposal, **8:**76
   Philadelphia, **8:**44
   Susquehanna River Basin, **8:**67, 68, 70
Pepsi and PepsiCo, **4:**21, **5:**146, **6:**27, **7:**26, 36, 161
Permitting:
   desalination plant, **8:**101, 112
   groundwater wells, **8:**137, 138
   wastewater, **4:**150
Perrier bottled water, **4:**21, 38, 40, 41, **5:**151
Persian Gulf War, **1:**110, 111
Persians, **3:**184
Persistent organic pollutants (POP), **7:**48, 60. *See also* Organic contaminants
Peru, **1:**46, 59–60, **2:**178–79
   cholera, **8:**359, 373
   conflicts concerning freshwater and mining, **8:**166
Pesticides, **5:**20, **7:**48, 60. *See also* Runoff, agricultural
PET. *See* Polyethylene terephthalate
Petroleum and fossil fuels:
   carbon footprint, **8:**85, 86
   case studies, **7:**90–92
   climate change caused by burning of, **6:**9, 40–43, 53, **7:**80, 84
   energy content, **7:**75
   impacts of contamination
      drinking water, **7:**88–89, **8:**63
      economic, **7:**64–65
      freshwater ecosystems, **7:**84–87
      health effects, **7:**51–52, 57
      human communities, **7:**87–88
      overview, **7:**92–93
      water quality, **7:**51, 73–74, 76–84
   mining process (*See* Mining, fossil fuels)
   oil spills, **7:**73, 74, 77, 87, 90
   origins of, **6:**4
   spills from hydraulic fracturing for natural gas, **8:**65, 66, 75, 76
   water consumption (water footprint), **7:**25, 26, 31, 73, 74–75
Pets, purchased food going to feed, **2:**77
Pharmaceutical contaminants, **7:**49, 51
Philippines:
   bottled water, **4:**40
   cholera, **1:**63
   conflict/cooperation concerning freshwater, **5:**11
   dams, **1:**71
      with Chinese financiers/developers/builders, **7:**333
      World Commission on Dams, **3:**161
   environmental concerns, top, **7:**287
   loss of tourism revenue due to water pollution, **7:**64
   mining spill, **7:**62, 65
   prices, water, **3:**69
   privatization, **3:**60, 61, 66, **4:**46

Pinchot, Gifford, **1:**80–81
Pluto, **3:**220
Pneumonia, **7:**259–63
Poland, **3:**66, 160–61, **4:**38
    fossil-fuel production, **7:**75
Poland Spring bottled water, **4:**21, **7:**162
Polar satellite, **3:**209–10
Poliomyelitis, **7:**272
Politics. *See* Government/politics
Pollution. *See* Environmental issues; Water quality, contaminants
Pollution prevention, **7:**65, **8:**24–25
Polychlorinated biphenyls (PCBs), **7:**48, 60
Polycyclic aromatic hydrocarbons (PAHs), **7:**89
Polyethylene terephthalate (PET), **4:**39, 41, **7:**158–59
Pomona (CA), **2:**138
Population issues:
    by continent, **2:**213–14
    diseases, projected deaths from water-related, **4:**12–13
    displaced people, **6:**145–46, **8:**148, 164 (*See also* Dams, social impacts, displaced people)
    drinking water access, **6:**63, 68
    expanding water-resources infrastructure, **1:**6
    food needs, **2:**63–64, 66–67
    growth
        0–2050, **2:**212
        2000–2020, **4:**10
    effects on water quality, **7:**53
    Millennium Development Goals (MDGs) affected by, **6:**63
    sanitation, **6:**63
    total and urban population data, **1:**246–50
    withdrawals, water, **1:**10, 12, 13, **3:**308–9
    See also Developing countries
Portugal, **1:**71
Poseidon Resources Corp., **2:**108, **8:**110–16
Poseidon Water Resources, **5:**61
Postel, Sandra, **1:**111
Potash, **8:**155
Poverty, **4:**40–41, **5:**123–24, **6:**58, **7:**61–62. *See also* Developing countries; Environmental justice
Power, measuring, **2:**308, **3:**326, **4:**333, **7:**349
Power generation. *See* Hydroelectric production; Nuclear power; Solar power; Wind power
Precipitation:
    acid rain, **7:**87
    Atlanta (GA), **6:**104
    China, **6:**87
    climate change, **1:**140–41, 146–47, **4:**159, 166, **6:**40, 87
    as "green water" in water footprint concept, **8:**85–86
    Las Vegas (NV), **6:**104
    Pacific Island developing countries, **3:**115–16, 124–25
    rainwater catchment/harvesting, **7:**120–21, **8:**25, 42, 48
    Seattle (WA), **6:**104
    snowfall/snowmelt, **1:**142, 147, **4:**160–61
    Standard Precipitation Index, **5:**93
    stocks and flows of freshwater, **2:**20, 22
    and use of term "withdrawal," **7:**222
    volume, and stormwater runoff production, **8:**39
Precision Fabrics Group, **1:**52
Pressure, measuring, **2:**307, **3:**325, **4:**332, **7:**348
Preston, Guy, **4:**50
Pricing, water:
    agricultural irrigation, **7:**111–12
    Australia, **7:**111–12, 118
    block, **1:**26, **4:**56
    bottled water, **4:**22–23
    climate change, **4:**180–81
    desalinated seawater, **8:**102, 106, 110–11, 113, 115
    households in different/cities/countries, **3:**304
    for hydraulic fracturing for natural gas, **8:**68
    Jordan, **1:**117
    from major water projects, **8:**104
    market approach, **1:**27, **7:**111
    off-take agreement, **8:**106
    peak-load, **1:**26
    privatization, **3:**69–71, 73, **4:**53–55
    rate structures (*See* Water rate structures)
    San Diego (CA), **8:**100
    seasonal, **1:**26
    take-or-pay contract, **8:**106, 112, 115, 117, 119
    tier, **1:**26
    twentieth-century water-resources development, **1:**24–28
    urban areas, **1:**25–27, **4:**124–25, 150
    and water policy reform, **7:**153
    See also Economy/economic issues, subsidies
Private goods, **3:**34
Privatization:
    business/industry, water risks that face, **5:**152–53
    conflict/cooperation concerning freshwater, **3:**xviii, 70–71, 79, **4:**54, 67
    defining terms, **3:**35
    drivers behind, **3:**58–59
    economic issues, **3:**70–72, **4:**50, 53–60
    environmental justice, **5:**131–33
    failed, **3:**70
    forms of, **3:**63–67, **4:**47, 48
    history, **3:**59–61
    opposition to, **3:**58
    overview, **3:**57–58, **4:**xvi, 45–46
    players involved, **3:**61–63
    principles and standards
        can the principles be met, **4:**48–49
        economics, use sound, **3:**80–81, **4:**53–60
        overview, **3:**79, **4:**47–48, **7:**106
        regulation and public oversight, government, **3:**81–82, **4:**60–73
        social good, manage water as a, **3:**80, **4:**49–53
    risks involved
        affordability questions, pricing and, **3:**69–73
        dispute-resolution process, weak, **3:**79
        ecosystems and downstream water users, **3:**77
        efficiency, water, **3:**77–78
        irreversible, privatization may be, **3:**79
        local communities, transferring assets out of, **3:**79
        monitoring, lack of, **3:**75–77
        overview, **3:**67–68, **8:**19–31
        public ownership, failing to protect, **3:**74–75
        underrepresented communities, bypassing, **3:**68
        water quality, **3:**78
    sanitation services, **5:**273–75
    summary/conclusions, **3:**82–83, **4:**73–74
    update on, **4:**46–47
    World Water Forum (2003), **4:**192, 193–94
Procter & Gamble, **5:**157, **6:**27
Productivity:
    agricultural, **2:**74–76
    land, **8:**86
    water, **3:**17–19
Progressive Habitat Development Alternative, **6:**135
Projections, review of global water resources:
    Alcamo et al. (1997), **2:**56–57
    analysis and conclusions, **2:**58–59
    data constraints, **2:**40–42
    defining terms, **2:**41
    Falkenmark and Lindh (1974), **2:**47–49
    Gleick (1997), **2:**54–55
    inaccuracy of past projections, **2:**43–44

Kalinin and Shiklomanov (1974) and De Mare (1976), **2:**46–47
L'vovich (1974), **2:**44–47
Nikitopoulos (1962, 1967), **2:**44
overview, **2:**39–40
Raskin et al. (1997, 1998), **2:**55–56
Seckler et al. (1988), **2:**57–58
Shiklomanov (1993, 1998), **2:**50–53
in 2002, **3:**xvii–xviii
World Resources Institute (1990) and Belyaev (1990), **2:**49–50
See also Sustainable vision for the Earth's freshwater
Public Citizen, **4:**69
Public goods, **3:**34
Public Limited Companies (PLC), **4:**72–73
Public participation:
    business/industry water management, **7:**37–38
    climate change adaptation, **6:**49
    drought management, **7:**116–17
    Great Lakes–St. Lawrence River Basin Water Resources Compact, **7:**169
    sustainable vision, **1:**82, **7:**24–25
    water decision making, **5:**150–51, **6:**96–97, **8:**11
    See also Education
Public perception:
    company public relations, **8:**23, 29–30
    environmental concerns around the world, top, **7:**285–88
    environmental concerns of U.S. public, top, **7:**282–84
    hydraulic fracturing for natural gas, **8:**66
    North American Water and Power Alliance, **8:**129
    satisfaction with air quality (*See* Air quality, satisfaction)
    satisfaction with water quality (*See* Water quality, satisfaction)
    terrorism, **5:**2–3
    water costs and desalination, **8:**99
    water risks of business/industry, **7:**26–27
Public-private partnerships, **3:**74–75, **4:**60–73, 193–94, **6:**92, **7:**40
    for desalination projects, **8:**110, 111–12, 119
    lessons learned, **8:**110, 111–12, 119
    risk allocation, **8:**105–6
    shared risk, **8:**19–31
Public Trust Doctrine, **5:**37
Puerto Rico, **3:**46, **5:**34
Pupfish, desert, **3:**142
Pure Life, **4:**40
Purified water, **4:**29

Q
Qatar, **7:**75
Qinghai-Tibetan Plateau, **6:**88
Quality of life (QOL), **3:**88–96, **6:**61
Quantitative measures of water availability/use, **2:**25

R
Race and environmental discrimination, **5:**118. *See also* Environmental justice
Radiative forcing, **3:**120
Radioactive contaminants, **7:**52, 73, **8:**72
Rail, Yuma clapper, **3:**142
Rainfall. *See under* Precipitation
Ralph M. Parsons Company, **8:**124, 130
Rand Water, **1:**95, 96
Rates:
    wastewater (*See* Wastewater, rates for)
    water (*See* Pricing, water; Water rate structures)
Raw or value-added resource, water traded as a, **3:**42–47
Reagan, Ronald, **2:**95

Rebates, for water conservation, **6:**110–12, **7:**119, 121
Reber Plan (San Francisco Bay Project), **8:**131–34
Reclaimed water:
    agricultural water use, **1:**28, 29, **2:**139, 142, 145–46, **7:**54, **8:**74
    Australia, **7:**114, 120–21
    blackwater, **7:**120
    California (*See* California, reclaimed water)
    costs, **2:**159, **8:**49
    defining terms, **2:**139
    environmental and ecosystem restoration, **2:**149–50
    food needs for current/future populations, **2:**87
    graywater (greywater), **7:**120, **8:**38, 48–49
    groundwater recharge, **2:**150–51
    health issues, **2:**152–56
    from hydraulic fracturing, **8:**74
    Israel, **1:**25, 29, **2:**138, 142
    Japan, **2:**139, 140, 158–59
    Namibia, **2:**152, 156–58
    overview, **1:**28–29, **2:**137–38, **7:**60, **8:**48
    potable water reuse, direct/indirect, **2:**151–52
    primary/secondary/tertiary treatment, **2:**138
    processes involved, **2:**140
    on roads, **8:**74
    summary/conclusions, **2:**159–61
    urban areas, **1:**25, **2:**146–49, **4:**151, **7:**114
    uses, wastewater, **2:**139, 141–42, **8:**49
Recreation:
    costs of poor water quality, **7:**64
    effects of water restriction policies, **7:**115–16
    tourism, **7:**64, **8:**26, 128, 155, 164
Red List of Threatened Species (IUCN), **7:**292–302
Red Sea:
    Dead Sea projects, **8:**144, 153–57
    desalination, proposed, **8:**154, 155–56
Refugees. *See* Displaced people
Regulatory bodies. *See* Law/legal instruments/regulatory bodies
Rehydration therapy, cholera and, **1:**57
Reliability, desalination and water-supply, **5:**73–74
Religious importance of water, **3:**40
Remediation, **8:**45, 46, 48
Renewable freshwater supply:
    by continent, **2:**215–17
    by country, **1:**235–40, **2:**197–202, **3:**237–42, **4:**261–66, **5:**221–27, **6:**195–201
        2011 update, **7:**215–20
        2013 update, **8:**221–26
    developing, **1:**28, **8:**48–50, 51, 52
    fossil groundwater, **6:**9–10
    globalization and intl. trade of water, **3:**39
    Overseas Development Assistance, **7:**273–77
        by donating country
            2004-2011, **8:**317–19
        by subsector
            2007-2011, **8:**320–22
Renewable resources, **6:**6–7
Reptiles:
    crocodiles, **7:**56
    extinct or extinct in the wild species, **7:**298
    threatened, **1:**291–96, **7:**56
    turtles, **3:**49, 50
Reservoirs:
    built per year, number, **2:**116
    climate change, **1:**142, 144–45
    environmental issues, **1:**75, 77, 91, **8:**128
    Martian ice, **3:**215–16
    number larger than 0.1 km by continent/time series, **2:**271–72
    orbit of Earth affected by, **1:**70
    sediment, **1:**91, **2:**127, **3:**136, 139, **4:**169
    seismic activity induced by, **1:**77, 97, **6:**144–45

Reservoirs (*continued*)
    total number by continent/volume, **2:**270
    twentieth-century water-resources development, **1:**6
    U.S. capacity, **1:**70
    U.S. volume, **2:**116
    volume larger than 0.1km by continent/time series, **2:**273–74
    See also Dams
Reservoirs, specific:
    Diamond Valley, **3:**13
    Imperial, **3:**136
    Itaipú, **1:**75
    Mesohora, **1:**144
    Occoquan, **2:**152
Restrooms and CII water use, **4:**135, 136
Reuse, water. *See* Reclaimed water
Revelle, Roger, **1:**149
Reverse osmosis (RO):
    bottled water treatment, **7:**159–60, 161
    and desalination, **2:**96, 102–3, **5:**51, 55, 57, 58, 60, 72
    costs, **8:**94
    energy requirements, **7:**161
    for industrial water treatment, **8:**28
"Rey" water (water footprint concept), **8:**86, 323, 332, 341, 350
Rhodesia, **5:**7
Rice, **2:**74–79
Right to water. *See* Human right to water; Nature's right to water
Risk assessment:
    dams and, **3:**153, **7:**137
    desalination projects, **8:**101–6, 118–19
    hydraulic fracturing for natural gas, **8:**77–78
Risk communication, **8:**11
Risk management:
    droughts and, **5:**99
    policy engagement to reduce, **8:**20
    and regulatory uncertainty, **8:**24
    risk allocation with desalination projects, **8:**115–16
    "shared risk" and the private sector, **8:**19, 25–27
Rivers:
    climate change, **1:**142, 143, 145, 148, **7:**2–10
    consequences of poor water quality, **7:**54–55
    dams' ecological impact on, **1:**77, 91
    deltas, **3:**xix
    diversion projects, large-scale, **1:**74, **8:**124–31, 136–44
    Federal Wild and Scenic Rivers Act of 1968, **1:**15, **8:**129
    floods, **5:**104, **7:**14–15
    flow rates, **2:**305–6, **3:**104, 324, **4:**168, 172, 331
    National Wild and Scenic Rivers Act of 1997, **2:**120, **5:**36
    Overseas Development Assistance, **7:**277
        by donating country 2004-2011, **8:**317–19
        by subsector 2007-2011, **8:**320–22
    pollution and large-scale engineering projects, **1:**6
    restoration, **2:**xix, 127, **3:**143–44, **8:**45–46, 48
    runoff, **1:**142, 143, 148, **2:**222–24, **4:**163–67, **5:**29
        Composite Runoff V1.0 database, **8:**274, 292
        effects of climate change, **7:**10
        limitations in data, **8:**274, 292
        monthly for major river basins, by basin name, **8:**292–309
        monthly for major river basins, by flow volume, **8:**274–91
        "natural," **8:**274, 292
        Rhine River Basin, **7:**10
    transboundary (*See* International river basins)
    wastewater dumping into, **6:**82
    See also Environmental flow; Stocks and flows of freshwater; Sustainable vision for the Earth's freshwater
Rivers, specific:
    Abang Xi, **1:**69
    Agrio, **7:**65
    Allier, **2:**125
    Amazon, **1:**75, 111, **2:**32, **7:**56, **8:**165
    American, **1:**16
    Amu Darya, **3:**3, 39–40, **7:**52
    Amur, **7:**131
    Apple, **2:**118
    Athabasca, **7:**87, 91–92
    AuSable, **2:**119
    Beilun, **7:**131
    Bhagirathi, **1:**16
    Boac, **7:**65
    Brahmaputra, **1:**107, 111, 118–19, 206–9, **6:**290
        watershed within China, **7:**131
    Butte Creek, **2:**120–23
    Cache, **8:**46, 48
    Carmel, **5:**74
    Cauvery, **1:**109, **2:**27, **7:**3, **8:**163
    Clyde, **2:**125–26
    Colorado (*See* Colorado River)
    Columbia, **3:**3, **7:**9, 18
        water transfer, proposed, **8:**126, 128, 129, 130–31
    Congo, **1:**75, 111, 156, **2:**31, 32–33
    Crocodile, **1:**123
    Danube, **1:**109, **2:**31, **5:**33, 111, **7:**5
    Elwha, **2:**117, 127–28
    Emory, **7:**84
    Euphrates, **1:**109–11, 118, **2:**33, **3:**182, 183–87, **6:**290
    Ganges, **1:**107, 111, 118–19, 206–9, **4:**81, **6:**290
        threatened/at risk species, **7:**56
        watershed within China, **7:**131
    Gila, **3:**139, 140
    Gordon, **2:**126–27
    Hai He, **6:**82, 90
    Han, **1:**109–10
    Har Us Nur, **7:**131
    Hsi/Bei Jiang, **7:**131
    Ili/Junes He, **7:**131
    Incomati, **1:**120–23
    Indus, **1:**16, 77, 111, **7:**131
    Irrawaddy, **7:**131, 132–33
    Jinsha, **6:**92
    Jordan, **1:**107, 109, 111, 115–16, **2:**31, **5:**33
        contamination, **8:**155
        diversion, effects on the Dead Sea, **8:**154–55
    Juma, **6:**90
    Kennebec, **2:**117, 123–25, **5:**34
    Kettle, **2:**119
    Kissimmee, **5:**32
    Kosi, **7:**6
    Kromme, **5:**32
    Laguna Salada, **3:**139
    Lamoille, **2:**128
    Lancang, **6:**88, **7:**130–32 (*See also* Rivers, specific, Mekong)
    Lerma, **3:**77
    Letaba, **1:**123
    Limpopo, **1:**120–23
    Logone, **5:**32
    Loire, **2:**117, 125
    Lower Snake, **2:**131–34
    Luvuvhu, **1:**123
    Mahaweli Ganga, **1:**16, **5:**32
    Malibamats'o, **1:**93
    Manavgat, **1:**203, 204–5, **3:**45, 47
    Manitowoc, **2:**119

Maputo, **1:**121
McCloud, **5:**123
Meghna, **6:**290, **7:**131
Mekong, **1:**111, **7:**9, 13–15, 14, 130–32
Merrimack, **2:**118
Meuse, **5:**15
Milwaukee, **2:**117–19
Mississippi, **2:**32, 33, **3:**13, **5:**110, 111
    water transfer, proposed, **8:**127, 140, 141
Missouri, **3:**13, **8:**140–44
Mooi, **4:**51
Murray-Darling, **5:**33, 42
    decline in water flow due to drought, **7:**102
    ecological effects of drought, **7:**100–101
    location, **7:**99
    water management, **7:**110–11, 146–47
    water markets, **7:**110–14, 146
Murrumbidgee, **7:**113
Narmada, **5:**133
Naryn, **8:**166
Neuse, **2:**126
Niger, **1:**55, 111, **2:**31, 85, **7:**88
Nile, **1:**77, 111, **2:**26, 32–33, **3:**10–11, **5:**111
    canal in Sudan, **8:**144
    dams, **8:**164
    effects of water contamination on fisheries, **7:**62
    hydrology, **7:**10
    intl. agreements, **7:**5–6, 10–13, **8:**91
    oil spills, **7:**77
Nujiang, **6:**89
Ob, **7:**131
Okavango, **1:**111, 119, 121–24
Olifants, **1:**123, **7:**64
Orange, **1:**93, 98–99, 111, **7:**52
Orontes, **1:**111, 115
Pamehac, **5:**34
Paran, **1:**111, **3:**13
Patauxent, **5:**34
Po, **5:**111
Prairie, **2:**118
Puerco, **7:**52
Pu-Lun-To, **7:**131
Red/Song Hong, **7:**131
Rhine, **5:**111, **7:**5, 10
Rhone, **3:**45, 47
Rio Grande, **1:**111, **5:**41, **7:**4, 8, **8:**125
Rogue, **2:**119, 128
Sabie, **5:**32
Sacramento, **2:**120–23, **4:**164, 167, 169, **5:**34, 111
    delta region, **8:**133
        effects of dams on fish, **8:**128
        and the Reber Plan, **8:**132
St. Lawrence, **7:**165, 167–69
Salween (Nu), **7:**131, 132
San Joaquin, **4:**164, 169, **8:**132, 133
Senegal, **1:**111, **2:**85
Shingwedzi, **1:**123
Sierra Nevada, **4:**164
Silala/Siloli, **8:**165–66
Snake, **2:**131–34, **3:**3, **8:**129
Songhua, **6:**83
Spöl, **5:**33
Sujfun, **7:**131
Susquehanna, **8:**67, 68, 70
Suzhou, **5:**32
Syr Darya, **3:**3, 39–40, **7:**52
Tarim, **5:**32, **7:**131
Temuka, **5:**33
Theodosia, **5:**34
Tigris, **1:**69, 111, 118, **2:**33, **3:**182, 187–90, **6:**290
Tumen, **7:**131
Vaal, **1:**95, **7:**52, **8:**27–28

Vakhsh, **8:**166
Volga, **1:**77
Wadi Mujib, **5:**33
Waitaki, **5:**33
White Salmon, **2:**119
Wind, **8:**73
Xingu, **8:**165
Yahara, **2:**118
Yalu, **7:**131
Yangtze, **5:**133, **6:**81, 86, 88, 91, 143–44 (*See also* Dams, specific, Yangtze River)
    threatened/at risk species, **7:**56
Yarlung Sangpo/Siang, **7:**134
Yarmouk, **1:**109, 115–16
Yellow, **5:**5, 15–16, 111, **6:**86, 91
Zambezi, **1:**111
Zhang, **6:**90
Zhujiang, **6:**86
See also China, dams, construction overseas; Dams, by continent and country; Lesotho Highlands project
The Road not Taken (Frost), **3:**1
Roads:
    permeable pavement, **8:**42
    reduction of water percolation due to, **7:**53
    runoff from, **7:**53, 77
    use of reclaimed water on, **8:**74
Roaring Springs/Global Beverage Systems, **4:**39
Rodenticides, **5:**20
Rome, ancient, **1:**40, **2:**137, **3:**184
Roome, John, **1:**99
Roosevelt, Franklin, **1:**69
Roundworm, **7:**61–62
Runoff:
    agricultural, **5:**128, 305–7, **7:**46, 48, 49–50, **8:**50, 155
    construction, **8:**76
    effects of climate change, **6:**43, **7:**10
    from hydraulic fracturing for natural gas, **8:**76–77
    "natural," **8:**274, 292
    river (*See* Rivers, runoff)
    from roads and parking lots, **7:**53, 77
    stormwater, **7:**53, 76, 77–78, **8:**39, 41–44
Rural areas:
    development and the World Water Forum, **4:**203
    drinking water, **6:**70–71 (*See also* Drinking water, access, by country)
    sanitation services, **6:**70–71 (*See also* Sanitation services, access by country)
Russell, James M., III, **1:**196
Russia:
    dams, **1:**75, 77
    environmental concerns, top, **7:**287, 288
    fossil-fuel production, **7:**75
    groundwater, **5:**125
    hydroelectric production, **1:**71, **7:**129
    irrigation, **4:**296
    and the Kyrgyzstan-Tajikistan conflict over Rogun Dam, **8:**166
    Siberia, reversal of rivers' flows, proposed, **8:**129
    threatened/at risk species, **1:**77
    water policy reform, **7:**149
    See also Soviet Union, former
Rwanda, **1:**62, **7:**11, 264
RWE/Thames, **5:**162

S
SABMiller, **7:**36
Safeway Water, **4:**39
Saint Lucia, **7:**264
St. Petersburg (FL), **2:**146–47
Salinization:
    climate change, **4:**168, 169–70, **7:**8, 54
    continental distribution, **2:**268

Salinization (*continued*)
    by country, **2:**269
    ecological effects, **7:**47
    from fossil-fuel production, **7:**76
    groundwater, **4:**87, **7:**55, **8:**148–49
    salt concentrations of different waters, **2:**21, 94
    soil fertility, **2:**73–74
    See also Desalination
Salmon, **1:**77, **2:**117, 120–21, 123, 128, 132, 133, **3:**3
    at risk due to dams, **8:**128–29
Salt. *See* Brine; Salt water
Salton Sea:
    air-quality monitoring, **6:**137
    background of, **6:**129
    Bureau of Reclamation, **6:**135
    California water transfers, **6:**129–31
    Colorado River inflows, **6:**129, 132
    Imperial Irrigation District, **6:**130
    inflows, **6:**129–33
    location of, **6:**127–28
    restoration of, **6:**131–37, **8:**144
    salinity of, **6:**128, 137
    seismic activity, **6:**135
Salton Sea Authority (SSA), **6:**134
Salt water, **6:**5, **8:**72
Samoa, **3:**118
Samosata, **3:**184–85
Samsat, **3:**184–85
San Francisco Bay, **3:**77, **4:**169, 183, **5:**73, **8:**132
San Francisco Bay Project (Reber Plan), **8:**131–34
San Francisco Chronicle, **4:**24
Sanitation services:
    access by country, **1:**256–60, **3:**261–69, **5:**247–55,
        6:221–29, **7:**241–50
        urban and rural, 1970-2008, **8:**252–53, 255–62
        urban and rural, 2011, **8:**252–53, 263–67
    access by region, MDG progress on, **3:**270–72,
        5:256–61, **6:**67–70, 233–35, **7:**254–56
        2013 update, **8:**271–73
    childhood mortality and, **6:**58, **7:**57–58, 61–62
    costs of, **6:**73
    developing countries, **1:**263–64, **7:**62
    diarrhea reduction through, **6:**75, **7:**58
    economic return on investments in, **7:**63–64
    education services affected by, **6:**58, **7:**61
    environmental justice, **5:**127–31
    falling behind, **1:**39–42, **5:**117, 124
    funding of, **6:**73
    as a human right, **8:**271
    importance of, **6:**58
    "improved," use of term, **7:**241, 254, **8:**252–53, 271
    inadequate, **6:**58
    intl. organizations, recommendations by, **2:**10–11
    investment in infrastructure projects with private
        participation, **5:**273–75
    lack of, statistics, **7:**52, **8:**1, 3
    limitations in data and reporting, **6:**61–62, **7:**242,
        254–55, **8:**253, 271–72
    maternal health affected by, **6:**58
    nongovernmental organization resources for, **6:**74
    Overseas Development Assistance, **4:**282–83,
        7:273–77
        by donating country
            2004-2011, **8:**317–19
        by subsector
            2007-2011, **8:**320–22
    poverty eradication and, **6:**58
    prioritizing of, **6:**75
    rural areas, **6:**70–71
    twentieth-century water-resources development,
        3:2
    urban areas, **6:**70–71
    well-being, measuring water scarcity and, **3:**96–98
    within-country disparities in, **6:**77
    women and access to water, **5:**126 (*See also*
        Women, responsibility for water
        collection)
    World Health Organization, **4:**208, **6:**62
    World Water Forum (2003), **4:**202, 205
    See also Health, water issues; Human right to
        water; Millennium Development Goals;
        Soft path for water; Well-being, measuring
        water scarcity and
San Jose/Santa Clara Wastewater Pollution Control
    Plant, **2:**149–50
San Pellegrino bottled water, **4:**21
Santa Barbara (CA), **5:**63–64, **8:**102–3, 119
Santa Rosa (CA), **2:**145–46
Sapir, Eddie, **4:**69
Sargon of Assyria, **1:**110
Sasol, **7:**39, **8:**27–28
Saudi Arabia:
    desalination, **2:**94, 97, **8:**93
    Disi Aquifer, **8:**157
    dracunculiasis, **1:**52
    fossil-fuel production, **7:**75
    groundwater, **3:**50
    intl. river basin, **2:**33
    pricing, water, **1:**24
Save the Children Fund, **4:**63
Save Water and Energy Education Program (SWEEP),
    4:114
Saving Water Partnership (SWP), **6:**109
SCA, **6:**27
Schistosomiasis, **1:**48, 49, **7:**58, 272
School of Environmental Studies (SOES), **2:**167, 171
Scientific American, **3:**89
Seagram Company, **3:**61
Sea-level rise, **3:**124, **4:**169–70, **7:**8, 54
Seattle (WA):
    description, **6:**103
    per-capita water demand, **6:**104–6
    population growth, **6:**104
    precipitation, **6:**104
    temperature, **6:**104
    wastewater rate structure, **6:**118–19
    water conservation, **6:**109, 110–12
    water rate structures, **6:**115–17
    water-use efficiency, **6:**109, **8:**38
Sedimentation:
    dams/reservoirs and, **1:**91, **2:**127, **3:**136, 139, **4:**169
    due to deforestation, **8:**26
    ecological effects, **7:**46, 53, 85
    of wetlands, **7:**56
Seismic activity:
    caused by filling reservoirs, **1:**77, 97, **6:**144–45
    and floods, **5:**106
    Salton Sea, **6:**135
    San Andreas fault, **6:**135
    Three Gorges Dam, **6:**144–45
Seljuk Turks, **3:**184, 188
Senegal, **1:**55
Serageldin, Ismail, **1:**166
Serbia, **7:**333
Services, basic water. *See* Drinking water, access;
    Employment; Health, water issues; Human
    right to water; Municipal water; Sanitation
    services
Servicio Nacional de Meterologia e Hidrologia, **2:**179
Sewer systems, condominial, **3:**6. *See also* Sanitation
    services
Shad, **2:**123
Shady, Aly M., **1:**174
Shale, **7:**75, 78–79, **8:**63, 67, 68
Shaping the 21st Century project, **3:**91
Shellfish. *See under* Invertebrates

Shigella, **7:**57
Shoemaker, Eugene, **1:**196
Showerheads, **4:**109, 114, **6:**106, **7:**28
Shrimp, **3:**49, 50, 141, 142
Siemens, **7:**133
Sierra Club, **1:**81
Singapore:
    access to water, strict, **2:**41
    conflict/cooperation concerning freshwater, **1:**110
    desalination, **2:**108, **5:**51
    disputes with Malaysia, **1:**22
    toilets, energy-efficient, **1:**22
    water-use efficiency, **4:**58–60
Sinkholes, **8:**155
Skanska, **3:**167
Slovakia, **1:**109, 120
Slovenia, **1:**71
SNC, **1:**85
Snow. *See under* Precipitation
Snow, John, **1:**56–57
Social goods and services, **3:**36–37, 80, **4:**49–53
Société de distribution d'eau de la Côte d'Ivoire (SODECI), **4:**66
Société pour l'aménagement urbain et rural (SAUR), **4:**66
Socioeconomic issues, **5:**94, **7:**29–30
Soft Energy Paths , **3:**xiii
Soft path for water:
    definition of, **6:**13, 101, **8:**37
    description of, **6:**12–14, **7:**150
    economies of scale in collection/distribution, **3:**8
    efficiency of use, definitions/concepts
        agriculture, **3:**19–20
        businesses, **3:**22–24
        conservation and water-use efficiency, **3:**17
        maximum practical/cost-effective savings, **3:**23
        municipal scale, **3:**20–22
        overview, **3:**16–17
        poem, **5:**219
        productivity and intensity, water, **3:**17–19
        social objectives, establishing, **3:**17
    emerging technologies, **5:**23–24
    end-use technology, simple, **3:**8–9
    how much water is really needed, **3:**4
    and integrated water resource management, **8:**27
    moving forward, **3:**25–29
    myths about
        cost-effective, efficiency improvements are not, **3:**12–15
        demand management is too complicated, **3:**15–16
        market forces, water demand is unaffected by, **3:**9
        opportunities are small, efficiency, **3:**9
        real, conserved water is not, **3:**10–11
        risky, efficiency improvements are, **3:**11–12
    overview, **3:**30, xviii
    redefining the energy problem, **3:**xiii
    sewer systems, condominial, **3:**6
    user participation, **3:**5, 6
    vs. hard path, **3:**3, 5–7, **6:**13–14
    See also Sustainable vision for the Earth's freshwater
Soil:
    changes, **1:**141–42, 148, **4:**167
    climate change and moisture, **4:**167
    compaction, **7:**83
    degradation by type/cause, **2:**266–67
    dust storms, **7:**105–6
    erosion, **7:**46, 79, 105–6, **8:**75
    food needs for current/future populations, **2:**71, 73–74
    hard path for meeting water-related needs, **3:**2

moisture in, as "green water" in water footprint concept, **8:**85–86
Solar power:
    carbon footprint, **8:**85
    Dead Sea project, proposed, **8:**154
    desalination and, **2:**105–6
    as flow-limited resource, **6:**6–7
    water consumption (water footprint), **7:**25
Solar radiation powering climate, **1:**138
Solomon Islands, **8:**168
Solon, **5:**5
Somalia, **5:**106, **8:**167
Sonoran Desert, **3:**142
South Africa:
    bottled water, **4:**22
    conflict/cooperation concerning freshwater, **1:**107, 119–21, 123–24, **5:**7, 9
    conflicts concerning freshwater 2011-2012, **8:**168
    dams, **1:**81, **3:**161, **7:**52
    Development Bank of South Africa, **1:**95, 96
    drinking water, access to, **7:**145
    environmental flow, **5:**32, 35, 37, 42
    fossil-fuel production, **7:**75
    human right to water, **2:**9, **4:**211
    hydrology, **1:**156–58
    introduced/invasive species, **7:**48
    legislation and policy
        Apartheid Equal Rights Amendment (ERA), **1:**158–59
        Constitution and Bill of Rights, **1:**159–60, **2:**9
        General Agreement on Tariffs and Trade, **3:**49
        National Water Conservation Campaign, **1:**164–65
        review process for, **1:**160–64
        water policy reform, **7:**145–46
        White Paper on Water Supply, **1:**160
    loss of tourism revenue due to water pollution, **7:**64
    mining, **7:**65
    privatization, **3:**60, **4:**49–51
    sanitation services, **7:**145
    South African Department of Water Affairs and Forestry, **1:**96
    threatened/at risk species, **7:**56
    water conservation, **8:**27–28
    See also Lesotho Highlands project
South America:
    aquifers, transboundary, **7:**3
    bottled water, **4:**18, 289, 291, **5:**163, 281, 283
    cholera, **1:**266, 270, 271
    dams, **1:**75, **3:**293
    drinking water, **1:**253, **3:**257, **5:**243, **7:**236–37
    environmental flow, **5:**34
    groundwater, **4:**86
    hydroelectric production, **1:**278
    irrigation, **1:**299, **2:**80, 259, 265, **3:**289, **4:**296, **5:**299, **6:**327, 333
    mortality rate, under-5, **7:**260–61
    population data, total/urban, **1:**248
    privatization, **3:**60
    renewable freshwater supply, **1:**238, **2:**200–201, 217, **3:**240, **4:**264, **5:**224
        2011 update, **7:**218
        2013 update, **8:**224
    reservoirs, **2:**270, 272, 274
    river basins, **6:**289, 308–11
    rivers, transboundary, **7:**3
    runoff, **2:**23
    salinization, **2:**268
    sanitation services, **1:**258, **3:**266, **5:**252, **7:**246–47
    threatened/at risk species, **1:**293
    water availability, **2:**217

South America (*continued*)
  withdrawals, water, **1:**243, **2:**207–8, **3:**247–48, **4:**271–72, **5:**232–33
    by country and sector, **7:**225–26, **8:**231–32
  See also Latin America
Southeastern Anatolia Project (GAP):
  archaeology in the region, **3:**183
  Euphrates River, developments on the, **3:**183–87
  overview, **3:**181–83
  summary/conclusions, **3:**190–91
  Tigris River, developments along the, **3:**187–90
Southern Bottled Water Company, **4:**39
South Sudan, **8:**164
Soviet Union, former:
  air quality, satisfaction, 2012, **8:**391
  cholera, **1:**58
  climate change, **1:**147
  dams, **1:**70, **3:**293
  environmental movement, **1:**15
  intl. river basin, **2:**29, 31
  irrigation, **1:**301, **2:**263, 265, **3:**289, **4:**296
  renewable freshwater supply, **4:**266
    2011 update, **7:**220
    2013 update, **8:**226
  water quality, satisfaction, 2012, **8:**391
  withdrawals, water, **1:**244, **2:**211, **3:**250–51
  See also Russia
Spain:
  agriculture, **4:**89
  conflict/cooperation concerning freshwater, **5:**5
  dams
    hydroelectric production, **1:**71
    Three Gorges Dam, **1:**89
    World Commission on Dams, **3:**161
  environmental flow, **5:**33
  globalization and intl. trade of water, **3:**45, 47
  groundwater, **4:**89
  mining, **7:**65
Sparkling water, **4:**30, **7:**158. *See also* Bottled water
Special Climate Change Fund (SCCF), **6:**51–52
Spectrometer, neutron, **3:**213
Spectroscopy, telescopic, **5:**175
Spiritual issues. *See* Religious importance of water
Spragg, Terry, **1:**203–5, **8:**135
Spring water, **4:**30, **7:**161–62, 163. *See also* Bottled water
Sri Lanka, **1:**69, **2:**86, **3:**161–62
  dams, **5:**134, **7:**334
  environmental flow, **5:**32
  floods, **5:**106
Starbucks, **5:**163, **7:**26
State Environmental Protection Administration. Chile (SEPA), **6:**80–81, 94
Stationarity, **6:**45
Statoil, **6:**23
Stewardship, water, **8:**22, 26
Stock-limited resources, **6:**6–7
Stocks and flows of freshwater:
  flows of freshwater, **2:**22–24
  hydrologic cycle, **2:**20–27
  major stocks of water on Earth, **2:**21–22
  overview, **2:**19–20
  rivers (*See* Rivers, runoff)
  summary/conclusions, **2:**36–37
  transboundary agreement strategies, **7:**8
  See also International river basins
Stone & Webster Company, **2:**108, **5:**61
Storage volume relative to renewable supply (S/O), **3:**102
Storm frequency/intensity, changes in, **1:**142–43, **4:**161–63
Stormwater runoff, **7:**53, 76, 77–78
  and hydraulic fracturing, **8:**76–77
  sustainable practices, **8:**39, 41–44

Streams, **5:**305–7
  effects of poor water quality on, **7:**54–55
  impacts of fossil-fuel extraction/processing, **7:**85, 87
  restoration, **8:**45–46
Strong, Maurice, **1:**88
Structure of Scientific Revolutions (Kuhn), **1:**193
Stunting, **6:**58
Sturgeon, **1:**77, 90, **2:**123
S2C Global Systems, **8:**135–36
Submillimeter Wave Astronomy Satellite (SWAS), **3:**219–20
Sub-Saharan Africa. *See* Africa, sub-Saharan
Subsidies. *See* Economy/economic issues
Substitutes, **6:**8–9
Sudan, **1:**55, **2:**26, **5:**7, 13
  dams with Chinese financiers/developers/builders, **7:**133, 334–35
  and the Nile River Basin, **7:**11, **8:**144, 164
  schistosomiasis and dam construction, **7:**58
  water shortages in refugee camps, **8:**164
Suez Canal, **8:**153
Suez Lyonnaise des Eaux, **3:**61–63, **4:**46
Supervisory Control and Data Acquisition (SCADA), **5:**16
Supply. *See* Renewable freshwater supply
Supply-chain management policies and programs, **6:**24
Supply-side development. *See* Twentieth-century water-resources development
Surface water:
  as "blue water" in water footprint concept, **8:**86, 89–90, 323, 332, 341, 350
  in China, **6:**81
  effects of climate change, **6:**43
  "produced" water from hydraulic fracturing, **8:**72
  See also Lakes; Rivers; Streams
Sustainability reports. *See* Global Reporting Initiative
Sustainable Asset Management (SAM) Group, **3:**167
Sustainable vision for the Earth's freshwater:
  agriculture, **1:**187–88
  business/industry, **7:**23, **8:**19–22, 30, 31
  climate change, **1:**191
  conflict/cooperation concerning freshwater, **1:**190
  criteria, **1:**17–18, **8:**37
  diseases, water-related, **1:**186–87
  ecosystems water needs identified and met, **1:**188–90
  funding, need for sustainable, **8:**8, 9–10
  human needs, basic, **1:**185–86
  overview, **1:**183–84, **8:**37
  public participation/perception, **1:**82, **7:**24–25
  stormwater management, **8:**41–44
  sustainable water jobs, **8:**35–59
  See also Soft path for water; Twenty-first century water-resources development
Swaziland, **1:**121
Sweden, **1:**52, 96, **3:**162
Switzerland, **1:**89, 171, **3:**162, **5:**33, 35
S&W Water, LLC, **5:**61
Sydney Morning Herald, **5:**66
Synthesis Report (2001), **5:**136
Syria:
  conflicts and the role of water, **1:**109, 110–11, 116, 118
  climate change, **8:**149–51
  overview and context, **8:**147–49, 165
  strategic attacks on infrastructure, **8:**149
  2011-2012, **8:**165
  dams, **7:**335
Systems Research, **2:**56

T
Taenia solium, **4:**8
Tahoe-Truckee Sanitation Agency, **2:**152
Tajikistan, **5:**9, **7:**335, **8:**166
Tampa Bay (FL). *See under* Desalination, plants
Tanzania, **1:**63, **7:**11, 335, **8:**168
Tapeworm, pork, **4:**8
Target Corporation, **6:**27
Tar sands, **7:**75, 78, 85, 87, 91–92
Tear Fund, **5:**131
Technical efficiency, **4:**103–4
Technology development, **5:**219, 220, **6:**60, **7:**67
Technology transfer, **8:**10–11
Temperature, measuring, **2:**307, **3:**325, **4:**332, **7:**348
Temperature rise, global, **1:**138, 145, **3:**120–23, **4:**159, 166. *See also* Climate change *listings;* Greenhouse effect; Greenhouse gases
Tennant Method and environmental flow, **5:**38
Tennessee Valley Authority, **1:**69–70, 145, **7:**307
Terrorism, **2:**35, **5:**1–3
   chemical/biologic attacks, vulnerability to, **5:**16–22
   cyberterrorism, **5:**16, **6:**152, **7:**176
   defining terms, **5:**3–5
   detection and protection challenges, **5:**23
   early warning systems, **5:**23–24
   environmental terrorism, **5:**3–5
   overview, **5:**1–2, **7:**176
   physical access, protection by denying, **5:**22–23
   poisoning of water supply, **8:**163, 165, 167
   policy in the U.S., security, **5:**23, 25
   public perception/response, **5:**2–3
   response plans, emergency, **5:**24–25
   summary/conclusions, **5:**25–26
   water infrastructure attacks, **5:**15–16, **8:**149, 167
   in water-related conflict, **5:**5–15, **7:**176, **8:**165 (*See also* Water Conflict Chronology)
   and water treatment, reducing vulnerability, **5:**2
Texas, **3:**74–75
   Austin, **1:**22
   hydraulic fracturing for natural gas, **8:**67–68
   North American Water and Power Alliance, **8:**126
Thailand, **4:**40, **5:**33, 106, 134
   arsenic in groundwater, **7:**59
   dams with Chinese financiers/developers/builders, **7:**335–36
   drought, **7:**132
   and the Mekong River, **7:**14, 130
   and the Salween River, **7:**132
Thames Water, **3:**63, **5:**62
Thatcher, Margaret, **1:**106, **3:**61
Thirsty for Justice, **5:**122
Threatened/at risk species:
   Colorado River, **3:**134, 142
   by country, **2:**291–97
   dams, **1:**77, 83, 90, **2:**120, 123
   extinct in the wild, freshwater animal species, **7:**292–302
   fish (*See* Fish, threatened/at risk species)
   proportion of species at risk in U.S., **4:**313–16
   Red List, **7:**292
   by region, **1:**291–96, **7:**56
   twentieth-century water-resources development, **3:**3
   water transfers, **3:**39–40
   See also Extinct species
Three Affiliated Tribes, **5:**123
Three Gorges Dam, **6:**139–49
   chronology of events, **1:**85–87, **6:**147–48
   climatic change caused by, **6:**146–47
   costs of, **6:**141–42
   dimensions of, **6:**140–41
   displaced people, **1:**78, 85, 90, **5:**134, 151
   economic issues, **1:**16, **6:**141–42
   financial costs of, **6:**141–42
   fisheries, **6:**142–43
   flood protection benefits, **6:**144
   funding of, **1:**86–89, **6:**141–42
   geological instability caused by, **6:**144–45
   history, **6:**140, **7:**133
   hydroelectric production, **1:**84, **6:**140
   impacts of, **1:**89–92, **6:**142–43
   largest most powerful ever built, **1:**84
   military targeting of, **6:**146
   opposition to, **1:**91–93
   overview, **6:**148–49, **7:**129
   population relocation and resettlement caused by, **6:**145–46
   river sediment flow effects, **6:**143–44
   seismicity caused by, **6:**144–45
   shipping benefits of, **6:**144
   size of, **6:**140
   storage capacity of, **6:**140
   threats to, **6:**139
Time, measuring, **2:**304, **4:**329, **7:**345
Timor Leste, **5:**9, **7:**264
Togo, **1:**55, **7:**336
Toilets, **1:**21–22, **3:**4, 118, **4:**104, 109, 113–14, **6:**106, 110
   low-flush, employment to install, **8:**39
Tonga, **3:**46, 118
Touré, A. T., **1:**53
Tourism, **7:**64, **8:**26, 128, 155, 164
Toxic waste dumps, **5:**119, 124
Toxic Wastes and Race in the United States, **5:**119
Trachoma, **1:**48, **7:**272
Trade. *See* Globalization and international trade of water
Traditional planning approaches, **1:**5. *See also* Projections, review of global water resources; Twentieth-century water-resources development
Transfers, water, **1:**27–28, 74–75, **3:**39–40
   Alaska water shipments, proposed, **8:**133, 135–36
   bag technology, **1:**200–205, **8:**135
   Disi Water Conveyance Project, **8:**157
   Garrison Diversion Project, **8:**140
   from Gulf of California to the Salton Sea, **8:**144
   Las Vegas Pipeline Project, **8:**136–40
   Missouri River diversion project, **8:**140–44
   North American Water and Power Alliance (NAWAPA), **1:**74, **8:**124–31
   Reber Plan (San Francisco Bay Project), **8:**131–34
   Red Sea-Dead Sea projects, **8:**144, 153–57
   by tanker, **8:**130, 135, 136
   See also Dams
Transparency International, **8:**12
Transparency issues, **8:**12–14, 112
Transpiration loss of water into atmosphere, **1:**141, **2:**83, **4:**159–60
Transportability, **6:**7–8
Transportation:
   canals, **8:**128, 144, 153, 154
   energy costs, **7:**161–62, 163
   by tanker, **8:**130, 135, 136
Treaties. *See* Law/legal instruments/regulatory bodies; United Nations
Trichuriasis, **1:**48, **7:**272
Trinidad and Tobago, **2:**108, **5:**72, **7:**264
Trout Unlimited, **2:**118, 123, 128
True Alaska Bottling Company, **8:**135–36
Trypanosomiasis, **7:**272
Tuna, **3:**49, 50
Tunisia, **2:**142, **5:**33, **7:**336
Turkey:
   bag technology, water, **1:**202–5, **8:**135
   conflict/cooperation concerning freshwater, **1:**110, 118, **5:**8

Turkey (*continued*)
    conflicts concerning freshwater, and Syria, **8:**148
    dams with Chinese financiers/developers/
        builders, **7:**336
    environmental concerns, top, **7:**287
    globalization and intl. trade of water, **3:**45–47
    terrorism, **5:**22
Turkish Antiquity Service, **3:**183. *See also* Southeastern Anatolia Project
Turtles, **3:**49, 50
Tuvalu, **5:**136
Twentieth-century water-resources development:
    Army Corps of Engineers and Bureau of Reclamation, U.S., **1:**7–8
    benefits of, **3:**2
    capital investment, **1:**6–7
    drivers of, three major, **1:**6
    end of
        alternatives to new infrastructure, **1:**17–18
        demand, changing nature of, **1:**10–14
        economics of water projects, **1:**16–17
        environmental movement, **1:**12, 15–16
        opposition to projects financed by intl. organizations, **1:**17
        overview, **1:**9–10
    shift in paradigm of human water use, **1:**5–6
        government, reliance on, **1:**7–8
        limitations to, **1:**8–9, **3:**2–3
        problems/disturbing characteristics of current situation, **1:**1–2
    summary/conclusions, **1:**32
    supply-side solutions, **1:**6
Twenty-first century water-resources development:
    agriculture, **1:**23–24
    alternative supplies, **1:**28, **8:**48–50, 51, 52
    desalination, **1:**29–32
    efficient use of water, **1:**19–20
    governance issues, **8:**6–16
    industrial water use, **1:**20–21
    overview, **1:**18–19
    Pacific Island developing countries, **3:**121–23
    paradigm shifts, **1:**5–6, **8:**19, 22
    pricing, water, **1:**24–28
    reclaimed water, **1:**28–29
    residential water use, **1:**21–23
    summary/conclusions, **1:**32–33
    See also Soft path for water; Sustainable vision for the Earth's freshwater
Typhoid, **1:**48, **7:**57, 58
Typhus, **1:**48

U
Uganda, **1:**55, **4:**211
    conflicts concerning freshwater, 2011-2012, **8:**167
    dams with Chinese financiers/developers/
        builders, **7:**336
    and the Nile River Basin, **7:**11
    wastewater treatment by the Akivubo Swamp, **7:**63
Ultraviolet radiation, bottled water treatment, **7:**159, 161
Unaccounted for water, **3:**305, 307, **4:**59
Underground storage tanks (UST), **7:**77
Undiminished principle and the human right to water, **5:**37
Unilever, **5:**149, **6:**24, **7:**35
United Arab Emirates (UAE), **5:**68–69, **7:**75
United Kingdom:
    environmental concerns, top, **7:**287, 288
    See also specific countries
United Nations:
    Agenda 21, **1:**18, 44, **3:**90
    arsenic in groundwater, **2:**167, 172
    Children's Fund (UNICEF), **1:**52, 55, **2:**167, 172, 173, **6:**60, 62
        collaboration with UN-Water, **8:**7
        data collection by, **7:**230, 241, **8:**236, 252
        Group DANONE aid, **7:**41
    Commission on Human Rights, **2:**5
    Commission on Sustainable Development, **2:**10, **3:**90
    Committee on Economic, Social, and Cultural Rights, **5:**117, 137
    Comprehensive Assessment of the Freshwater Resources of the World (1997), **1:**42–43
    Conference on International Organization (1945), **2:**5
    conflict/cooperation concerning freshwater, **1:**107, 114, 118–19, 124, 210–30, **2:**36, **8:**7–8
    data, strict access to water, **2:**41–42
    Declaration on the Right to Development (1986), **2:**8–10
    Development Programme (UNDP), **1:**52, 82, 171, **2:**172, 173, **3:**90, **4:**7, **5:**100, **8:**7
    diseases, water-related, **5:**117
    dracunculiasis, **1:**52, 55
    drinking water, **1:**40, 251
    droughts, **5:**100
    Earth Summit (1992), **3:**38, 88, 101
    Economic Commission for Asia and the Far East (UNECAFE), **7:**13
    Educational, Scientific, and Cultural Organization (UNESCO), **8:**7, 11, 14
    environmental justice, **5:**137–38
    Environment Programme (UNEP), **1:**137, **3:**127, 164, **7:**34, **8:**7
    Food and Agriculture Organization (FAO), **2:**64, 67, **5:**126
        AQUASTAT database, **4:**81–82, **7:**215, 221, **8:**221, 310
        collaboration with UN-Water, **8:**7
        Syria, **8:**148
    food needs for current/future populations, **2:**64, 66, 67
    Framework Convention on Climate Change (UNFCC), **3:**126, **6:**48–51
    Global Water Partnership, **1:**165, 166, 171, 175, **5:**183, **6:**73
    greenhouse gases, **3:**126
    groundwater, **2:**167, 172, **4:**80–81
    Human Poverty Index, **3:**87, 89, 90, 109–11
    human right to water, **2:**3, 5–9, 14, **4:**208, 214, **5:**117
        formal recognition, **7:**251, **8:**268
    Industrial Development Organization, **7:**278–81
    Inter-agency Group for Child Mortality Estimation, **7:**264
    intl. river basins assessment, **7:**2–3
    intl. water transfer policy, **8:**130
    public participation and sustainable water planning, **1:**82
    Summit for Children (1990), **1:**52, **2:**14
    Universal Declaration of Human Rights, **2:**4–10, **4:**208
    UN-Water, **8:**7, 8, 12
    watercourses, uses of (*See* Convention of the Law of the Non-Navigational Uses of International Watercourses)
    well-being, measuring water scarcity and, **3:**90, 96, 109–11
    World Water Council, **1:**172, 173, 175
    See also Law/legal instruments/regulatory bodies, International Covenant on Economic, Social, and Cultural Rights; League of Nations; Millennium Development Goals
United States:
    Alaska water shipments, proposed, **8:**133, 135–36
    bottled water (*See under* Bottled water)
    budgets, U.S. federal agency water-related, **7:**303–7

business/industry, water risks, **5:**162–63
cholera, **1:**56, 266, 270, 271
climate change, **1:**148, **7:**153
Colorado River Basin, intl. agreements, **7:**6, 8–9, 15–16, **8:**91
conflict/cooperation concerning freshwater, **1:**110, 111, **5:**6–9, 11–12, 24
dams, **1:**69–70, **3:**293, **7:**52
desalination, **2:**94–95, 97, **5:**58–63
diseases, water-related, **4:**308–12
dracunculiasis, **1:**52
drinking water, **1:**253, **3:**257, **5:**242
    access, **3:**280–88, **7:**236, **8:**36–37
droughts, **5:**93, **6:**44, **8:**130
economic productivity of water, **4:**321–24
environmental concerns, **6:**339–41
environmental concerns of the public, top, **7:**282–84, 287, 288
environmental flow, **5:**34, 36–37
environmental justice, **5:**119–20, 122–23
floods, **4:**305–7
food needs for current/future populations, **2:**68–69
fossil-fuel production, **7:**75, 78–79, 82, 85, 88, 90
General Agreement on Tariffs and Trade, **3:**50
Great Lakes Basin, intl. agreements, **7:**165–69
groundwater, **3:**2, **4:**82, 86, 96, **5:**125
    arsenic in, **7:**59
human right to water, **4:**213
hydroelectric production, **1:**71, 278, **7:**129
introduced/invasive species, **7:**48
irrigation, **1:**299, **2:**265, **3:**289, **4:**296, **5:**299, **7:**16
Las Vegas Pipeline Project, **8:**136–40
meat consumption, **2:**79–80
Missouri River diversion project, **8:**140–44
mortality rate, under-5, **7:**261
North American Free Trade Agreement, **3:**47–48, 51–54, **8:**130
North American Water and Power Alliance, **8:**124–31
pesticides, **5:**305–7
population data/issues, **1:**248, **2:**214
precipitation changes, **1:**146–47
privatization, **3:**58–60
radioactive contaminants, **7:**52
renewable freshwater supply, **1:**238, **2:**200, 217, **3:**240, **4:**264, **5:**224, **6:**83–84
    2011 update, **7:**218
    2013 update, **8:**224
reservoirs, **2:**270, 272, 274
runoff, **2:**23
salinization, **2:**268
sanitation services, **1:**258, **3:**266, 272, **5:**251, **7:**246
terrorism, **5:**21–23, 25
threatened/at risk species, **1:**293, **4:**313–16, **7:**56
transboundary waters, **7:**3, **8:**124–31
unemployment statistics, **8:**35
usage estimates, **1:**245, **8:**125
water availability, **2:**217, **8:**36–37
water footprint, **8:**87–88, 89, 90, 91
water industry revenue/growth, **5:**303–4
water policy reform
    background, **7:**143–44
    key steps to, **7:**151–54
    need for, **7:**143, 150–51, 154
well-being, measuring water scarcity and, **3:**92
withdrawals, water, **1:**10, 11, 12, 13, 243, **2:**207, **3:**247, 308–12, **4:**271, 317–20, **5:**232
    1900-2005, **8:**142
    2005, **7:**225, **8:**231
United Utilities, **3:**63
United Water Resources, **3:**61, 63
United Water Services Atlanta, **3:**62, **4:**46
Universidad de San Augustin, **2:**179
University of California at Santa Barbara (UCSB), **3:**20–22
University of Kassel, **2:**56
University of Michigan, **3:**183
Upper Atmosphere Research Satellite (UARS), **1:**196
Uranium, **7:**74
Urban areas:
    drinking water access in, **6:**70–71, **7:**97, 115–17
        (*See also* Drinking water, access, by country)
    droughts, **5:**98, **7:**114–21
    floods, **5:**104
    future demands in, **6:**102–4
    municipal water, **1:**29, **4:**29, **5:**73, **6:**101
    pricing, water, **1:**25–27, **4:**124–25, **7:**118
    privatization, **3:**76
    reclaimed water, **1:**25, **2:**146–49, **7:**114
    sanitation services in, **6:**70–71 (*See also* Sanitation services, access by country)
    soft path for meeting water-related needs, **3:**20–22, **7:**150
    stormwater runoff management techniques, **8:**42
    water rate structure (*See* Water rate structures)
    water use in, **6:**101–2, **7:**115–17
Urbanization, **5:**98, **7:**53
Urfa, **3:**185
Urlama, **5:**5
U.S. Filter Company, **3:**63
U.S. National Water Assessment, **5:**112
User fees and environmental flows, **5:**41–42
Utah:
    North American Water and Power Alliance, **8:**124–31
    Snake Valley, **8:**139
Utilities and risks that face business/industry, **5:**162–63
Uzbekistan, **1:**52, **4:**40, **7:**336, **8:**166

V
van Ardenne, Agnes, **4:**196
Varieties of Environmentalism (Guha & Martinez-Alier), **5:**123
Vegetation:
    Colorado River, **3:**134, 139–42
    gardens, **1:**23, **4:**122–23, **7:**115, 116, **8:**42
    green roofs, **8:**42
    lawns, **1:**23, **4:**122–23, **7:**115, 116
    planter boxes, **8:**42
    rain gardens, **8:**42
Vehicles, impacts of truck traffic, **8:**75
Velocity, measuring, **2:**305, **3:**323, **4:**330, **7:**346
Venezuela, **7:**75
Veolia, **6:**93
Veolia Environnement, **5:**162
Vermont Natural Resources Council, **2:**128
Vibrio, **7:**47. *See also* Cholera
Vibrio cholerae, **1:**56, 57, 58, **7:**57. *See also* Cholera
Vietnam, **4:**18, **5:**134, 163
    arsenic in groundwater, **7:**59
    dams with Chinese financiers/developers/builders, **7:**130, 336–37
    and the Mekong River Basin, **7:**14, 130
Viking, **3:**214, **5:**177
Virginia, **2:**152, **8:**54
Virgin Islands, U.S., **3:**46
Visalia (CA), **1:**29
Vision 21 process, **2:**3
Vivendi, **3:**61–64, 70, **4:**47
Voith and Siemens, **1:**85
Volume, measuring, **2:**303–4, **3:**321–22, **4:**328–29, **7:**344–45. *See also* Stocks and flows of freshwater

W
Waggoner, Paul, **1:**149
Waimiri-Atroari people, **5:**134
Wales, **7:**63. *See also* United Kingdom
Wall Street Journal, **1:**89
Warfare, **5:**4–5. *See also* Conflict/cooperation concerning freshwater; Terrorism; Water Conflict Chronology
Warming, global, **1:**138. *See also* Climate change listings; Greenhouse effect; Greenhouse gases
Washing machines, **1:**23, **4:**114–16, **5:**219, 220
Washington. *See* Seattle
Washington, D. C., green roofs initiative, **8:**44
Waste management:
    hazardous/toxic waste landfills, **5:**119, 124
    Overseas Development Assistance, **7:**277
        by donating country
            2004-2011, **8:**317–19
        by subsector
            2007-2011, **8:**320–22
    See also Wastewater
Wastewater:
    business/industry effluent, **7:**26, 28–29, 51, 55–56, 73–74
    disposal associated with hydraulic fracturing, **8:**72, 74–75, 76
    dumped into rivers, **6:**82
    from hydraulic fracturing for natural gas, **8:**64, 65, 66, 71–75
    human waste disposal, **7:**52–53
    pits for storage, **8:**72
    rates for, **6:**118–19
    treatment of, **1:**6, **2:**138, **5:**153, 159–60, **6:**27, 92
        energy considerations, **8:**74
        expansion and improvement, **7:**65
        inappropriate, in municipal treatment plants, **8:**74
        overwhelmed by stormwater runoff, **7:**53, **8:**39
        for reuse, **8:**49
        reverse osmosis, **8:**28
    volume produced per day, U.S., **8:**49
    See also Reclaimed water; Sanitation services
Wasting, **6:**58
Water:
    "blue," in water footprint concept, **8:**86, 89–90, 323, 332, 341, 350
    bottled (*See* Bottled water)
    embedded/embodied/virtual/indirect, **3:**18–19, **6:**7, **8:**83, 91 (*See also* Water footprint)
    footprint (*See* Water footprint)
    in goods, **6:**335–38
    "green," in water footprint concept, **8:**85–86, 90, 323, 332, 341, 350
    "grey," in water footprint concept, **8:**86, 323, 332, 341, 350
    lack of substitutes for, **6:**8–9, 13
    origins of, **1:**93–98, **3:**209–12, **6:**5
    pricing (*See* Pricing, water)
    produced, **6:**23, **7:**73, 76, 79
    rights (*See* Water rights)
    right to (*See* Human right to water; Nature's right to water)
    running out of, **6:**4–5
    stocks of, **6:**6
    units/data conversions/constants, **2:**300–309, **3:**318–27, **4:**325–34, **5:**319–28, **7:**341–50, **8:**395–404
    vs. oil, **6:**3–9, 14
Water (Vizcaino, et al.), **5:**219
WaterAid, **5:**131
Water allocation:
    instream, preserving/restoring, **5:**29–30 (*See also* Environmental flow)
    transboundary water agreements, **8:**14–15, 166
    transboundary waters and climate change, **7:**7, 13
    volumetric systems, **4:**95–97
Water-based diseases, **7:**58. *See also* Diseases, water-related
Waterborne diseases, **1:**47–49, 274–75, **4:**8, **7:**57–58. *See also* Diseases, water-related
Water Conflict Chronology, **1:**108–9, 125–30, **2:**35, 182–89, **3:**194–206, **4:**xvii–xviii, 238–56, **5:**5–15, 190–213, **6:**151–93
    cited by the media, **8:**173
    overview and use of terms, **8:**160
    2011 update, **7:**1, 175–205
    2013 update, **8:**173–208
    website, **7:**175, **8:**147, 160
Water conservation:
    Atlanta (GA), **6:**108–9
    Australia, **7:**106, 109, 114–21
    California commercial/industrial water use
        background to CII water use, **4:**132–33
        calculating water conservation potential, methods for, **4:**143–47
        current water use in CII sectors, **4:**133–38
        data challenges, **4:**139–40, 148–50, 152–53
        defining CII water conservation, **4:**132
        evolution of conservation technologies, **4:**149
        overview, **4:**131–32
        potential savings, **4:**140–43
        recommendations for CII water conservation, **4:**150–53
        summary/conclusions, **4:**153–54
        water use by end use, **4:**138–39
    California residential water use
        abbreviations and acronyms, **4:**126–27
        agricultural water use, **4:**107
        current water use, **4:**105–6, **8:**38
        data and information gaps, **4:**108–9
        debate over California's water, **4:**102–3
        defining conservation and efficiency, **4:**103–5
        economics of water savings, **4:**107–8
        indoor water use
            dishwashers, **4:**116
            end uses of water, **4:**112–13
            faucets, **4:**117
            leaks, **4:**117–18
            overview, **4:**109
            potential savings by end use, **4:**111–12
            showers and baths, **4:**114
            summary/conclusions, **4:**118
            toilets, **4:**113–14, **8:**39
            total use without conservation efforts, **4:**110
            washing machines, **4:**114–16
        outdoor water use
            current use, **4:**119–20
            existing efforts/approaches, **4:**120–21
            hardware improvements, **4:**122–23
            landscape design, **4:**122–23, **8:**38
            management practices, **4:**121–22
            overview, **4:**118–19
            rate structures, **4:**124–25
            summary/conclusions, **4:**125–26
        overview, **4:**101–2
    cost analysis, **8:**100
    description of, **6:**106
    indoor, **6:**110–12, **8:**38
    Las Vegas (NV), **6:**107–8, 110–12
    outdoor, **6:**112, **8:**38
    rainwater catchment, **7:**120–21, **8:**25, 42, 48
    rebates and incentives for, **6:**110–12, **7:**119, 121
    South Africa, **8:**27–28
    Tampa Bay (FL), **8:**107
    technology to monitor, **8:**39
    U.S., **8:**142
Water Efficient Technologies, **6:**107

Water footprint:
    calculation, **8:**85–86
    California vs. U.S., **8:**88–90, 91
    concept, **8:**83–85, 323, 332, 341, 350
    conclusion, **8:**90–92
    findings, **8:**87–90
    limitations in data and reporting, **8:**323–24, 332–33, 341–42, 350–51
    and nexus thinking, **8:**85–86
    per-capita, of national consumption, by country 1996-2005, **8:**323–31
    per-capita, of national consumption, by sector and country 1996-2005, **8:**332–40
    total, of national consumption, by country 1996-2005, **8:**341–49
    total, of national consumption, by sector and country 1996-2005, **8:**350–58
Water Footprint Network (WFN), **8:**85, 87
Water in Crisis: A Guide to the World's Fresh Water Resources (Gleick), **2:**300, **3:**318
Water industry. *See* Business/industry, water risks; Economy/economic issues
Water Integrity Network, **8:**12
Water landscape, **6:**36
Water market and water trading, **3:**47–48, **7:**110, 111–14, 146, **8:**102. *See also* Pricing, water
Water performance reporting, **6:**28–31
A Water Policy for the American People, **3:**16, **4:**103
Water & Process Technologies, **5:**159
Water quality:
    acidification (*See* Acidification)
    bottled water, **4:**17, 25–26, 31–32, 37–40, **7:**159–60, 161
    business/industry, water risks that face, **5:**146–49, **7:**26
    China, **6:**80–82
    climate change, **4:**167–68, **6:**44, **7:**7
    community-level consequences of poor, **7:**61–62
    contaminants
        assimilation, and "grey water" in water footprint concept, **8:**86, 323, 332, 341, 350
        emerging, **7:**48–49
        fecal, **7:**52–53, 57–58, **8:**155
        from fossil-fuel extraction/processing, **7:**88–89, **8:**63–64
        methane, **8:**69, 70–71, 73
        organic, **7:**278–81, **8:**155
        overview, **7:**46–47, **8:**3
        pathogenic organisms, **7:**47
        See also specific contaminants
    droughts, **5:**98, 102
    ecological consequences of poor, **7:**54–57
    economic/social consequences of poor, **7:**62–65
    environmental justice, **5:**127–29
    floods, **5:**109
    groundwater, **4:**83, 87, **8:**68
    Guidelines for Drinking-Water Quality, **4:**26–27, 31
    human health consequences of poor, **7:**57–60
    impacts of fossil-fuel extraction/processing, **7:**51, 73–74, 76–93, **8:**66
    monitoring technology, **8:**49
    overview, **7:**45
    pollution prevention, **7:**65, **8:**24–25, 52
    privatization, **3:**78
    salinity issues (*See* Desalination; Salinization)
    satisfaction
        countries with most and least, **8:**392–93
        by country, **7:**289–91
        limitations on data collection by poll, **8:**387, 389, 392
        by region
            2012, **8:**389–91
        sub-Saharan Africa, by country, **7:**290
            2012, **8:**387–88
    sedimentation (*See* Sedimentation)
    temperature/thermal pollution, **7:**46–47, 51, 53, 85
    three-tier classification of impacts, **7:**52
    transfer of water by tanker, **8:**136
    water quantity consequences of poor, **7:**60
    See also Desalination; Drinking water, access; Environmental flow; Salinization
Water rate structures, **3:**303
    Atlanta (GA), **6:**115–16
    average price, **6:**117
    benefits of, **6:**112, 114
    consumption charges, **6:**117
    flat, **6:**114
    inclining block, **6:**114
    Las Vegas (NV), **6:**115–16
    seasonal, **1:**26, **6:**114
    Seattle (WA), **6:**115–17
    summary of, **6:**119
    uniform, **6:**114
    wastewater, **6:**118–19
Water reporting:
    by companies, **6:**18, 20 (*See also* Corporate reporting)
    inconsistency in, **6:**33
    performance, **6:**28–31
    recommendations for, **6:**37–38
    by sector, **6:**32–35
Water Resources Policy Committee, **3:**16
Water rights, **7:**111, 112, 113, **8:**138, 156
Water risk, corporate. *See* Business/industry, water risks
Watersheds, **6:**10–11. *See also* International river basins
    impact of hydraulic fracturing for natural gas, **8:**67–68
    restoration, **8:**45
Water Supply and Sanitation Collaboration Council, **7:**230, 241, **8:**236, 252
Water use:
    company reporting on, **6:**18, 20 (*See also* Corporate reporting)
    consumptive/nonconsumptive, **6:**7, **8:**84, 87 (*See also* Water footprint)
    definitions and use of term, **1:**12, **8:**84
    direct vs. indirect, **3:**18–19, **6:**7, **8:**83 (*See also* Water footprint)
    domestic (*See* Water footprint, per capita, of national consumption, by sector and country)
    efficiency (*See* Water-use efficiency)
    estimates, **1:**46, 246
    fossil-fuel extraction/processing and energy production, **7:**25, 26, 31, 73–84
    global, statistics, **8:**87
    increases in, **6:**1
    industrial, percent used for, **7:**74
    institutional, **6:**102
    measurement, **2:**25, **7:**30–31
    process water use and CII water use, **4:**134–36
    restriction policy in urban areas, **7:**115–17
    vs. Gross Domestic Product, **8:**142
    See also Withdrawals, water
Water-use efficiency, **1:**19–20, **3:**77–78, **4:**xvi–xvii, 58–60, **5:**153, 157, **6:**106–9
    age of homes and, **6:**112
    agriculture, **3:**4, 19–20, **7:**109–10, 146, **8:**50–54
    Atlanta (GA), **6:**108–9
    Australia, **7:**118–21
    business/industry, **7:**35, 37, 39, **8:**28

Water-use efficiency (*continued*)
    jobs created by, **8:**39
    Las Vegas (NV), **6:**107–8
    legislation and policy, **7:**28, 118–19, 153–54
    measurement, **7:**30–31
    overview of techniques to increase, **8:**37–38
    Seattle (WA), **6:**109, **8:**38
    Tampa Bay (FL), **8:**107
    U.S. policy, **7:**153–54
    See also Soft path for water; Sustainable vision for the Earth's freshwater; Twenty-first century water-resources development
Waynilad Water, **4:**46
WCD. *See* World Commission on Dams
Weather Underground group, **5:**20
Websites, water-related, **1:**231–34, **2:**192–96, **3:**225–35
    short documentaries and films, **7:**174
    Water Conflict Chronology, **7:**175, **8:**147, 160
Weight of water, measuring, **2:**309, **3:**327, **4:**334, **7:**350
Well-being, measuring water scarcity and:
    Falkenmark Water Stress Index, **3:**98–100
    multifactor indicators
        Human Poverty Index, **3:**87, 89, 90, 109–11
        Index of Human Insecurity, **3:**107, 109
        International Water Management Institute, **3:**108, 197
        overview, **3:**101
        vulnerability of water systems, **3:**101–4
        Water Poverty Index, **3:**110–11
        Water Resources Vulnerability Index, **3:**105–6
    overview, **3:**xviii–xix, 87–88
    quality-of-life indicators, **3:**88–96, **6:**61
    single-factor measures, **3:**96–98, 101–3
    summary/conclusions, **3:**111
Well water, **4:**28, 30. *See also* Groundwater
Western Pacific Region. *See* Pacific Region, Western
Wetlands, **1:**6, **3:**141–43, **5:**111, **6:**86, 88
    degradation, **7:**56, 63
    ecosystem services by, **7:**56, 63
    loss, **8:**4, 132
    restoration, **8:**48
Wetlands, specific:
    Amazon River, **7:**56
    Ciénega de Santa Clara, **3:**141–43
    El Doctor, **3:**141
    El Indio, **3:**141, 142
    Nakivubo, **7:**63
    Rio Hardy, **3:**141
    San Francisco Bay, **8:**132
Wheat, **2:**75, **4:**89
Whipworm, **7:**61
White, Gilbert, **3:**16
Wildlife:
    Colorado River, **1:**77, **3:**134
    National Fish and Wildlife Foundation, **2:**124
    reserves, **7:**47–48
    World Wildlife Fund International, **3:**157
    See also Birds; Mammals
Williams, Ted, **2:**119
Wind power, **2:**105, **7:**25
Wintu people, **5:**123
Wisconsin, **2:**117–18
Withdrawals, water:
    conflict/cooperation concerning, **1:**112, **8:**24
    by country and sector, **1:**241–44, **2:**203–11, **3:**243–51, **4:**267–75, **5:**228–36, **6:**202–10, **7:**223–29
        2013 update, **8:**227–33
    definitions and use of term, **1:**12, **7:**221–22, **8:**84, 227–28
    Great Lakes Basin restrictions, **7:**168, 169
    gross national product
        China, **3:**316–17

        Hong Kong, **3:**313–15
        U.S., **3:**310–12
    hydraulic fracturing for natural gas, **8:**65, 66, 67–69
    Las Vegas area, proposed, **8:**137, 138, 139
    population in the U.S., **1:**10, 12, 13
    soft path for meeting water-related needs, **3:**23–24
    threatened/at risk species, **3:**3
    total/per-capita, **1:**10, 11
    See also Groundwater, monitoring/management problems; Projections, review of global water resources
Wolf, Aaron, **2:**28
Wolff, Gary, **3:**xiv
Women:
    effects of poor water quality on, **7:**61
    and environmental justice, **5:**126, 134
    increased exposure to contaminated water, **7:**89
    responsibility for water collection, **7:**61, 89
    violence to, during water collection, **8:**168
World Bank:
    arsenic in groundwater, **2:**172–73
    business/industry, water risks that face, **5:**147
    dams, **1:**82–83, **7:**133, 134, 138
    Development Research Group, **7:**278
    diseases, water-related, **4:**9
    displaced people, dams and, **1:**78
    dracunculiasis, **1:**52
    effects of climate change on Middle East and Northern Africa, **8:**149
    Global Burden of Disease assessment, **7:**270
    Global Water Partnership, **1:**165, 171, **5:**183
    human needs, basic, **1:**44, 47
    human right to water, **2:**10–11
    Lesotho Highlands project, **1:**96, 99
    opposition to projects financed by, **1:**17
    overruns, water-supply projects, **3:**13
    privatization, **3:**59, 70, **4:**46
    Red Sea-Dead Sea projects, **8:**154, 155
    sanitation services, **5:**273
    self-review of dams funded by, **1:**175–76
    Southeastern Anatolia Project, **3:**190–91
    Three Gorges Dam, **1:**85, 88
    World Water Council, **1:**173
World Business Council for Sustainable Development (WBCSD), Global Water Tool, **7:**33, 34
World Climate Conference (1991), **1:**149
World Commission on Dams (WCD):
    data and feedback from five major sources, **3:**150
    environmental flow, **5:**30
    environmental justice, **5:**134, 135
    findings and recommendations, **3:**151–53
    goals, **3:**150–51, **7:**136
    organizational structure, **3:**149–50, **7:**136
    origins of, **1:**83, 177–79, **7:**136
    overview, **3:**xix, **7:**136–37
    priorities/criteria/guidelines, **3:**153–58, **7:**137–38, 139
    reaction to the report
        conventions, intl., **3:**166
        development organizations, intl., **3:**164–65
        funding organizations, **3:**158, 162–64, 167–69, **7:**138–39
        governments, **3:**170–71, **7:**139–40
        industry/trade associations, intl., **3:**169–70, **7:**139
        national responses, **3:**159–62, **7:**138–39
        nongovernmental organizations, **3:**157, **7:**138
        overview, **3:**155–56, **7:**138
        private sector, **3:**166
        regional groups, **3:**165
    rights and risk assessment, **3:**153, **7:**137
    Southeastern Anatolia Project, **3:**191
    summary/conclusions, **3:**171–72

World Conservation Union (IUCN), **1:**82–83, 121, 177, **3:**164. *See also* International Union for Conservation of Nature
World Council on Sustainable Development, **5:**158
World Court, **1:**109, 120
World Food Council, **2:**14
World Fund for Water (proposed), **1:**174–75
World Health Assembly, **1:**52
World Health Organization (WHO):
    arsenic in groundwater, **2:**166, 167, 172
    bottled water, **4:**26–27
    childhood mortality, data, **7:**257
    cholera, **1:**61, 271 (*See also* Cholera, cases reported to the WHO; Cholera, deaths reported to the WHO)
    collaboration with UN-Water, **8:**7
    desalination, **5:**75
    diseases, water-related, **4:**9, **5:**117
    dracunculiasis, **1:**52, 55
    drinking water, **3:**2, **4:**2, 208, **6:**211
        access to, data, **7:**230, **8:**236
    Global Burden of Disease assessment, **7:**270
    human needs, basic, **1:**44
    human right to water, **2:**10–11
    Joint Monitoring Programme, **6:**60, 73, **7:**230, 241, **8:**236, 252
    reclaimed water, **2:**154, 155
    sanitation services, **1:**256, **3:**2, **4:**2, 208, **6:**62
        access to, data, **7:**241, **8:**252
    unaccounted for water, **3:**305
    well-being, measuring water scarcity and, **3:**90, 91
World Health Reports, **4:**8, 9
World Meteorological Organization (WMO), **1:**137, **5:**100, **8:**7
World Resources Institute, **2:**27–28, 49–50, **8:**2
World Trade Organization (WTO), **3:**48–50
Worldwatch Institute, **2:**28
World Water Council (WWC), **1:**172–76, **3:**165, **4:**192–93, **5:**183
World Water Development Report, **8:**6
World Water Forum:
    2000, **3:**xviii, 58, 59, 90, 173
    2003
        background to, **4:**192–94
        Camdessus Report, **4:**195–96, 206
        efficiency and privatization, lack of attention given to, **4:**192
        focus of, **4:**191
        human right to water, **4:**212
        Millennium Development Goals, **4:**6, 7
        Ministerial Statement, **4:**194–95, 200–204
        NGO Statement, **4:**192, 198, 205–6
        overview, **4:**xv
        successes of, **4:**191–92
        Summary Forum Statement, **4:**196–97
        value of future forums, **4:**192
    2006, **5:**186–88
World Wildlife Fund International, **3:**157
Wyoming, Pavillion gas field, **8:**71, 72, 73

X
Xeriscaping, **1:**23, **4:**123–24
Xylenes, **8:**72

Y
Yangtze! Yangtze!, **1:**92
Yeates, Clayne, **1:**194–95
Yemen, **1:**53, 55, **8:**164–65

Z
Zambia, **1:**63, **4:**211, **5:**9, 32
    dams with Chinese financiers/developers/builders, **7:**133, 337–38
Zimbabwe, **1:**107, **5:**7, 140, **7:**338
Zombie water projects:
    Alaskan water shipments, **8:**133, 135–36
    Las Vegas Pipeline Project, **8:**136–40
    Missouri River diversion project, **8:**140–44
    North American Water and Power Alliance (NAWAPA), **1:**74, **8:**124–31
    Reber Plan (San Francisco Bay Project), **8:**131–34
    water transfer to the Salton Sea, **8:**144
Zuari Agro-Chemical, **1:**21

Made in the USA
Columbia, SC
11 August 2018